SERIES ON SEMICONDUCTOR SCIENCE AND TECHNOLOGY

Series Editors

Series on Semiconductor Science and Technology

---

1. M. Jaros: *Physics and applications of semiconductor microstructures*
2. V. N. Dobrovolsky and V. G. Litovchenko: *Surface electronic transport phenomena in semiconductors*
3. M. J. Kelly: *Low-dimensional semiconductors*
4. P. K. Basu: *Theory of optical processes in semiconductors*
5. N. Balkan: *Hot electrons in semiconductors*
6. B. Gil: *Group III nitride semiconductor compounds: physics and applications*
7. M. Sugawara: *Plasma etching*
8. M. Balkanski, R. F. Wallis: *Semiconductor physics and applications*
9. B. Gil: *Low-dimensional nitride semiconductors*
10. L. Challis: *Electron-phonon interactions in low-dimensional structures*
11. V. Ustinov, A. Zhukov, A. Egorov, N. Maleev: *Quantum dot lasers*
12. H. Spieler: *Semiconductor detector systems*
13. S. Maekawa: *Concepts in spin electronics*
14. S. D. Ganichev, W. Prettl: *Intense terahertz excitation of semiconductors*
15. N. Miura: *Physics of semiconductors in high magnetic fields*
16. A. Kavokin, J. J. Baumberg, G. Malpuech, F. P. Laussy: *Microcavities*

Physics of Semiconductors in High Magnetic Fields

Noboru Miura

Institute for Solid State Physics, University of Tokyo

OXFORD
UNIVERSITY PRESS

Great Clarendon Street, Oxford OX2 6DP

Oxford University Press is a department of the University of Oxford.
It furthers the University's objective of excellence in research, scholarship,
and education by publishing worldwide in

Oxford New York

Auckland Cape Town Dar es Salaam Hong Kong Karachi
Kuala Lumpur Madrid Melbourne Mexico City Nairobi
New Delhi Shanghai Taipei Toronto

With offices in

Argentina Austria Brazil Chile Czech Republic France Greece
Guatemala Hungary Italy Japan Poland Portugal Singapore
South Korea Switzerland Thailand Turkey Ukraine Vietnam

Oxford is a registered trade mark of Oxford University Press
in the UK and in certain other countries

Published in the United States
by Oxford University Press Inc., New York

First published 2008

British Library Cataloguing in Publication Data
Data available

Library of Congress Cataloging in Publication Data
Data available

Printed in Great Britain
on acid-free paper by
Biddles Ltd. www.biddles.co.uk

ISBN 978–0–19–851756–6 (Hbk)

1 3 5 7 9 10 8 6 4 2

PREFACE

The recent progress of magnet technology has enabled us to study solid state physics in very high magnetic fields. Steady fields up to about 40–45 T are available with hybrid magnets in several facilities in the world. Using pulsed fields, 50–90 T can be easily produced with a long duration of 10–100 ms, and we can now obtain a field even in the megagauss range (above 100 T) by special techniques, and these high fields are routinely employed for various measurements in solid state physics.

High magnetic fields quantize the electronic states in substances, and the effects of the quantization give rise to many new phenomena. In semiconductors, the Landau quantization effect is particularly significant because of the small effective masses and high mobilities. Thus, semiconductor physics is one of the areas where high magnetic fields are most useful.

Recently, remarkable progress in semiconductor technology has made it possible to grow many new systems that have never before been available. The development of techniques such as molecular beam epitaxy (MBE), metal organic chemical vapor deposition (MOCVD), or various fine processes have created low dimensional electron systems. In low dimensional systems, the quantization effect by magnetic fields is even more enhanced, combined with the quantization by the potentials built in the samples. The integer and fractional quantum Hall effects are among the most remarkable manifestations of the quantization effects. A variety of quantum phenomena have been observed in artificial new low dimensional systems.

The author has devoted himself to developing techniques for generating very high pulsed magnetic fields up to several megagauss (several hundred tesla) by various kinds of approaches, and applied them to various solid state experiments for many years. Many new results have been obtained using such high fields in a wide range of physics. Many interesting phenomena have also been observed by collaborating scientists who come from outside including overseas to use our facilities. Among them, typical results on semiconductor physics will be described in each chapter of this book.

There have been many excellent textbooks on semiconductor physics. There have also been many valuable proceedings of international conferences on semiconductor physics in high magnetic fields as listed at the end of the book. However, there have not been so many books in the form of a monograph sharply devoted to the physics of semiconductors in high magnetic fields. The aim of this book is to give a review of recent progress in this area with a particular emphasis on the new phenomena observed in the very high field range, mainly in the pulsed field range above 20 or 30 T up to a few megagauss.

This book is partly based on the notes for a series of lectures which the author gave at the graduate course of the University of Tokyo, with the additional descriptions of new phenomena observed in high magnetic fields. In order to make the entire book consistent as a monograph, considerable effort has been directed to enriching the introductory part of each chapter. Therefore, the book will hopefully have the dual character of textbook and monograph giving a review of the latest progress. In the last chapter, the experimental techniques for generating pulsed high magnetic fields for the data acquisition are described.

The author strongly wishes that this book will serve as a guide for young scientists, students, and engineers to increase their interest in high magnetic fields.

Noboru Miura
Tokyo, Japan
September 2007

CONTENTS

1

INTRODUCTION

Magnetic fields quantize the energy states of the conduction bands and the valence bands in semiconductors. In high magnetic fields, the quantization of energy states in semiconductors becomes very prominent and the radius of the cyclotron motion of conduction electrons or the electron wave-function extension is much reduced. The quantization of the electronic states and the modification of the electronic wave function cause many new quantum phenomena in high magnetic fields. Electrons in most semiconductors possess high mobility or long relaxation time between scattering. Today, electron mobility of the best GaAs crystals reaches higher than 10^7 $\mathrm{cm}^2/\mathrm{Vs}$ at low temperatures, which corresponds to a mean free path of mm. Moreover, in most cases, the effective masses of electrons and holes are smaller than the free electron mass. These features are most favorable for observing the effects of quantization.

The recent advances in semiconductor technology have enabled us to fabricate high quality samples in which electrons and holes are confined in artificial quantum potentials embedded in them. They are heterostructure interfaces, superlattices, quantum wells, quantum wires, and quantum dots. In these samples, the low dimensional electrons confined in quantum potentials show unique properties in magnetic fields. For instance, the two-dimensional electron systems in quantum wells or heterostructures are fully quantized when magnetic fields are applied in a perpendicular direction to the system. Therefore, under the unique situation of full quantization, many new quantum phenomena have been observed. The integer and fractional quantum Hall effects are the highlights of these. When fields are applied parallel to the two-dimensional systems, on the other hand, a variety of phenomena take place originating from the interplay between the magnetic field potential and the artificially formed quantum potentials. In quantum dots or quantum wires, phenomena originating from the potential interplay are also observed, and they are extremely useful for the characterization of new quantum devices. Thus the physics of low dimensional electron systems in high magnetic fields is one of the most important subjects of semiconductor physics.

Figure 1.1 shows the cyclotron energy $\hbar\omega_c$, the cyclotron radius r_c, and the spin Zeeman splitting $g\mu_B B$ as a function of magnetic field B. The cyclotron energy and the Zeeman splitting of free electrons increases as a linear function of magnetic field, and reaches 11.6 meV (136 K) at 100 T. As $\hbar\omega_c$ is inversely proportional to the effective mass m^* which is smaller than the free electron mass m in most semiconductors, $\hbar\omega_c$ is even larger than that of free electrons. In Fig. 1.1,

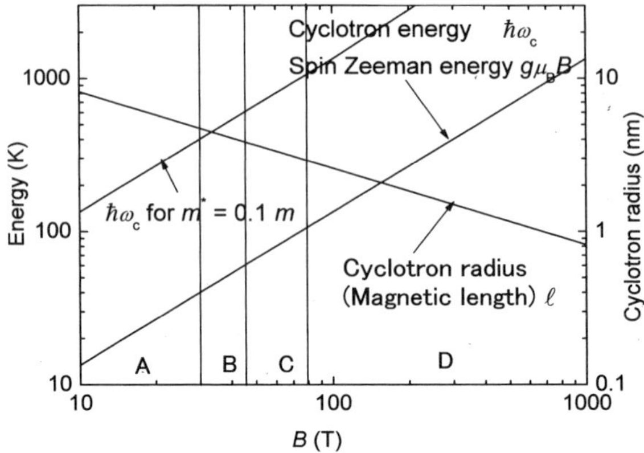

FIG. 1.1. Cyclotron energy $\hbar\omega_c$ and the spin Zeeman splitting $g\mu_B B$ for a free electron in a unit of K, and the cyclotron radius of the ground state r_c which is the same as the magnetic length l, as a function of magnetic field B. The range of the magnetic fields available with the present technology is shown: A. Superconducting magnet, B. Water-cooled or hybrid magnet, C. Non-destructive long pulse magnet, D. Destructive short pulse magnet.

the line for an effective mass $m^* = 0.1m$ is also shown. In high enough magnetic fields, $\hbar\omega_c$ becomes larger than various excitation energies in semiconductors, such as the longitudinal optical (LO) phonon energy $\hbar\omega_o$, the band gap energy $\mathcal{E}_g$, the plasmon energy $\mathcal{E}_p$, the binding energies of excitons or impurity states, *etc.* Under such unusual circumstances, we can observe new phenomena that have never been observed in the lower field range. The cyclotron radius of the ground state l, on the other hand, decreases with increasing magnetic field in proportion to $1/\sqrt{B}$. It is reduced to 2.6 nm at 100 T and 0.81 nm at 1000 T, and becomes smaller than various characteristic lengths in semiconductors. In other words, very high magnetic fields by themselves create electronic states of nano-scale structure. For example, it may become smaller than the wave-function extension of excitons or impurity states. This situation also causes various interesting phenomena in a new regime. In artificial quantum structures, it can become smaller than the wave-function extension of electrons confined in the quantum potential. In very high magnetic fields, l can be reduced to a very small value: even smaller than the lattice constant of crystals. The effective mass approximation that is quite adequate to treat the electronic states in a magnetic field may break down in extremely high magnetic fields. It is an interesting question what will happen under such conditions.

Today, as high field magnet technology develops, the maximum field available for experiments has increased more and more, and we can study semiconductor physics in very high magnetic fields. Table 1.1 shows various techniques for

Table 1.1 Different techniques for generating high magnetic fields and the maximum fields generated by them.

Technique	Maximum field (T)	Duration
Electromagnet	3	DC
Supeconducting magnet	20	DC
Water cooled magnet	33	DC
Hybrid magnet	45	DC
Wire-wound pulse magnet	90	1–100 ms
Helical pulse magnet	70	$\sim$100 μs
Single turn coil technique	300	$\sim$ μs
Electromagnetic flux compression	620	$\sim$ μs
Explosive driven flux compression (Bellow type)	200	$\sim$ μs
Explosive driven flux compression (Cylindrical type)	> 1000	$\sim$ μs

producing high magnetic fields. In the 1960s, the Francis Bitter National High Magnetic Field Laboratory (FBNHMFL) was built at the MIT, and steady high magnetic fields were generated were generated by using a large DC generator of 10 MW and a large water cooling system for the magnets heated by the large current (water-cooled magnet). High magnetic fields up to about 20 T were generated and the fields were employed for many different experiments, but the major application was in the field of semiconductor physics. In those days, the FBNHMFL was one of the main centers of semiconductor research in the world. In later years, several high magnetic field facilities have been built in Europe and Japan as well, and the maximum field has increased. The technology of superconducting magnets using Nb-Ti alloy or Nb_3Sn wires also made great progress in the 1960s. By using commercial superconducting magnets, it has become possible to generate high magnetic fields (around 15–17 T) easily in ordinary laboratories. The maximum field now available with superconducting magnets is almost 21 T. However, because of the limited critical field and the critical current of presently available superconducting materials, in order to obtain higher fields, we have to use the water-cooled magnets in large facilities as mentioned above. At present, the maximum field obtained with such water-cooled magnets is 33 T at the new National High Magnetic Field Laboratory (NHMFL) in Tallahassee succeeding the FBNHMFL in USA. In most large DC field facilities, hybrid magnets are constructed combining a water-cooled magnet with a superconducting magnet set outside. The hybrid magnet can generate higher magnetic fields than simple water-cooled magnets. The highest field now at Tallahassee is 45 T.

For generating a field above this field level, pulse magnets are employed. By supplying a large pulse current from a capacitor bank, a wire-wound solenoid type magnet can generate magnetic fields up to 60–90 T or so, within a duration time of 1–100 ms. For pulse magnets, it is not necessary to install a large power supply which consumes an enormous electric power, nor a large water cooling system. The use of pulse magnets extended the available field range significantly.

Together with the progress of fast data acquisition equipments, researches using pulse magnets have become more and more popular since the late 1970s.

However, there is a limitation to the maximum field that can be generated by pulse magnets. The main reason is a large magnetic stress (Maxwell stress) exerted on the magnet due to the high magnetic fields. The Maxwell stress is proportional to the square of the field, and exceeds the mechanical strength of the magnet. Thus the magnets are inevitably destroyed if the field exceeds some limit. A great deal of effort is being made worldwide to generate the highest possible field non-destructively. So far, the maximum non-destructive field is limited to nearly 80 T. A number of laboratories in the world are now competing in the generation of higher magnetic fields aiming at 100 T.

Above this field level, we have to employ destructive techniques, explosive method, electromagnetic flux compression, the single turn coil technique, *etc.* The first attempt of the destructive technique was made at Los Alamos in USA and Alzamas-16 in the former Soviet Union (now known as Sarov in Russia) by using chemical explosives to compress the magnetic flux. Generation of very high magnetic fields up to 600 T or even higher has been reported. Although the explosive techniques can generate extremely high magnetic fields, the use of the generated fields is not so easy in ordinary experimental rooms because of the destructive nature. Electromagnetic flux compression or the single-turn coil technique is more suitable for the generation of megagauss fields, as they use capacitor banks as an energy source. At the Institute for Solid State Physics (ISSP) of the University of Tokyo, these two techniques have been developed since 1970 so that they may be applied to solid state experiments. By electromagnetic flux compression, magnetic fields up to 622 T have been generated. This is the highest field which has ever been generated by indoor experiments.

By these techniques mentioned above, fields above 100 T can now be obtained. As $B = 100$ T corresponds to 1 MG (megagauss) in cgs units, magnetic fields higher than 100 T are often called "megagauss fields". Except for the single turn coil technique among different megagauss generating tecniques, samples are also destroyed at every shot. The rise time or the duration time of the field is a few microseconds. Although the time is very short, the present measuring techniques enable us to obtain high accuracy data with good signal to noise ratio in these high fields. Thus generated megagauss fields have been conveniently employed in many different kinds of experiments and many new phenomena have been observed in a variety of substances.

More details of these techniques are given in Chapter 7. The state of art of the magnet technology world-wide is seen in a recently published reference [1].

This book is devoted to describing the fundamental physics of semiconductors in high magnetic fields and to introduce interesting new phenomena actually observed in recently available very high magnetic fields. The topics include transport phenomena, cyclotron resonance and related phenomena, magneto-optical spectroscopy, and diluted magnetic semiconductors. Each chapter includes mostly experimental data obtained in these very high magnetic fields, up to 40–60 T with long pulse magnets, up to ∼260 T with the single turn coil technique, and up to ∼600 T with electromagnetic flux compression, in addition to typical basic data of high field physics of semiconductors. In many of the figures in this book, readers will be impressed by the scale of the magnetic fields usually represented on the horizontal axis. In the following chapters, you will be invited to exploit this fantastic world of high magnetic fields.

2

ELECTRONIC STATES IN HIGH MAGNETIC FIELDS

In this chapter, we overview the electronic states in high magnetic fields for mobile electrons and bound electrons. This provides an important basis for understanding the contents of the subsequent chapters. Readers who are already familiar with these basics can skip this chapter and jump to Chapter 3.

2.1 Free electrons in magnetic fields

First, let us consider the motion of free electrons in magnetic fields and their kinetic energy. The Hamiltonian of free electrons is given by

$$\mathcal{H} = \frac{1}{2m}\boldsymbol{p}^2, \tag{2.1}$$

where the momentum $\boldsymbol{p}$ is represented by

$$\boldsymbol{p} = m\boldsymbol{v}. \tag{2.2}$$

The energy eigenvalue of Eq. (2.1) is

$$\mathcal{E} = \frac{p^2}{2m}. \tag{2.3}$$

When a magnetic field H is applied, free electrons conduct a cyclotron motion. The classical kinetic equation for electrons in an electric field E and a magnetic flux density $B = \mu_0 H$ is

$$m\ddot{\boldsymbol{r}} = -e\boldsymbol{E} - e(\boldsymbol{v} \times \boldsymbol{B}). \tag{2.4}$$

If $E = 0$ and B is parallel to the z-axis, this leads to

$$\begin{aligned} m\frac{d^2x}{dt^2} &= -\mathrm{e}Bv_y, \\ m\frac{d^2y}{dt^2} &= +\mathrm{e}Bv_x. \end{aligned} \tag{2.5}$$

The solution of (2.5) is represented as follows:

$$\begin{cases} x(t) = X + \xi \\ y(t) = Y + \eta, \end{cases}, \tag{2.6}$$

with constants X and Y, and

$$\begin{cases} \xi(t) = r_c \cos\omega_c(t - t_0) \\ \eta(t) = r_c \sin\omega_c(t - t_0) \end{cases} . \tag{2.7}$$

Here, (X, Y) is the center of the cyclotron motion, and (ξ, η) is the relative coordinate around the center.

$$\omega_c = \frac{eB}{m} \tag{2.8}$$

is the frequency of the cyclotron motion called the cyclotron frequency. The kinetic energy of the electron in the plane perpendicular to $\boldsymbol{B}$ is given by

$$\mathcal{E}_\perp = \frac{1}{2}mv^2, \tag{2.9}$$

and the radius of the orbit is

$$r_c = \frac{mv}{eB}. \tag{2.10}$$

In quantum mechanics, the energy of the cyclotron motion is quantized. The magnetic flux density B and electric fields are represented by a vector potential $\boldsymbol{A}$ and scalar potential V as

$$\boldsymbol{B} = \nabla \times \boldsymbol{A} = rot\boldsymbol{A}, \tag{2.11}$$

and

$$\boldsymbol{E} + \frac{\partial \boldsymbol{A}}{\partial t} = -\nabla V. \tag{2.12}$$

It should be noted that $\boldsymbol{A}$ and V cannot be uniquely determined to give the same $\boldsymbol{B}$ and $\boldsymbol{E}$. We can easily show that when $\boldsymbol{A}$ and V are the potentials that satisfy (2.11) and (2.12), $\bar{\boldsymbol{A}}$ and $\bar{V}$ with a gauge transformation

$$\begin{aligned} \bar{\boldsymbol{A}} &= \boldsymbol{A} + \nabla\lambda \\ \bar{V} &= V - \frac{\partial \lambda}{\partial t} \end{aligned} \tag{2.13}$$

also give the same values of $\boldsymbol{B}$ and $\boldsymbol{E}$. This provides unique phenomena in connection with the gauge invariance in a magnetic field.

Putting the Lagrangian as

$$\mathcal{L} = \frac{mv^2}{2} + eV - e(\boldsymbol{v} \cdot \boldsymbol{A}), \tag{2.14}$$

we obtain a kinetic equation

$$\frac{d}{dt}\frac{\partial \mathcal{L}(q_i, \dot{q}_i)}{\partial \dot{q}_i} - \frac{\partial \mathcal{L}(q_i, \dot{q}_i)}{\partial q_i} = 0, \tag{2.15}$$

from which Eq. (2.4) is derived [GT-2]. The Hamiltonian is

$$\mathcal{H}(p_i, q_i) = \sum_i (\dot{q}_i p_i - \mathcal{L}). \tag{2.16}$$

From the kinetic equation

$$p_i = \frac{\partial \mathcal{L}}{\partial \dot{q}_i}, \tag{2.17}$$

we can see that

$$\boldsymbol{p} = m\boldsymbol{v} - e\boldsymbol{A}. \tag{2.18}$$

From (2.14) and (2.16) we obtain the Hamiltonian

$$\mathcal{H} = \frac{1}{2m}(\boldsymbol{p} + \mathrm{e}\boldsymbol{A})^2 - \mathrm{e}V. \tag{2.19}$$

The equations of motion for p_i and q_i are

$$\dot{p}_i = -\frac{\partial \mathcal{H}}{\partial q_i}, \quad \dot{q}_i = \frac{\partial \mathcal{H}}{\partial p_i}. \tag{2.20}$$

Equation (2.18) indicates that the momentum $\boldsymbol{p}$ is not proportional to the velocity $\boldsymbol{v}$ when $\boldsymbol{A}$ is non-zero.

Let us consider now the quantum mechanical meaning of the center coordinates X and Y of the cyclotron motion in the presence of a magnetic field in the z-direction [2]. Suppose a magnetic field is applied parallel to the z-axis. Defining the quasi-momentum by

$$\boldsymbol{P} = \boldsymbol{p} + e\boldsymbol{A}, \tag{2.21}$$

the velocity is given by

$$\frac{P_x}{m} = v_x, \quad \frac{P_y}{m} = v_y. \tag{2.22}$$

The Hamiltonian is

$$\mathcal{H} = \frac{1}{2m}P^2. \tag{2.23}$$

The following commutation relations can be readily proven,

$$[P_x, x] = \hbar/i, \quad [P_y, y] = \hbar/i, \quad [P_x, y] = 0, \quad [P_y, x] = 0. \tag{2.24}$$

It should be noted that P_x and P_y do not commute with each other.

$$[P_x, P_y] = \frac{\hbar^2}{il^2} = -im\hbar\omega_c, \tag{2.25}$$

where

$$l = \sqrt{\frac{\hbar}{eB}}. \tag{2.26}$$

l is called the magnetic length, and it is one of the fundamental physical parameters which has a dimension of length. As will be shown in (2.66), it represents the radius of the cyclotron orbit of the ground state. If we define the coordinates (ξ, η) and (X, Y) by

$$\begin{cases} \xi = \dfrac{1}{m\omega_c} P_y = \dfrac{l^2}{\hbar} P_y, \\ \eta = -\dfrac{1}{m\omega_c} P_x = -\dfrac{l^2}{\hbar} P_x, \end{cases} \tag{2.27}$$

and

$$\begin{cases} x = X + \xi, \\ y = Y + \eta, \end{cases} \tag{2.28}$$

we can see that (X, Y) and (ξ, η) have the same character as the center coordinates and the relative coordinates defined by (2.6) classically. The commutation relations among these coordinates are

$$[\xi, \eta] = -il^2 = -i\frac{\hbar}{eB}, \tag{2.29}$$

$$[X, Y] = il^2 = i\frac{\hbar}{eB}, \tag{2.30}$$

$$[\xi, X] = [\eta, X] = [\xi, Y] = [\eta, X] = 0. \tag{2.31}$$

(2.32)

The Hamiltonian is

$$\mathcal{H} = \frac{\hbar^2}{2ml^4}(\xi^2 + \eta^2). \tag{2.33}$$

It should be noted that this Hamiltonian does not contain the center coordinates X and Y. Thus the energy is independent of the X and Y. However, from the commutation relation above, the center coordinates are quantized and there is an uncertainty relation between their x and y components:

$$\Delta X \cdot \Delta Y = 2\pi l^2. \tag{2.34}$$

As long as (2.11) holds, we can choose any arbitrary $\boldsymbol{A}$ for a given $\boldsymbol{B}$. However, it is convenient in most cases to take either one of the two gauges, the Landau gauge or the symmetric gauge. When $\boldsymbol{B} \parallel \boldsymbol{z}$, the vector potential with the Landau gauge is given by

$$\boldsymbol{A} = [0,\ Bx,\ 0]. \tag{2.35}$$

On the other hand, with the symmetric gauge,

$$\boldsymbol{A} = \left[-\frac{1}{2}By,\ \frac{1}{2}Bx,\ 0\right]. \tag{2.36}$$

In the Landau gauge, the Hamiltonian is

$$\mathcal{H} = \frac{p_x^2}{2m} + \frac{1}{2m}(p_y + eBx)^2 + \frac{p_z^2}{2m}. \tag{2.37}$$

From the Hamiltonian above, the equations of motion are derived:

$$\begin{cases} \dot{p}_x = -\dfrac{eB}{m}(p_y + eBx), & \dot{p}_y = 0, \quad \dot{p}_z = 0. \\ \dot{x} = \dfrac{p_x}{m}, \quad \dot{y} = \dfrac{1}{m}(p_y + eBx), & \dot{z} = \dfrac{p_z}{m}. \end{cases} \tag{2.38}$$

Therefore, p_y and p_z are constants of motion. That is, we can put

$$p_y = \text{const} = p_y^0, \quad p_z = \text{const} = p_z^0. \tag{2.39}$$

The equation of motion for x becomes

$$m\ddot{x} = -\omega_c(p_y^0 + eBx). \tag{2.40}$$

The solution is

$$x = r_c \cos(\omega_c t + \alpha) - \frac{p_y^0}{eB}, \tag{2.41}$$

$$\dot{y} = \frac{1}{m}(p_y^0 + eBx) = \frac{eB}{m} r_c \cos(\omega_c t + \alpha), \tag{2.42}$$

$$y = r_c \sin(\omega_c + \alpha) + Y. \tag{2.43}$$

with a constant Y. When we put

$$X = -\frac{p_y^0}{eB} = -\frac{m\dot{y}}{eB} + x, \tag{2.44}$$

the above equations give

$$(x - X)^2 + (y - Y)^2 = r_c^2, \tag{2.45}$$

$$r_c^2 = \frac{\dot{x}^2 + \dot{y}^2}{\omega_c^2}. \tag{2.46}$$

X and Y can be regarded as the center of the circular motion. Since

$$\begin{aligned} (x - X)^2 &= \frac{\dot{y}^2}{\omega_c^2}, \\ (y - Y)^2 &= r_c^2 - (x - X)^2 = \frac{\dot{x}^2 + \dot{y}^2}{\omega_c^2} - \frac{\dot{y}^2}{\omega_c^2}, \\ &= \frac{\dot{x}^2}{\omega_c^2}, \end{aligned}$$

we obtain

$$Y = y - \frac{\dot{x}}{\omega_c} = y - \frac{p_x}{m\omega_c} = y - \frac{p_x}{eB}. \tag{2.47}$$

The kinetic energy of the circular motion in the plane perpendicular to B is obtained as

$$\mathcal{E}_\perp = \frac{1}{2}m(\dot{x}^2 + \dot{y}^2) = \frac{1}{2}mr_c^2\omega_c^2. \tag{2.48}$$

In the z-direction, the kinetic energy is independent of B,

$$\mathcal{E}_z = \frac{1}{2}m\dot{z}^2 = \frac{(p_z^0)^2}{2m}. \tag{2.49}$$

2.2 Landau levels

Now we consider the quantization of the energy levels [3]. From the Hamiltonian with the Landau gauge,

$$\mathcal{H} = \frac{1}{2m}(\boldsymbol{p}+e\boldsymbol{A})^2 = \frac{p_x^2}{2m} + \frac{(p_y+eBx)^2}{2m} + \frac{p_z^2}{2m}. \tag{2.50}$$

The Schrödinger equation is

$$\mathcal{H}\Psi = \mathcal{E}\Psi. \tag{2.51}$$

Putting

$$\Psi = e^{ik_z z} f(x,y), \tag{2.52}$$

we obtain

$$\mathcal{E} = \mathcal{E}_z + \mathcal{E}_\perp, \qquad \mathcal{E}_z = \frac{\hbar^2}{2m}k_z^2, \tag{2.53}$$

$$\frac{1}{2m}(P_x^2+P_y^2)f(x,y) = \mathcal{E}_\perp f(x,y), \tag{2.54}$$

$$\left[\frac{1}{2m}p_x^2 + \frac{1}{2m}(p_y+eBx)^2\right] f(x,y) = \mathcal{E}_\perp f(x,y). \tag{2.55}$$

We can separate the function as

$$f(x,y) = e^{ik_y y}\varphi(x). \tag{2.56}$$

Then the equation for $\varphi(x)$ is

$$\left[-\frac{\hbar^2}{2m}\frac{\partial^2}{\partial x^2} + \hbar\omega_c k_y x + \frac{m\omega_c^2}{2}x^2\right]\varphi = \left(\mathcal{E}_\perp - \frac{\hbar^2 k_y^2}{2m}\right)\varphi. \tag{2.57}$$

As in (2.44), we define

$$X = -k_y l^2, \tag{2.58}$$

and we put $x = x' - k_y l^2 = x' + X$. Then,

$$\left[-\frac{\hbar^2}{2m}\frac{\partial^2}{\partial x'^2} + \frac{m\omega_c^2}{2}x'^2\right]\varphi(x) = \mathcal{E}\varphi(x). \tag{2.59}$$

The solution of this one-dimensional differential equation is

$$\varphi(x) = \Phi_N\left(\frac{x-X}{l}\right) \qquad N = 0,1,2, \quad \cdots, \tag{2.60}$$

where

$$\Phi_N(x) = C_N \exp\left(-\frac{x^2}{2}\right) H_N(x), \tag{2.61}$$

is a Harmonic oscillator function with the Hermite polynominals,

$$H_N(x) = (-1)^N e^{x^2} \frac{\partial^N}{\partial x^N}(e^{-x^2}). \tag{2.62}$$

The first few lowest order functions are

$$H_0(x) = 1, \quad H_1(x) = 2x, \quad H_2(x) = 4x^2 - 2, \cdots. \tag{2.63}$$

The quantized energy is

$$\mathcal{E}_\perp = \left(N + \frac{1}{2}\right)\hbar\omega_c. \tag{2.64}$$

From (2.44) and (2.47), the radius r_c of the cyclotron motion for the N-th state is given by

$$r_c^2 = (x - X)^2 + (y - Y)^2 = (m\omega_c)^{-2}[p_x^2 + (p_y + eBx)^2] = \frac{2\hbar(N + \frac{1}{2})}{m\omega_c}. \tag{2.65}$$

Thus

$$r_c = (2N + 1)^{\frac{1}{2}} l. \tag{2.66}$$

l is the magnetic length defined by (2.26). We can see that r_c is inversely proportional to $\sqrt{B}$. Note that r_c is the extension of the wave function $\Phi_N(x)$ in (2.61). It is an interesting point that r_c and l are independent of any material parameters like m^*, uniquely determined by B. For $B = 10$ T, the radius of the ground state ($N = 0$) $r_c = l$ is 8.11 nm, and for $B = 100$ T, it is just 2.57 nm. Decreasing the wave function extension to such small values as nanometer size brings about many interesting features, as will be seen in the following chapters. The quantized levels are called Landau levels.

The energy quantization (2.64) can also be obtained in a more elegant way by using the commutation relations (2.32). If we define

$$\begin{cases} a = -\dfrac{1}{\sqrt{2}l}(\eta + i\xi), \\ a^\dagger = -\dfrac{1}{\sqrt{2}l}(\eta - i\xi), \end{cases} \tag{2.67}$$

the Hamiltonian (2.33) is written as

$$\mathcal{H} = \hbar\omega_c\left(a^\dagger a + \frac{1}{2}\right). \tag{2.68}$$

As a and $a^\dagger$ satisfy the following commutation relation,

$$[a, a^\dagger] = 1, \tag{2.69}$$

they are regarded as annihilation and creation operators for the eigenfunctions. From the properties of these operators and the Hamiltonian (2.33), the quantized energy (2.64) is readily derived.

Next, we consider the density of states in the Landau levels. Suppose the electrons are confined in a box with dimensions (L_x, L_y, L_z). From the boundary conditions that the wave function should be zero at the edges of the box, there is a restriction for k_y.

$$k_y = \frac{2\pi}{L_y} \times N, \qquad N = \text{integer}. \tag{2.70}$$

Hence

$$X = -k_y l^2 = -\frac{2\pi l^2}{L_y} \times N = -\frac{h}{eBL_y} \times N. \tag{2.71}$$

Since the center of the orbit should be within the box,

$$0 \leq |X| \leq L_x.$$

$$N \leq \frac{L_x L_y}{2\pi l^2}. \tag{2.72}$$

Thus we can see that each Landau level has a $1/(2\pi l^2)$-fold degeneracy per unit area. This degeneracy is with respect to the center coordinate X. Namely, it is a result of the fact that the all levels with different X have the same energy if the quantum number N is the same.

Next, we need to consider the energy of motion in the z-direction. However, before go into the three-dimensional (3D) case, let us consider first the two-dimensional electron system. As will be shown in Section 2.5, there are electron systems where the electron motion is restricted within the $x-y$ plane and not allowed in z-direction. Such systems are two-dimensional. The number of states of each Landau level in a two-dimensional system is obtained from (2.72) as

$$N_{2D}(B) = \frac{e}{h} B = \frac{1}{2\pi l^2}. \tag{2.73}$$

Using the values of the physical constants, Eq. (2.73) is represented as

$$N_{2D}(B) = 2.41 \times 10^{10} B(T)[\text{cm}^{-2}]. \tag{2.74}$$

The density of states of 2D electrons $\mathcal{D}_{2D}(\mathcal{E})$ is a series of delta functions as shown in Fig. 2.1. Note that $\mathcal{D}_{2D}(\mathcal{E})$ in zero field is a constant value $\mathcal{D}_{2D}^{B=0}(\mathcal{E}) = 2m/(4\pi\hbar^2)$, so that the product of $\mathcal{D}_{2D}^{B=0}(\mathcal{E})$ and the spacing of the Landau levels $\hbar\omega_c$ (the hatched area in Fig. 2.1) is just equal to the number of states in each Landau level $N_{2D}(B)$. It implies that the uniformly distributed density of states is bunched into the delta functions by the magnetic field, the total number being unchanged. $N_{2D}(B)$ is a function of the magnetic field but independent of any material parameters. The coefficient of the field dependence is represented only by fundamental physical parameters such as the effective masses. This is the underlying background of the interesting properties of 2D electron systems such as the quantum Hall effect. The electron motion is fully quantized by the

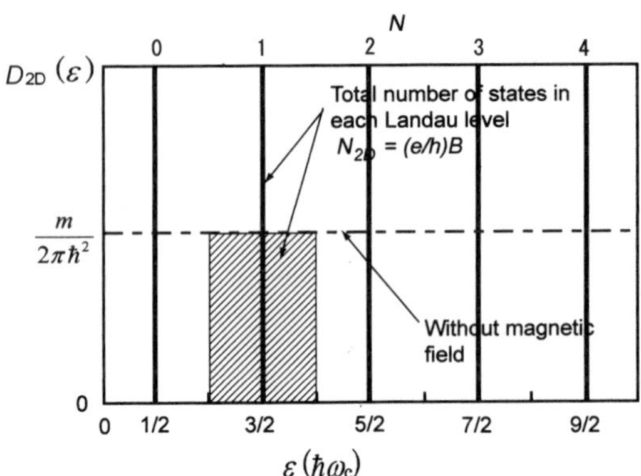

FIG. 2.1. Density of states of Landau levels for a 2D system. The dot-broken line and the solid line show the density of states without and with magnetic fields.

quantum potential and the magnetic field. In a real crystal, however, there is scattering of electrons, so that the Landau levels are broadened. The density of states $\mathcal{D}_{2D}(\mathcal{E})$ is actually not delta-function like and have finite widths. It should be noted that there is the following relation between the broadened density of states and $N_{2D}(B)$:

$$\int_{\text{Landau level}} \mathcal{D}_{2D}(\mathcal{E})d\mathcal{E} = N_{2D}(B). \tag{2.75}$$

In a three-dimensional space, there is another degree of freedom of the motion in the direction parallel to the magnetic field (z-direction), which is not affected by magnetic fields. The kinetic energy in the z-direction is

$$\mathcal{E}_z = \frac{\hbar^2 k_z^2}{2m}. \tag{2.76}$$

There is also a spin Zeeman energy due to spins of electrons. Adding all these energies, the total energy of electrons in magnetic fields is expressed by

$$\mathcal{E} = (N + \frac{1}{2})\hbar\omega_c + \frac{\hbar^2}{2m}k_z^2 \pm g\mu_B H. \tag{2.77}$$

The density of states for the energy of the one-dimensional z-direction motion is

$$\mathcal{D}_z(\mathcal{E})d\mathcal{E}_z = \frac{L_z}{2\pi}dk_z, \tag{2.78}$$

per each spin. Therefore, we obtain

$$\mathcal{D}_z(\mathcal{E})\frac{L_z}{2\pi}\frac{1}{\left(\frac{\partial \mathcal{E}_z}{\partial k_z}\right)} = \frac{L_z}{2\pi}\frac{m}{\hbar^2 k_z}.$$

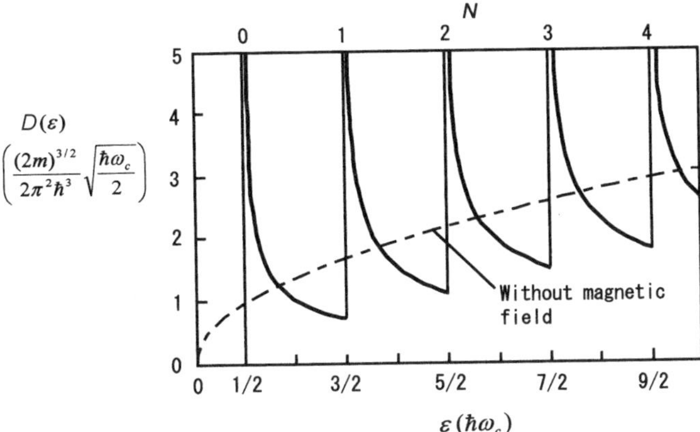

FIG. 2.2. Density of states of Landau levels (3D system). The dot-broken line and the solid line show the density of states without and with magnetic fields.

Since

$$k_z = \frac{\sqrt{2m\mathcal{E}_z}}{\hbar} = \frac{\sqrt{2m[\mathcal{E} - (N + \frac{1}{2})\hbar\omega_c]}}{\hbar},$$

$\mathcal{D}_z(\mathcal{E})$ is derived as

$$\begin{aligned}\mathcal{D}_z(\mathcal{E}) &= \frac{L_z\hbar}{2\pi\sqrt{2m}}\frac{1}{\sqrt{\mathcal{E} - (N + \frac{1}{2})\hbar\omega_c}} \times \frac{m}{\hbar^2} \\ &= \frac{L_z\sqrt{2m}}{4\pi\hbar\sqrt{\mathcal{E} - (N + \frac{1}{2})\hbar\omega_c}}.\end{aligned} \tag{2.79}$$

Adding the degeneracy of the Landau levels shown in Eq. (2.73), the total density of states is represented as

$$\mathcal{D}(\mathcal{E}) = \frac{V}{8\pi^2}\frac{(2m)^{\frac{1}{2}}}{\hbar l^2}\sum_{N=0}^{N_{max}}\frac{1}{\sqrt{\mathcal{E} - (N + \frac{1}{2})\hbar\omega_c}}. \tag{2.80}$$

The function $\mathcal{D}(\mathcal{E})$ is illustrated in Fig. 2.2.

In the Landau gauge, the electron states are characterized by the quantum numbers (N, X, k_z). The energy does not depend on the center coordinate X, although it is a quantum number. On the other hand, Y is left undetermined from the uncertain relation (2.34). This comes from the fact the wave function is plane-wave-like as shown by Eq. (2.56). The angular momentum

$$L_z = (\boldsymbol{r} \times \boldsymbol{p})_z = \frac{1}{2}m\omega_c[r_c^2 - (X^2 + Y^2)] \tag{2.81}$$

is also left uncertain in the Landau gauge.

We now turn to the symmetric gauge. $\boldsymbol{A} = \left[-\frac{1}{2}By, \frac{1}{2}Bx, 0\right]$. The Hamiltonian is

$$\begin{aligned}\mathcal{H} &= \frac{p^2}{2m} + \frac{1}{2}\omega_c(xp_y - yp_x) + \frac{m}{8}\omega_c^2(x^2+y^2) \\ &= \frac{p^2}{2m} + \frac{1}{2}\omega_c L_z + \frac{m}{8}\omega_c^2\rho^2, \qquad \rho^2 = x^2+y^2. \end{aligned} \tag{2.82}$$

Here,

$$L_z = xp_y - yp_x$$

is the z component of the angular momentum $\boldsymbol{L}$. Taking the cylindrical coordinate (ρ, ϕ, z), solution of the Schrödinger equation

$$\mathcal{H}\Psi = \mathcal{E}\Psi \tag{2.83}$$

is

$$\Psi = Ce^{iM\phi}e^{ik_z z}\rho^{|M|}e^{-\frac{\rho^2}{4l^2}}L^{|M|}_{N+|M|}(\rho^2/2l^2). \tag{2.84}$$

Here, L^α_N is the associated Laguerre function represented by

$$L^\alpha_N(x) = \frac{(\alpha+N)_N}{N!}\sum_{r=0}^{N}\frac{[-N]_r}{[\alpha+1]_r}\frac{x^r}{r!} = \sum_{r=0}^{N}\begin{pmatrix}\alpha+N \\ N-r\end{pmatrix}\frac{(-x)^r}{r!}.$$

The eigenenergy in the $x-y$ plane is

$$\mathcal{E}_\perp = \left(N + \frac{M+|M|}{2} + \frac{1}{2}\right)\hbar\omega_c. \tag{2.85}$$

For $M<0$, the energy becomes the same as in Eq. (2.64) being independent of $|M|$. The energy is degenerate with respect to various values of $M<0$. This is in contrast to the case of the Landau gauge where the degeneracy is with respect to X. Since $\hbar M$ is the eigenvalue of L_z, M is a quantum number, in other words, the representation is diagonal for L_z. Also, from

$$X^2 + Y^2 = 2l^2\left[\frac{l^2}{2\hbar^2}(P_x^2+P_y^2) - \frac{1}{\hbar}L_z\right] \tag{2.86}$$

$$= 2l^2\left(N + \frac{|M|-M}{2} + \frac{1}{2}\right), \tag{2.87}$$

we can see that the representation is diagonal for X^2+Y^2 as well. Namely, the centers of the wave function are located on the coaxial circles whose diameters are expressed by (2.87).

The degeneracy of the states with the same energy is calculated as follows. The probability density $\rho|\Psi(\rho)|^2$ expected from Eq. (2.84) takes a maximum at a radius $\rho = (2|M|+1)^{\frac{1}{2}}l$. Let the radius of the system be R. Then the states with M which satisfy $\rho < R$ exist in the system.

$$(2|M|+1)^{\frac{1}{2}}l < R,$$

$$|M| \leq \frac{1}{2}\left(\frac{R}{l}\right)^2.$$

The number of states per unit area is therefore

$$\mathcal{D} = \frac{|M|}{\pi R^2} \leq \frac{1}{2\pi l^2}/(\pi R^2) = \frac{1}{2\pi l^2}, \tag{2.88}$$

which is the same as in the case of the Landau gauge.

It is reasonable that both the Landau gauge and the symmetric gauge result in the same energies and the same density of states, although the quantum numbers are different. The two gauges are converted by the gauge transformation (2.13). Generally, when the gauge is transformed by $\overline{\boldsymbol{A}} = \boldsymbol{A} + \nabla\lambda$, the wave function is transformed by adding a phase factor,

$$\Psi \rightarrow \overline{\Psi} = \exp\left(-\frac{ie\lambda}{\hbar}\right)\Psi. \tag{2.89}$$

For the transformation from the Landau gauge to the symmetric gauge, $[0, Bx, 0] \rightarrow [-\frac{1}{2}By, \frac{1}{2}Bx, 0]$,

$$\lambda = \frac{B}{2}xy.$$

The wave function can be transformed by the following relation:

$$\overline{\Psi} = \exp\left(-i\frac{eB}{2\hbar}xy\right)\Psi = \exp\left(-i\frac{xy}{2l^2}\right)\Psi. \tag{2.90}$$

2.3 Bloch electrons in magnetic fields

2.3.1 *Effective mass approximation*

In the previous section we have considered free electrons for which no potential is exerted except the static electric and magnetic fields. In solid state crystals, conduction electrons or holes in semiconductors suffer from the periodic potential, $U(\boldsymbol{r})$. The periodic potential creates energy bands. This is the difference between free electrons and conduction electrons or holes in semiconductors. Conduction electrons moving through periodic potential in crystals are called Bloch electrons. Their wave function is represented as [4],

$$\varphi_{n\boldsymbol{k}}(x) = e^{i\boldsymbol{k}\cdot\boldsymbol{r}}u_{n\boldsymbol{k}}(\boldsymbol{r}), \tag{2.91}$$

where $\boldsymbol{k}$ is the wave vector, n is the index of the band. $u_{n\boldsymbol{k}}$ is the Bloch function which has a periodicity of the lattice potential and oscillates near the nucleus of each atom in the crystal. The exponential function $e^{i\boldsymbol{k}}$ is responsible for the free electron-like characters of band electrons and the Bloch function part bears the characteristics of each band, reflecting atomic wave functions. In the vicinity of the band extrema, the energy of the Bloch electrons is represented by

$$\mathcal{E} = \frac{\hbar^2 k^2}{2m^*}, \tag{2.92}$$

with the effective mass m^*.

In most cases, the free electron model is applicable to the motion of the Bloch electrons simply by replacing m with m^*. This replacement is justified by the $\boldsymbol{k}\cdot\boldsymbol{p}$ perturbation theory [5,6] and the effective mass theory [7]. First, we consider the Bloch electrons in the absence of magnetic fields. In the $\boldsymbol{k}\cdot\boldsymbol{p}$ perturbation theory, $\varphi_{n\boldsymbol{k}}(x)$ is expanded by a linear combination of

$$\phi_{n\boldsymbol{k}}(x) = \sqrt{\frac{\Omega}{V}} e^{i\boldsymbol{k}\cdot\boldsymbol{r}} u_{n0}(\boldsymbol{r}).$$

the difference between $\varphi_{n\boldsymbol{k}}(x)$ and $\phi_{n\boldsymbol{k}}(x)$ is that the Bloch part functions is at $\boldsymbol{k} = 0$ in the latter. We assume that the functions

$$\phi_{n0}(x) = \sqrt{\frac{\Omega}{V}} u_{n0}(\boldsymbol{r})$$

are the exact solutions at $\boldsymbol{k} = 0$ for the Hamiltonian,

$$\mathcal{H} = \frac{p^2}{2m} + U(\boldsymbol{r}). \tag{2.93}$$

When we operate $\mathcal{H}$ on $\varphi_{n\boldsymbol{k}}(x)$, we obtain

$$\begin{aligned}\mathcal{H}\phi_{n\boldsymbol{k}}(\boldsymbol{r}) &= e^{i\boldsymbol{k}\cdot\boldsymbol{r}} \left(\frac{p^2}{2m} + U(\boldsymbol{r}) + \frac{\hbar}{m}\boldsymbol{k}\cdot\boldsymbol{p} + \frac{\hbar^2 k^2}{2m}\right) u_{n0}(\boldsymbol{r}) \\ &= e^{i\boldsymbol{k}\cdot\boldsymbol{r}} \left(\mathcal{E}_n(\boldsymbol{k}=0) + \frac{\hbar^2 k^2}{2m} + \frac{\hbar}{m}\boldsymbol{k}\cdot\boldsymbol{p}\right) u_{n0}(\boldsymbol{r}).\end{aligned} \tag{2.94}$$

Regarding the term containing $\boldsymbol{k}\cdot\boldsymbol{p}$ as a perturbation, the energy is obtained to the second order perturbation,

$$\begin{aligned}\mathcal{E}_n(\boldsymbol{k}) &= \mathcal{E}_n(0) + \frac{\hbar^2 k^2}{2m} + \sum_{\mu} \left(\frac{\hbar}{m}\right)^2 \frac{\boldsymbol{k}\cdot\boldsymbol{p}_{n\mu}\boldsymbol{p}_{\mu n}\cdot\boldsymbol{k}}{\mathcal{E}_n(0) - \mathcal{E}_\mu(0)} \\ &= \mathcal{E}_n(0) + \frac{\hbar^2}{2}\boldsymbol{k}\cdot\widehat{m}^{-1}\cdot\boldsymbol{k}.\end{aligned} \tag{2.95}$$

Here, the tensor

$$\widehat{m}^{-1} = \frac{1}{m}\boldsymbol{I} + \frac{2}{m}\sum_{\mu} \frac{\boldsymbol{p}_{n\mu}\boldsymbol{p}_{\mu n}}{\mathcal{E}_n(0) - \mathcal{E}_\mu(0)}$$

is the reciprocal effective mass tensor. Such a model treating the term of $\boldsymbol{k}\cdot\boldsymbol{p}$ as a perturbation is called the $\boldsymbol{k}\cdot\boldsymbol{p}$ approximation or $\boldsymbol{k}\cdot\boldsymbol{p}$ theory.

When the system is spherically symmetric,

$$\widehat{m}^{-1} = \begin{pmatrix} \frac{1}{m^*} & 0 & 0 \\ 0 & \frac{1}{m^*} & 0 \\ 0 & 0 & \frac{1}{m^*} \end{pmatrix}.$$

The energy is represented by

$$\mathcal{E}(\boldsymbol{k}) = \mathcal{E}(0) + \frac{\hbar k^2}{2m^*}. \tag{2.96}$$

m^* is called the effective mass. When $\widehat{m}^{-1}$ is diagonal but anisotropic, we have

$$\mathcal{E}(\boldsymbol{k}) = \mathcal{E}(0) + \frac{\hbar^2}{2m_x^*}k_x^2 + \frac{\hbar^2}{2m_y^*}k_y^2 + \frac{\hbar^2}{2m_z^*}k_z^2. \tag{2.97}$$

Next, we consider the case when magnetic fields are present. By the analogy from the free electron case where the substitution , $\boldsymbol{p} \to \boldsymbol{p} + e\boldsymbol{A}$ is made, with a substitution

$$\hbar\boldsymbol{k} \to \boldsymbol{p} + e\boldsymbol{A}, \tag{2.98}$$

for a given energy dispersion $\mathcal{E}_n(\boldsymbol{k})$, we obtain the following Schrödinger equation.

$$\mathcal{E}_n\left(\frac{1}{i}\nabla + \frac{e}{\hbar}\boldsymbol{A}\right) f(\boldsymbol{k}) = \mathcal{E} f(\boldsymbol{r})$$

If $\mathcal{E}$ has a parabolic form,

$$\mathcal{E}(\boldsymbol{k}) = \mathcal{E}(0) + \frac{\hbar k^2}{2m^*}.$$

$$\frac{1}{2m^*}\left(\frac{1}{i}\nabla + \frac{e}{\hbar}\boldsymbol{A}\right)^2 f(\boldsymbol{k}) = \mathcal{E} f(\boldsymbol{r}).$$

It is known that such a substitution is appropriate in most cases. The substitution (2.98) is called the Landau-Peierls substitution.

The validity of the Landau-Peierls substitution is shown by Luttinger and Kohn as the effective mass theory [7]. In the effective mass theory, the wave function φ is expanded as a series of the functions

$$\phi_{n\boldsymbol{k}}(\boldsymbol{r}) = \sqrt{\frac{\Omega}{V}} u_{n0}(\boldsymbol{r}) e^{i\boldsymbol{k}\cdot\boldsymbol{r}}$$

which form an orthonormal set,

$$\varphi = \sum_{n'} \int d\boldsymbol{k}' A_{n'}(\boldsymbol{k}') \phi_{n'\boldsymbol{k}'}.$$

The function $\phi_{n\boldsymbol{k}}$ has an orthogonal property

$$(\phi_{n\boldsymbol{k}}, \phi'_{n'\boldsymbol{k}'}) = \delta(\boldsymbol{k}' - \boldsymbol{k})\delta_{nn'}.$$

The Hamiltonian in the presence of the magnetic field is

$$\mathcal{H} = \frac{1}{2m}(\boldsymbol{p} + e\boldsymbol{A})^2 + U(\boldsymbol{r}), \tag{2.99}$$

and

$$\mathcal{H}\varphi = \mathcal{E}\varphi$$

is an equation to be solved. Interband interaction terms $< \phi_{n'\boldsymbol{k}'}|\mathcal{H}|\phi_{n\boldsymbol{k}} >$ can be taken into the effective mass, and it can be shown that the wave function can have a form

$$\varphi(\boldsymbol{r}) = \sum_n F_n(\boldsymbol{r})\phi_{n0}(\boldsymbol{r}). \tag{2.100}$$

Here, the function $F_n(\boldsymbol{r})$ is an eigenfunction of the following Schrödinger equation for the n-th band:

$$\mathcal{E}_n\left(\frac{1}{i}\nabla + \frac{e}{\hbar}\boldsymbol{A}\right)F_n(\boldsymbol{r}) = \mathcal{E}F_n(\boldsymbol{r}). \tag{2.101}$$

The function $F_n(\boldsymbol{r})$ represents the effect of the magnetic field and is called the envelope function. When $\mathcal{E}$ has a parabolic dispersion, for example, such as

$$\mathcal{E} = \frac{\hbar^2 k^2}{2m^*},$$

then

$$\varphi(\boldsymbol{r}) \quad = \quad f(\boldsymbol{r}) \quad \times \quad u_n(\boldsymbol{r}). \tag{2.102}$$

In the Landau gauge, $f(\boldsymbol{r})$ is an eigenfunction of the same expression as (2.56) with $m \to m^*$, so that it is a harmonic oscillator function. The energy is given

$$\mathcal{E} = \left(N + \frac{1}{2}\right)\hbar\omega_c, \qquad \omega_c = \frac{eB}{m^*}. \tag{2.103}$$

Thus we see that the Landau levels in the conduction band, which has a parabolic energy dispersion, have the same energy as free electrons except that the free electron mass is replaced by the effective mass. The wave function is a product of the slowly varying envelope function standing for the cyclotron motion and the rapidly oscillating Bloch part standing for the periodic motion in the crystal.

If the energy band is not entirely parabolic, or if a few bands are degenerate, the solutions are more complicated, but we can obtain the Landau levels by properly treating the dispersion or the degeneracy, as shown in later sections.

The essential point of the effective mass approximation is that the effect of the periodic potential of the crystal lattice is squeezed into the effective mass, treating the Schrödinger equation without the periodic potential. It should be

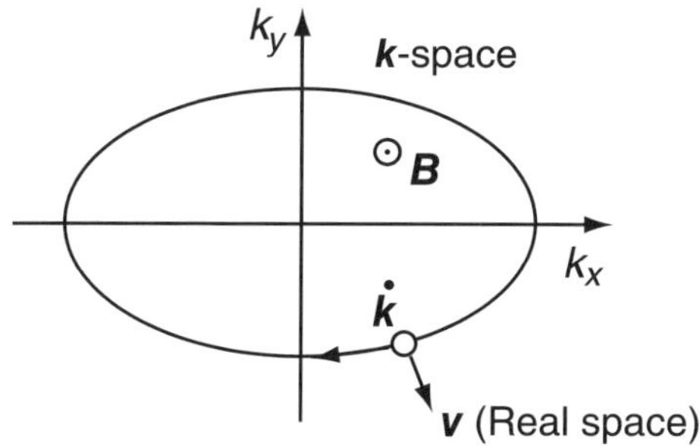

FIG. 2.3. Motion of a Bloch electron in (k_x, k_y) plane.

noted that the effective mass approximation [7] is valid not only for the electronic states in the presence of magnetic fields, but also for those in the presence of electric potential. Therefore, it can be applicable for treating impurity states, or electronic states in the presence of artificially introduced quantum potentials in heterostructures, as long as the extension of the envelope wave function is sufficiently larger than the period of the lattice potential.

2.3.2 *Semiclassical treatment*

The quantization of the energy levels can also be derived by a semiclassical approach [3]. The motion of Bloch electrons with a wave vector $\boldsymbol{k}$ in an external force $\boldsymbol{F}$ is given by an acceleration theorem,

$$\hbar \frac{d\boldsymbol{k}}{dt} = \boldsymbol{F}. \tag{2.104}$$

If the external force is the Lorentz force, it is

$$\hbar \frac{d\boldsymbol{k}}{dt} = -e\boldsymbol{v} \times \boldsymbol{B}. \tag{2.105}$$

Let the magnetic field direction be along the z-axis. Then the motion of an electron with an energy $\mathcal{E}$ is along a periodic orbit on the equi-energy surface in the (k_x, k_y) plane as shown in Fig. 2.3, since in the magnetic field, electrons gain no energy. Since $\dot{\boldsymbol{k}}$ and $\boldsymbol{v}$ are perpendicular to each other, the electron motion in real space is also periodic and obtained by rotating the vector in k-space by $\pi/2$ and multiplying $eB/\hbar$.

The quantum condition is

$$\oint \boldsymbol{p} \cdot d\boldsymbol{q} = 2\pi\hbar(N + \gamma), \tag{2.106}$$

where N is an integer and γ is a phase factor which is 1/2 for free electrons. If we denote the position vector of electrons in the (x, y) plane in real space as $\boldsymbol{\rho}$, by integrating (2.105) we obtain

$$\hbar \boldsymbol{k} = -e\boldsymbol{\rho} \times \boldsymbol{B}, \tag{2.107}$$

apart from a constant. By substituting $\hbar \boldsymbol{k}$ with $(\boldsymbol{p} + e\boldsymbol{A})$, we obtain

$$\oint \boldsymbol{p} \cdot d\boldsymbol{q} = \oint (\hbar \boldsymbol{k} - e\boldsymbol{A}) \cdot d\boldsymbol{q} = -e \oint (\boldsymbol{\rho} \times \boldsymbol{B} + \boldsymbol{A}) \cdot d\boldsymbol{q}. \tag{2.108}$$

Here the integral of the first term is

$$\oint \boldsymbol{\rho} \times \boldsymbol{B} \cdot d\boldsymbol{q} = -\boldsymbol{B} \cdot \oint \boldsymbol{\rho} \times d\boldsymbol{\rho} = -2\Phi, \tag{2.109}$$

where Φ is the flux contained in the orbit, and that of the second term is

$$\oint \boldsymbol{A} \cdot d\boldsymbol{q} = \int rot \boldsymbol{A} \cdot d\boldsymbol{\sigma} = \int \boldsymbol{B} \cdot d\boldsymbol{\sigma} = \Phi, \tag{2.110}$$

where $d\boldsymbol{\sigma}$ is the area element of real space. The integration constant involved in (2.107) diminishes in the above integration. Thus we obtain

$$\oint \boldsymbol{p} \cdot d\boldsymbol{q} = \frac{e}{\hbar}\Phi. \tag{2.111}$$

From (2.106) and (2.111),

$$\Phi = \frac{2\pi\hbar}{e}(N + \gamma) = (N + \gamma)\Phi_0, \tag{2.112}$$

where $\Phi_0 = h/e$ is the unit flux. Putting the area of the orbit in $\boldsymbol{k}$ space S_k, the quantization condition (2.112) by multiplying $(eB/\hbar)^2$.

$$BS_k = \frac{h}{e}(N + \gamma)(eB/\hbar)^2, \tag{2.113}$$

Thus

$$S_k = (N + \gamma)\frac{2\pi eB}{\hbar}, \tag{2.114}$$

Such semiclassical treatment of the electron motion is useful when considering the electron dynamics of the Fermi surface in a magnetic field.

2.4 Landau levels in semiconductors

2.4.1 *Two-band model*

The simplest approximation to treat energy bands and Landau levels in actual semiconductors within the framework of the $\boldsymbol{k} \cdot \boldsymbol{p}$ approximation is the two-band model. In this model we consider only a single conduction band and a single valence band which are faced with each other with a relatively small band gap, as shown in Fig. 2.4. In such a case, the effect of the other bands is considered

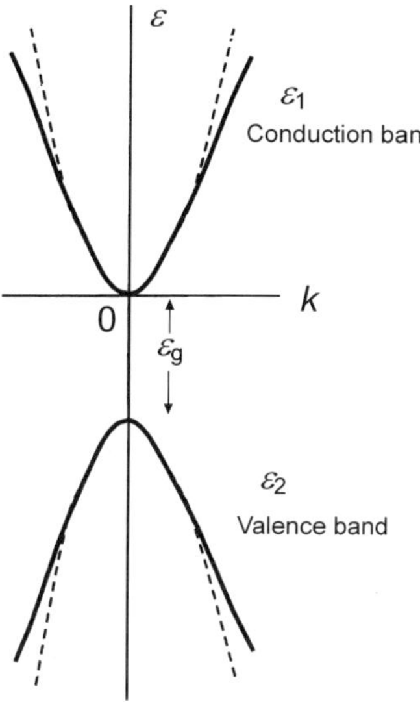

FIG. 2.4. Energy dispersion in the two-band model. Broken lines indicate the dispersion of parabolic bands.

to be negligibly small in comparison to the mixing of the two bands. The matrix elements for the $\boldsymbol{k} \cdot \boldsymbol{p}$ Hamiltonian $\mathcal{H}$ are written as

$$(\mathcal{H}) = \begin{pmatrix} \mathcal{E}_1 & \mathcal{H}_{12} \\ \mathcal{H}_{21} & \mathcal{E}_2 \end{pmatrix} = \begin{pmatrix} \frac{1}{2m}\hbar^2 k^2 & \frac{\hbar}{m} p_{12} \cdot k \\ \frac{\hbar}{m} p_{12}^* \cdot k & -\frac{1}{2m}\hbar^2 k^2 - \mathcal{E}_g \end{pmatrix}. \tag{2.115}$$

Within the approximation $\hbar^2 k^2/2m \ll \mathcal{E}_g$, the eigenenergies of the two bands are obtained as

$$\mathcal{E}_{1,2} = -\frac{\mathcal{E}_g}{2} \pm \frac{1}{2}\sqrt{\mathcal{E}_g^2 + \frac{\hbar^2}{m^2} p_{12}^2 k^2}. \tag{2.116}$$

For small k near the band edge, we obtain the following approximate expression by expanding the square root function,

$$\begin{aligned} \mathcal{E}_1 &\simeq \tfrac{1}{2}\left[-\mathcal{E}_g + \mathcal{E}_g\left(1 + \frac{\hbar^2 p_{12}^2}{2m^2 \mathcal{E}_g^2} k^2\right)\right] \\ &\simeq \frac{\hbar^2 p_{12}^2}{4m^2 \mathcal{E}_g^2} k^2. \end{aligned} \tag{2.117}$$

If we define the band edge mass $m^*(0)$ by

$$\mathcal{E}_1 \simeq \frac{\hbar^2}{2m^*(0)}k^2, \tag{2.118}$$

for small k, we get

$$\frac{1}{m^*(0)} \simeq \frac{\hbar^2 p_{12}^2}{2m^2\mathcal{E}_g}. \tag{2.119}$$

Thus for a small k, the energy dispersion is parabolic with respect to k just like free electrons with the effective mass $m^*(0)$. For an arbitrary k, replacing a parameter involving p_{12}^2 by $m^*(0)$ using (2.119), we obtain

$$\mathcal{E}_{1,2} = \frac{1}{2}\left[-\mathcal{E}_g \pm \sqrt{\mathcal{E}_g^2 + \frac{2\hbar^2\mathcal{E}_g}{m^*(0)}k^2}\right], \tag{2.120}$$

for the conduction band and the valence band. Equation (2.120) indicates that as k increases, the energy deviates from the parabolic dispersion, and the effective mass appears to become heavier. Such a deviation from the parabolic energy dispersion of the energy band is called non-parabolicity. When magnetic fields are applied, the term $\hbar^2k^2/2m^*(0)$ is replaced by the quantized form $(N+\frac{1}{2})\hbar\omega_c(0)$ and the following expression is obtained for the Landau levels,

$$\mathcal{E}_1(N) = \frac{1}{2}\left[-\mathcal{E}_g + \sqrt{\mathcal{E}_g^2 + 4\mathcal{E}_g\left(N+\frac{1}{2}\right)\hbar\omega_c(0)}\right]. \tag{2.121}$$

We can see that, for a small energy or low magnetic fields, the energy of the Landau levels is

$$\mathcal{E}_1(N) \simeq \left(N+\frac{1}{2}\right)\frac{\hbar eB}{m^*(0)} \equiv \left(N+\frac{1}{2}\right)\hbar\omega_c(0), \tag{2.122}$$

with the band edge effective mass $m^*(0)$, but as the energy is increased, the energy deviates from the linear relation with B and the effective mass becomes larger. It should be noted that the relation (2.121) is rewritten as

$$\mathcal{E}_1(N)\left(1+\frac{\mathcal{E}_1(N)}{\mathcal{E}_g}\right) = \left(N+\frac{1}{2}\right)\hbar\omega_c(0). \tag{2.123}$$

This is the equation to express the non-parabolicity of the energy band in terms of the two-band model.

The energy of the other band $\mathcal{E}_1(N)$ has of course a symmetric form. The two-band model is best applicable to PbTe, PbSe or Bi, which have a narrow energy gap between non-degenerate conduction and valence bands.

2.4.2 *Conduction bands*

Although the two-band model sometimes works very well in narrow gap semiconductors, it is too simple in most cases. Generally, it is necessary to take account

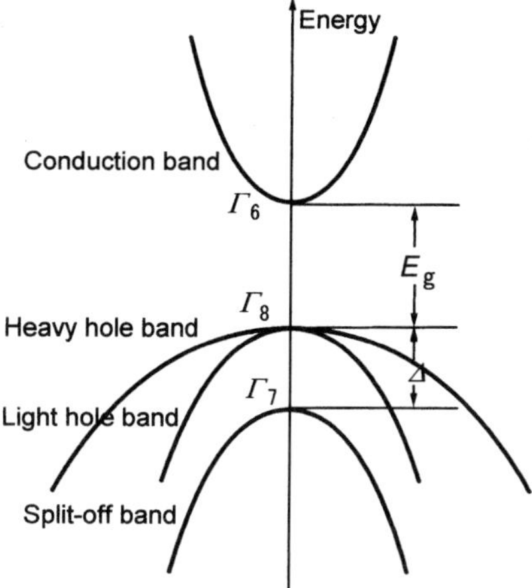

FIG. 2.5. Energy band structure in diamond, Ge, Si and zinc blende crystals.

of the interaction with a number of other bands nearby. In diamond, Ge, Si with the diamond structure, and III-V or II-VI compounds with the zinc blende structure, the valence band has a degeneracy as shown in Fig. 2.5.

The simple two-band model can be improved by taking account of the entire valence bands and their degeneracy for the interband matrix element for $\boldsymbol{k} \cdot \boldsymbol{p}$. Such an approach was first made by Roth, Lax, and Zwerdling [8], and Lax *et al.* [9] based upon the Kane model [5, 6]. By taking account of the interband matrix element of the momentum operator including the spin-orbit interaction

$$\boldsymbol{\pi} = \frac{\boldsymbol{p}}{m} - \frac{\mu_B}{m}\nabla V \times \boldsymbol{S}, \tag{2.124}$$

Lax *et al.* obtained the energy of the n-th Landau level of the conduction band as [9]

$$\begin{aligned}\mathcal{E}_N^{\pm} &= \left(N + \frac{1}{2}\right)\hbar e B \\ &\times \frac{1}{m^*(0)}\frac{\mathcal{E}_g(\mathcal{E}_g+\Delta)}{3\mathcal{E}_g+2\Delta}\left(\frac{2}{\mathcal{E}_N^{\pm}+\mathcal{E}_g} + \frac{1}{\mathcal{E}_N^{\pm}+\mathcal{E}_g+\Delta}\right) \\ &\pm \frac{1}{2}g^*(0)\mu_B B\frac{\mathcal{E}_g(\mathcal{E}_g+\Delta)}{\Delta}\left(\frac{1}{\mathcal{E}_N^{\pm}+\mathcal{E}_g} - \frac{1}{\mathcal{E}_N^{\pm}+\mathcal{E}_g+\Delta}\right),\end{aligned} \tag{2.125}$$

where $\mathcal{E}_g$ is the band gap, Δ is the spin-orbit splitting (the energy gap between the Γ_8 and the Γ_7 bands), $m^*(0)$ and $g^*(0)$ are the effective mass and the g-factor at the band edge ($\mathcal{E} = 0$). If we define the energy dependent effective mass $m^*(\mathcal{E})$ and the g factor $g^*(\mathcal{E})$ as

$$\frac{1}{m^*(\mathcal{E})} = \frac{1}{m^*(0)} \frac{\mathcal{E}_g(\mathcal{E}_g + \Delta)}{3\mathcal{E}_g + 2\Delta} \left(\frac{2}{\mathcal{E} + \mathcal{E}_g} + \frac{1}{\mathcal{E} + \mathcal{E}_g + \Delta} \right), \tag{2.126}$$

and

$$g^*(\mathcal{E}) = g^*(0) \frac{\mathcal{E}_g(\mathcal{E}_g + \Delta)}{\Delta} \left(\frac{1}{\mathcal{E} + \mathcal{E}_g} - \frac{1}{\mathcal{E} + \mathcal{E}_g + \Delta} \right), \tag{2.127}$$

then (2.125) can be written as follows [9]:

$$\mathcal{E} = \left(N + \frac{1}{2} \right) \frac{\hbar e B}{m^*(\mathcal{E})} \pm \frac{1}{2} g^*(\mathcal{E}) \mu_B B. \tag{2.128}$$

Johnson and Dickey also obtained a similar formula [10].

Regarding the band edge values of the effective mass and the g-factor, Roth *et al.* derived expressions as

$$\frac{1}{m^*(0)} = \frac{1}{m} + \frac{2}{m^2} \frac{3\mathcal{E}_g + 2\Delta}{3\mathcal{E}_g(\mathcal{E}_g + \Delta)} |< S|p_x|X >|^2, \tag{2.129}$$

$$g^*(0) = 2 - \frac{4}{m} \frac{\Delta}{3\mathcal{E}_g(\mathcal{E}_g + \Delta)} |< S|p_x|X >|^2, \tag{2.130}$$

where $< S|p_x|X >$ is the momentum matrix element between the S-like and X-like wave functions constituting the conduction and the valence bands. From (2.129) and (2.130), we can obtain a relation between $g^*(\mathcal{E})$ and $m^*(\mathcal{E})$ [8];

$$\frac{g^*(0) - 2}{m/m^*(0) - 1} = - \left(\frac{3}{2} \frac{\mathcal{E}_g}{\Delta} + 1 \right)^{-1}. \tag{2.131}$$

This relation is called Roth's relation, and demonstrates that both the effective mass and the g factor are greatly modified by the band mixing with the spin-orbit interaction in a similar manner. As can been seen in Table 2.1, Roth's relation holds well in actual semiconductor crystals with the zinc blende structure.

Usually, the band edge effective mass $m^*(0)$ is considerably smaller than the free electron mass m in most III-V and II-VI compounds. As the energy increases, $m^*(\mathcal{E})$ increases according to Eq. (2.126). The band edge effective g-factor $g^*(0)$ is negative except for some crystals as InP. As the energy increases, $g^*(\mathcal{E})$ shifts towards a positive value according to Eq. (2.127). These phenomena can be observed in high magnetic fields. The values of $m^*(\mathcal{E})$ and $g^*(\mathcal{E})$ gradually approach the free electron values m and 2, respectively, as the energy or magnetic field is increased.

2.4.3 *Valence bands*

The Landau levels in the valence band of Si, Ge and III-V are complicated because of the degeneracy of the uppermost band Γ_8, and the spin split off band Γ_7

Table 2.1 Comparison of the observed g-factor with that obtained from Roth's formula

Material	$\mathcal{E}_g$(eV)	Δ(eV)	$m^*(0)$	$g^*(0)_{calc}$	g^*_{obs}
CdTe	1.60	0.927	0.0963	–0.89	–1.59
GaAs	1.53	0.341	0.067	–0.07	–0.44
GaSb	0.814	0.752	0.041	–7.31	–9,25
InP	1.423	0.102	0.081	1.48	1.26
InAs	0.44	0.38	0.024	–13.2	–14.8
InSb	0.237	0.81	0.0139	–48.3	–51.3

as shown in Fig. 2.5. We have to take account of the matrix elements of the $\boldsymbol{k} \cdot \boldsymbol{p}$ perturbation between these bands. The Γ_8 band is four-fold degenerate including spins at $\boldsymbol{k} = 0$. Including spins, these valence bands consist of six bands altogether, which originate from the p-like Bloch functions X, Y, and Z with the same symmetry as x, y, and z, respectively. Considering the spin-orbit interaction and the symmetry of the zinc blende structure, the basis wave functions of the valence bands are represented as

$$
\begin{aligned}
\phi^{(\frac{3}{2})}_{\frac{3}{2}} &= \left|\frac{3}{2}, \frac{3}{2}\right\rangle = \frac{1}{\sqrt{2}}(X+iY)\alpha,\\
\phi^{(\frac{3}{2})}_{\frac{1}{2}} &= \left|\frac{3}{2}, \frac{1}{2}\right\rangle = \frac{i}{\sqrt{6}}[(X+iY)\beta - 2Z\alpha],\\
\phi^{(\frac{3}{2})}_{-\frac{1}{2}} &= \left|\frac{3}{2}, -\frac{1}{2}\right\rangle = \frac{1}{\sqrt{6}}[(X-iY)\alpha + 2Z\beta],\\
\phi^{(\frac{3}{2})}_{-\frac{3}{2}} &= \left|\frac{3}{2}, -\frac{3}{2}\right\rangle = \frac{i}{\sqrt{2}}(X-iY)\beta,\\
\phi^{(\frac{1}{2})}_{\frac{1}{2}} &= \left|\frac{1}{2}, \frac{1}{2}\right\rangle = \frac{1}{\sqrt{3}}[(X+iY)\beta + Z\alpha],\\
\phi^{(\frac{1}{2})}_{-\frac{1}{2}} &= \left|\frac{1}{2}, -\frac{1}{2}\right\rangle = \frac{i}{\sqrt{3}}[-(X-iY)\alpha + Z\beta].
\end{aligned}
\tag{2.132}
$$

Here, α and β denote up ($\uparrow$) and down ($\downarrow$) spin functions, respectively. The first four functions of (2.132) represent the wave functions of the uppermost Γ_8 band, and have angular momentum $J = \frac{3}{2}$ whose z components are $m_J = \frac{3}{2}, \frac{1}{2}, -\frac{1}{2}$, or $-\frac{3}{2}$. The last two functions represent the split-off band Γ_7 whose angular momentum is $J = \frac{1}{2}$ with $m_J = \frac{1}{2}$ and $-\frac{1}{2}$. When the spin-orbit splitting energy Δ is relatively large in comparison to the energy range of consideration, the energy levels can be obtained by taking account only of the first four bands (Γ_8 band). The $\boldsymbol{k} \cdot \boldsymbol{p}$ matrix on the basis of these wave functions is [7]

$$D_v = \begin{pmatrix} \frac{1}{2}P & L & M & 0 \\ L^* & \frac{1}{6}P + \frac{2}{3}Q & 0 & M \\ M^* & 0 & \frac{1}{6}P + \frac{2}{3}Q & -L \\ 0 & M^* & -L^* & \frac{1}{2}P \end{pmatrix}, \tag{2.133}$$

where the matrix elements are

$$\begin{aligned} P &= \frac{\hbar^2}{2m}[(A+B)(k_x^2+k_y^2)+2Bk_z^2], \\ Q &= \frac{\hbar^2}{2m}[B(k_x^2+k_y^2)+Ak_z^2], \\ L &= -\frac{i}{\sqrt{3}}\frac{\hbar^2}{2m}C(k_x-ik_y)k_z, \\ M &= \frac{1}{\sqrt{12}}\frac{\hbar^2}{2m}[(A-B)(k_x^2-k_y^2)-2iCk_xk_y]. \end{aligned} \tag{2.134}$$

In the absence of a magnetic field, diagonalizing (2.133) we can obtain the dispersion of the heavy hole band and the light hole band which are degenerate at $\boldsymbol{k} = 0$ as shown in Fig. 2.5. When a magnetic field is applied, the Landau levels are obtained by replacing $k_x \to (1/i)\nabla_x - eA_x$, $k_y \to (1/i)\nabla_y - eA_y$, and solve a four-dimensional Schrödinger equation

$$D_v F = \mathcal{E} F. \tag{2.135}$$

Luttinger derived a more easily accessible expression for the $\boldsymbol{k}\cdot\boldsymbol{p}$ Hamiltonian for the uppermost valence bands from a symmetry consideration [11]. Here cubic symmetry is assumed for Ge and Si and the set of parameters is reorganized from (P, Q, L, M) or (A, B, C) to $(\gamma_1, \gamma_2, \gamma_3, \kappa, q)$. The new set of parameters is called Luttinger parameters, and the formulation using these parameters is called the Luttinger model of the valence bands. Here γ_1, γ_2, and γ_3 represent effective masses and κ and q represent spin parameters determining spin splitting. Usually, q is negligibly small. Using these parameters, the Hamiltonian for the uppermost Γ_8 valence band with four-fold degeneracy for $B \parallel < 111 >$ is given by

$$D_v = \begin{pmatrix} D_{11} & -g_1 a^2 & -g_2 a^{\dagger 2} & -q/\sqrt{2} \\ -g_1 a^{\dagger 2} & D_{22} & 0 & g_2 a^{\dagger 2} \\ -g_2 a^2 & 0 & D_{33} & -g_1 a^2 \\ -q/\sqrt{2} & g_2 a^2 & -g_1 a^{\dagger 2} & D_{44} \end{pmatrix}. \tag{2.136}$$

Here

$$\begin{aligned} D_{11} &= (\gamma_1+\gamma_3)(a^\dagger a + \frac{1}{2}) + \frac{3}{2}\kappa + \frac{23q}{8}, \\ D_{22} &= (\gamma_1-\gamma_3)(a^\dagger a + \frac{5}{2}) - \frac{1}{2}\kappa - \frac{13q}{8}, \\ D_{33} &= (\gamma_1-\gamma_3)(a^\dagger a - \frac{3}{2}) + \frac{1}{2}\kappa + \frac{13q}{8}, \end{aligned}$$

$$D_{44} = (\gamma_1 + \gamma_3)(a^\dagger a + \frac{1}{2}) - \frac{3}{2}\kappa - \frac{23q}{8},$$
$$g_1 = \frac{(2\gamma_3 + \gamma_2)}{\sqrt{3}},$$
$$g_2 = \sqrt{\frac{2}{3}}(\gamma_3 - \gamma_2). \qquad (2.137)$$

$a^\dagger$ and a are the creation and annihilation operators as in (2.67). The solution gives the energy of Landau levels whose spacings are not uniform for small quantum number N. The calculation can only be done numerically, but if γ_2 and γ_3 are nearly the same, g_2 can be neglected and we can decompose the matrix into two (2×2) matrices.

$$D_v = \begin{pmatrix} D_a & 0 \\ 0 & D_b \end{pmatrix}, \qquad (2.138)$$

with

$$D_a = \hbar\omega_c \begin{pmatrix} (\gamma_1 + \bar{\gamma})(N + \frac{1}{2}) + \frac{3}{2}\kappa & -\sqrt{3}\bar{\gamma}a^2 \\ -\sqrt{3}\bar{\gamma}a^{\dagger 2} & (\gamma_1 - \bar{\gamma})(N + \frac{1}{2}) - \frac{1}{2}\kappa \end{pmatrix}, \qquad (2.139)$$

and

$$D_b = \hbar\omega_c \begin{pmatrix} (\gamma_1 - \bar{\gamma})(N + \frac{1}{2}) + \frac{1}{2}\kappa & -\sqrt{3}\bar{\gamma}a^2 \\ -\sqrt{3}\bar{\gamma}a^{\dagger 2} & (\gamma_1 + \bar{\gamma})(N + \frac{1}{2}) - \frac{3}{2}\kappa \end{pmatrix}. \qquad (2.140)$$

Here we put $\bar{\gamma} = \gamma_2 = \gamma_3$ and neglect q. $\gamma_2 - \gamma_3$ represents the anisotropy in the cubic symmetry and the approximation to set them zero is called the spherical approximation. The two (2×2) matrices are easily soluble using the following functions.

$$\psi_a = \begin{pmatrix} a_1 u_{N-2} \\ a_2 u_N \end{pmatrix}, \qquad (2.141)$$

$$\psi_b = \begin{pmatrix} b_1 u_{N-2} \\ b_2 u_N \end{pmatrix}, \qquad (2.142)$$

The energies for the a-set are;

$$\mathcal{E}_a^{\pm}(N) = \hbar\omega_c \Big[\gamma_1 N - (\frac{1}{2}\gamma_1 + \bar{\gamma}\frac{1}{2}\kappa)$$
$$\pm \{[\gamma N - (\gamma_1 - \kappa + \frac{1}{2}\bar{\gamma})]^2 + 3\bar{\gamma}^2 N(N-1)\}^2\Big], \quad (N \geq 2) \qquad (2.143)$$

$$\mathcal{E}_a(0) = \hbar\omega_c \left[\frac{1}{2}(\gamma_1 - \bar{\gamma}) - \frac{1}{2}\kappa\right],$$

$$\mathcal{E}_a(1) = \hbar\omega_c \left[\frac{3}{2}(\gamma_1 - \bar{\gamma}) - \frac{1}{2}\kappa\right], \tag{2.144}$$

and for the b-set,

$$\mathcal{E}_b^{\pm}(N) = \hbar\omega_c \Bigg[\gamma_1 N - (\frac{1}{2}\gamma_1 + \bar{\gamma}\frac{1}{2}\kappa) \pm \{[\bar{\gamma}N - (\gamma_1 - \kappa + \frac{1}{2}\bar{\gamma})]^2 + 3\bar{\gamma}^2 N(N-1)\}^2\Bigg], \quad (N \geq 2) \tag{2.145}$$

$$\begin{aligned} \mathcal{E}_b(0) &= \hbar\omega_c \left[\frac{1}{2}(\gamma_1 + \bar{\gamma}) - \frac{3}{2}\kappa\right], \\ \mathcal{E}_b(1) &= \hbar\omega_c \left[\frac{3}{2}(\gamma_1 + \bar{\gamma}) - \frac{3}{2}\kappa\right]. \end{aligned} \tag{2.146}$$

The Luttinger model has been used as a standard model to calculate the valence band of diamond and other zinc blende-type crystals. The energies of the Landau levels calculated by the Luttinger model are not uniformly spaced and the gaps between adjacent levels are irregular depending on N for small quantum number N. This effect is called the quantum effect in the Landau level structure of the valence band. Cyclotron resonance in the valence band at low temperatures thus shows a complicated spectrum as will be mentioned in Chapter 4. However, as N increases, the spacing between levels tends to be uniform, and the ordinary Landau levels can be well defined. At high temperatures, the quantum effect becomes less important, as most holes are populated in these uniformly spaced Landau levels with large N. In such a large N limit, the effective masses of light holes m_l^* and those of heavy holes m_h^* are well defined and obtained as

$$\left.\begin{matrix} m/m_l \\ m/m_h \end{matrix}\right) = \gamma_1 \pm (\gamma'^2 + 3\gamma''^2)^{1/2}, \tag{2.147}$$

where

$$\begin{aligned} \gamma' &= \gamma_3 + (\gamma_2 - \gamma_3)[3(\cos^2\theta - 1)/2]^2, \\ \gamma'' &= \frac{2}{3}\gamma_3 + \frac{1}{3}\gamma_2 + \frac{1}{6}(\gamma_2 - \gamma_3)[3(\cos^2\theta - 1)/2]^2. \end{aligned} \tag{2.148}$$

Lawaetz calculated the Luttinger parameters in most of the typical crystals from the fundamental parameters such as band gaps, ionicity and lattice constant [12].

2.4.4 *Pidgeon-Brown model*

Although the Luttinger model mentioned above gives a means to calculate a fairly accurate Landau level structure, it is not sufficient for narrow gap semiconductors where the non-parabolicity is significant because of the large band mixing between the conduction band and the valence bands. In the Luttinger model, the effect of the conduction band is involved only in the γ parameters treating it in the same way as for remote bands. Pidgeon and Brown calculated the Landau levels including the Γ_6 conduction band and the Γ_8 valence bands as well as the split-off Γ_7 valence band simultaneously in the $\boldsymbol{k}\cdot\boldsymbol{p}$ Hamiltonian [13]. Including the spin states, it becomes an (8×8) matrix. As in (2.132), the basis function was taken as follows,

$$\begin{aligned}
\phi_{1,0} &= |S\uparrow\rangle \\
\phi_{2,0} &= |S\downarrow\rangle \\
\phi_{3,0} &= \frac{1}{\sqrt{2}}\,|(X+iY)\uparrow\rangle \\
\phi_{4,0} &= \frac{i}{\sqrt{2}}\,|(X-iY)\downarrow\rangle \\
\phi_{5,0} &= \frac{1}{\sqrt{6}}\,|[(X-iY)\uparrow +2Z\downarrow]\rangle \\
\phi_{6,0} &= \frac{i}{\sqrt{6}}\,|[(X+iY)\downarrow -2Z\uparrow]\rangle \\
\phi_{7,0} &= \frac{i}{\sqrt{3}}|\,[-(X-iY)\uparrow +Z\downarrow]\rangle \\
\phi_{8,0} &= \frac{1}{\sqrt{3}}\,|[(X+iY)\downarrow +Z\uparrow]\rangle\,,
\end{aligned} \tag{2.149}$$

and solve the Schrödinger equation for the envelope function $f_j(\boldsymbol{r})$ defined by

$$\Psi(\boldsymbol{r}) = \sum_j f_j(\boldsymbol{r})\phi_{j,0}(\boldsymbol{r}). \tag{2.150}$$

Here $u_{j,0}$ indicates the Bloch function at $\boldsymbol{k} = 0$. The (8×8) matrix D can be decomposed to two (4×4) matrices D_a and D_b and is written as:

$$D \sim \begin{pmatrix} D_a & 0 \\ o & D_b \end{pmatrix}. \tag{2.151}$$

The remaining part can be treated just as a small perturbation. The Schrödinger equation then becomes

$$\begin{cases} D_a f_a = \mathcal{E}_a f_a, \\ D_b f_b = \mathcal{E}_b f_b. \end{cases} \tag{2.152}$$

As the momentum matrix elements

$$P = -i < S|p_x|X >= -i < S|p_y|Y >= -i < S|p_z|Z >$$

are involved explicitly in the matrix, a parameter set different from the Luttinger parameters was used. The parameters $(\gamma_1, \gamma_2, \gamma_3, \kappa)$ are now defined as

$$\gamma_1 = \gamma_1^L - \frac{2}{3}\frac{P^2}{\mathcal{E}_g},$$

$$\gamma_2 = \gamma_2^L - \frac{1}{3}\frac{P^2}{\mathcal{E}_g},$$
$$\gamma_3 = \gamma_3^L - \frac{1}{3}\frac{P^2}{\mathcal{E}_g},$$
$$\kappa = \kappa^L - \frac{1}{3}\frac{P^2}{\mathcal{E}_g}, \tag{2.153}$$

where the superscripts L indicate the Luttinger parameters. The solutions are obtained in the form

$$f_a = \begin{pmatrix} a_1 u_N \\ a_3 u_{N-1} \\ a_5 u_{N+1} \\ a_7 u_{N+1} \end{pmatrix}, \qquad f_b = \begin{pmatrix} b_2 u_N \\ b_6 u_{N-1} \\ b_4 u_{N+1} \\ b_8 u_{N-1} \end{pmatrix}. \tag{2.154}$$

The secular determinants to solve are obtained as follows,

$$\begin{vmatrix} a_{11} & i(sN)^{1/2}P & i[\frac{1}{3}s(N+1)]^{1/2}P & [\frac{2}{3}s(N+1)]^{1/2}P \\ -i(sN)^{1/2}P & a_{22} & -s[3N(N+1)]^{1/2}\gamma'' & -is[6N(N+1)]^{1/2}\gamma' \\ -i[\frac{1}{3}s(N+1)]^{1/2}P & -s[3N(N+1)]^{1/2}\gamma'' & a_{33} & is2^{1/2}[\gamma'(N+\frac{3}{2})-\frac{1}{2}\kappa] \\ +[\frac{2}{3}s(N+1)]^{1/2}P & is[6N(N+1)]^{1/2}\gamma' & -is2^{1/2}[\gamma'(N+\frac{3}{2})-\frac{1}{2}\kappa] & a_{44} \end{vmatrix} = 0, \tag{2.155}$$

for the a-set and

$$\begin{vmatrix} b_{11} & i(\frac{1}{3}sN)^{1/2}P & i[s(N+1)]^{1/2}P & [\frac{2}{3}sN]^{1/2}P \\ -i(\frac{1}{3}sN)^{1/2}P & b_{22} & -s[3N(N+1)]^{1/2}\gamma'' & -is2^{1/2}[\gamma'(N-\frac{1}{2})+\frac{1}{2}\kappa] \\ -i[s(N+1)]^{1/2}P & -s[3N(N+1)]^{1/2}\gamma'' & b_{33} & is[6N(N+1)]^{1/2}\gamma'] \\ +[\frac{2}{3}sN]^{1/2}P & is2^{1/2}[\gamma'(N-\frac{1}{2})+\frac{1}{2}\kappa] & -is[6N(N+1)]^{1/2}\gamma' & b_{44} \end{vmatrix} = 0, \tag{2.156}$$

for the b-set, where

$$s = \hbar\omega_c$$
$$a_{11} = \mathcal{E}_g - \mathcal{E}_a + s(N+1)$$
$$a_{22} = -s\left[(\gamma_1+\gamma')(N-\frac{1}{2}) + \frac{3}{2}\kappa\right] - \mathcal{E}_a$$
$$a_{33} = -s\left[(\gamma_1-\gamma')(N+\frac{3}{2}) - \frac{1}{2}\kappa\right] - \mathcal{E}_a$$
$$a_{44} = -s\left[\gamma_1(N+\frac{3}{2}) - \kappa\right] - \Delta - \mathcal{E}_a$$
$$b_{11} = \mathcal{E}_g - \mathcal{E}_a + sN$$
$$b_{22} = -s\left[(\gamma_1-\gamma')(N-\frac{1}{2}) + \frac{1}{2}\kappa\right] - \mathcal{E}_b$$
$$b_{33} = -s\left[(\gamma_1+\gamma')(N+\frac{3}{2}) - \frac{3}{2}\kappa\right] - \mathcal{E}_b$$
$$b_{44} = -s\left[\gamma_1(N-\frac{1}{2}) + \kappa\right] - \Delta - \mathcal{E}_b$$

and $\mathcal{E}_g$ and Δ are the band gap and the spin-orbit splitting energy, respectively. γ' and γ'' are the γ parameters for arbitrary direction of magnetic field in the $(1\bar{1}0)$ plane, as given in (2.148). In (2.155) and (2.156), higher order terms were neglected.

The Pidgeon-Brown model is very useful for calculating the Landau levels in most zinc blende type semiconductors, especially narrow gap semiconductors such as InAs, InSb, HgTe, HgSe, etc.

2.5 Low dimensional systems

The recent progress of the crystal growth and processing technology of semiconductors has enabled two-dimensional (2D), one-dimensional (1D) and zero-dimensional (0D) electron systems. In this section, we present a brief overview of low dimensional electron systems and the energy states in the presence of high magnetic fields.

2.5.1 *MOS-FET and heterostructure*

Since the early days of the study of two-dimensional electron systems in the 1960s, Si MOS-FET (metal oxide semiconductor - field effect transistor) have been extensively investigated [14,15]. The transistor has been used as one of the major semiconductor devices for electronics, but it also provides an excellent playground of electrons for research on a two-dimensional electron system. The structure of a Si MOS-FET is shown in Fig. 2.6. It consists of a p-type Si substrate, an oxide layer (SiO_2), and a gate electrode grown on top of it. When the gate voltage V_G is zero, no current is carried between the n-doped source and drain electrodes. With a positive bias V_G applied to the gate, the surface of Si near the interface with the oxide layer is negatively charged just like a capacitor across the oxide layer. In other words, electrons are induced on the surface of Si.

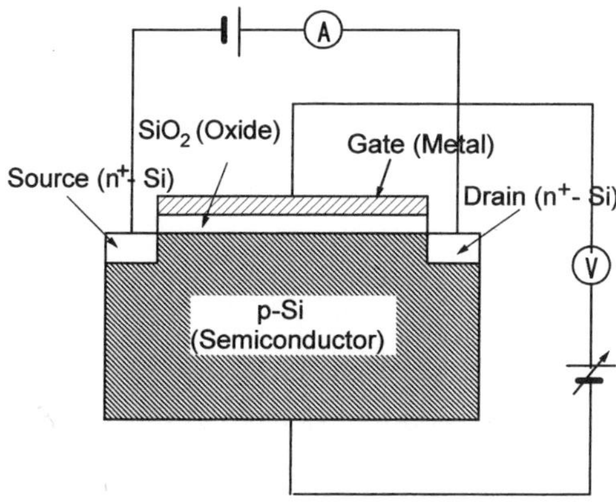

FIG. 2.6. Structure of Si MOS-FET.

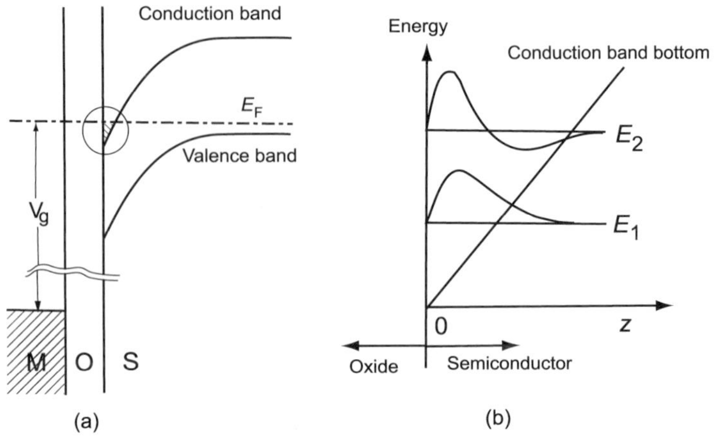

FIG. 2.7. Potential and the energy diagram of Si MOS-FET. (a) Energy potential diagram. (b) Simplified potential profile as represented by a triangular potential and the two-dimensional energy levels.

These electrons carry a current between the source and drain electrodes. As they can move around only on the surface of the Si, the electrons are 2D electrons. The energy profile of the MOS structure is shown in Fig. 2.7. The gate bias voltage V_G supplies the potential gradient in Si and bends the conduction and valence band edge. When a sufficiently large voltage is applied, the conduction band edge is lowered below the Fermi level, and 2D electrons are created. A great advantage of the Si MOS-FET is that the 2D electron density n in the inversion layer can be controlled by changing V_G. When the potential is approximated by a triangular shape as shown in Fig. 2.7 (b), the wave function of confined electrons in the potential is represented as

$$\phi = \exp[i(k_x x + k_y y)]\zeta_n(z). \tag{2.157}$$

Here $\zeta(z)$ is the Airy function and the quantized energy is given by

$$\begin{aligned}\mathcal{E}_N &= \left(\frac{\hbar^2 F^2}{2m_z^*}\right)^{1/3}\gamma_N + \frac{\hbar^2}{2m^{\parallel}}(k_x^2 + k_y^2),\\ \gamma_N &\sim \left(\frac{3\pi}{2}\right)^{2/3}\left(N + \frac{3}{4}\right)^{2/3}.\end{aligned} \tag{2.158}$$

Subbands are formed associated with each quantum level in a two-dimensional plane perpendicular to the potential. In Si, there are six valleys of conduction band minima in the (100) direction as shown in Fig. 2.8. As the quantized energy is inversely proportional to the effective mass in the direction of the quantized potential (perpendicular to the layer), the subbands formed in the two A valleys for which $m_z > m_{\parallel}$ are the lowest energy states. The two valleys are degenerate

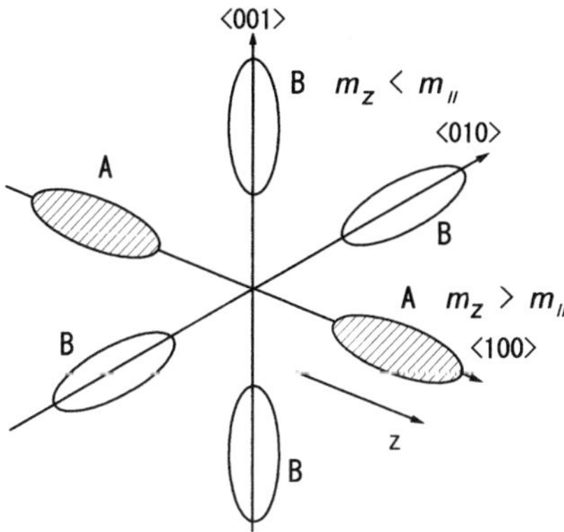

FIG. 2.8. Six valleys of the conduction band minima in Si.

in energy, but in the Shubnikov-de Haas effect a splitting is observed due to the interaction between the valleys (valley splitting) as mentioned in the next chapter. The lowest subband in the four B valleys is located about 20 meV higher than in A. So that in usual cases, it is not necessary to consider their presence. The second quantized levels in the A valleys are also located about 20 meV higher than the ground state.

When we apply magnetic fields, Landau levels are formed in each subband. The energy is then expressed as

$$\mathcal{E}_{Ni} = \left(N + \frac{1}{2}\right) \hbar\omega_c + \mathcal{E}_z^{(i)}. \tag{2.159}$$

A similar 2D system can be created in a heterostructure where two different semiconductors are put together adjacently. As an example of heterostructures, the energy profile of the interface between non-doped GaAs and n-doped AlGaAs ($Al_xGa_{1-x}As$) is shown in Fig. 2.9. At the interface, there is a step χ in the conduction band bottom due to the difference in the electron affinity between the two crystals, which is the difference in energy required to excite one electron from the bottom of the conduction band to the vacuum state. Likewise, a step exists in the valence band top. If n-type doping is done only in AlGaAs, electrons in AlGaAs near the interface fall into the GaAs, because the donor levels are above the bottom of the conduction band in GaAs. Therefore, a space charge layers is formed at the interface, which creates an electric field. As a result, the energy bands are bent as shown in Fig. 2.9, and a triangular-shaped potential is formed at the GaAs surface. Similar 2D levels are formed in the potential as in the case of Si-MOS. An advantage of such a structure is that the 2D electrons can move around on the surface of GaAs where there are no donors, so that they are free from the impurity scattering. Thus a high mobility is realized.

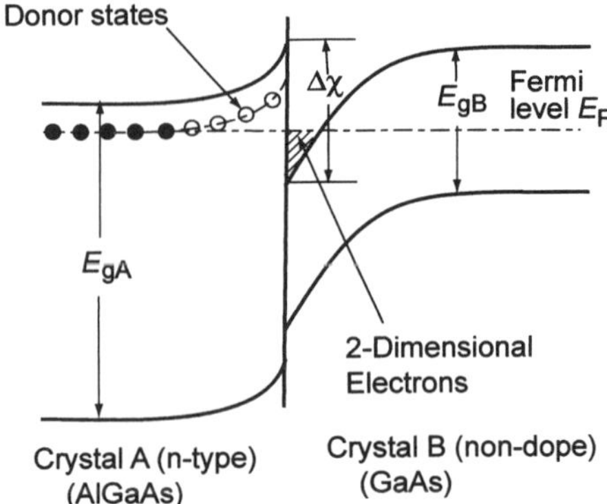

FIG. 2.9. Quantum potential profile in heterostructure interface. The case of the interface between n-doped-AlGaAs and non-doped GaAs is shown.

Such a structure with a doping in a remote area is called "modulation doping", and used for HEMT (high electron mobility transistor). In modulation-doped heterostructures, 2D electrons are created automatically at the heterointerface, but fabricating a gate electrode on top of the interface, we can control the carrier concentration by the gate voltage.

2.5.2 *Quantum well*

A quantum well is a structure of semiconductor layers sandwiched by other semiconductors which have a larger band gap, as shown in Fig. 2.10 (a). In quantum wells, electrons are confined by a square shaped quantum potential. A structure with just one well layer is a single quantum well, and a structure consisting of multiply stacked well layers is a multi-quantum well. The latter is useful for obtaining a sufficient absorption intensity in magneto-optical or cyclotron resonance experiments. In modulation doped quantum wells with donors doped in the barrier layer, electrons are transferred to the well. Due to the space charge layers at the interfaces, the internal electric fields are formed, so that the potential has a curvature as shown in Fig. 2.10 (b). Similarly to the case of heterostructure shown in Fig. 2.9, the barrier height is determined by the difference of the electron affinity between the two substances. Quantized energy levels are formed in quantum wells, depending on the shape of the quantum potentials. Electrons in the quantized level are 2D electrons that have a degree of freedom in the well layer. Therefore, the levels are called subbands.

A. Conduction band

First, let us consider the quantization in the conduction band with a square potential well, as in Fig. 2.9 (a). If we assume an infinite barrier height, the quantized electron energy perpendicular to the layers is represented in a simple

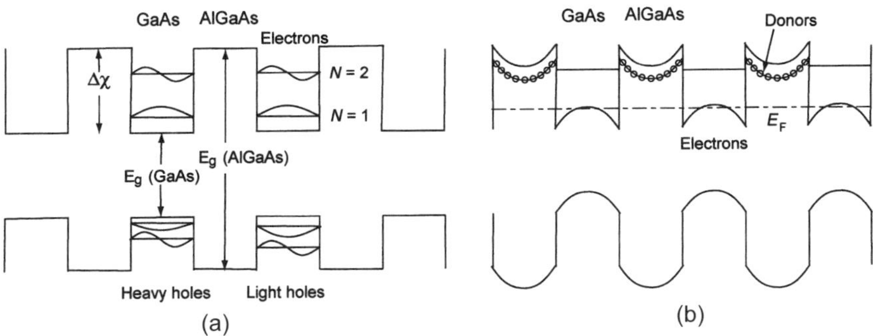

FIG. 2.10. Quantum potential profiles of (a) quantum well, and (b) modulation doped superlattice. Here the case of GaAs/AlGaAs system is shown as an example.

form,

$$\mathcal{E}_z = \frac{\hbar^2}{2m^*}\left(\frac{N\pi}{L_w}\right)^2, \tag{2.160}$$

where L_w represents the well width. Equation (2.160) gives a rough estimation of the quantized energy, but in most cases we have to consider the effect of the finite barrier height, because the wave function penetrates into the barrier layers. In order to obtain the accurate quantized energy, we have to calculate taking account of the penetration effect, most conveniently by the Kronig-Penny model.

When we apply magnetic fields perpendicular to the layer, the electron motion in the layer is quantized, and we obtain the energy of the 2D electrons as in (2.159). The magnetic quantization in the plane and quantization by a quantum potential in the perpendicular direction are basically independent. However, there is a small mixing between the Landau levels associated with the different subbands, because the valence bands have a character to mix the motion in the parallel and perpendicular direction to the layer and there is a mixing between the conduction band and the valence bands. This effect is particularly significant in narrow gap semiconductor quantum wells when the higher Landau levels of the ground subband cross the higher subbands. The Landau level-subband interaction becomes important when the magnetic field is tilted from the normal angle (see Section 4.11.2).

Another topic of interest concerning the electronic states of two-dimensional electron systems is the spin splitting. The spin splitting of two-dimensional electrons in inversion layers, quantum wells and heterostructures has been shown to depend on the electron concentration, the positions of the Fermi level, due to many-body effects [14, 16]. The g-factor is greatly modified from the three-dimensional value as expressed in Eq. (2.127), and shows an oscillatory behavior as a function of the filling factor of the Landau levels. Some topics related to this problem will be discussed in later chapters.

Apart from the many-body interactions in two-dimensional systems, the spin-orbit interaction in systems without inversion symmetry also brings about an additional modification of the spin splitting. It is known that if the system lacks spatial inversion symmetry, the spin degeneracy of two-dimensional electron states is lifted even when the external magnetic field is zero. Such effects concerning the zero-field spin splitting have been revealed in experiments of the Shubnikov-de Haas effect [17, 18] and electron spin resonance [19]. Two mechanisms are considered as origins of the zero-field spin splitting. One is the effect due to a k^3 term in conduction band Hamiltonian for bulk crystals lacking of inversion symmetry such as a zinc blende structure [20]. This term lifts the spin-degeneracy and is called Dresselhaus term. Another effect arises from an interface electric field existing in heterostructures. Rashba was the first to point out that if there is an electric field along the z-axis, the Hamiltonian has a term

$$\mathcal{H}_R = \alpha_R(\boldsymbol{\sigma} \times \boldsymbol{k}) \cdot \widehat{\boldsymbol{z}}, \tag{2.161}$$

where α_R is the Rashba parameter which is proportional to the electric field, $\boldsymbol{\sigma}$ is the Pauli spin matrices, and $\widehat{\boldsymbol{z}}$ is a unit vector along the z-axis (parallel to the electric field) [21, 22]. This term is called the Rashba term, and causes spin splitting in the absence of magnetic field. As there is inevitably an electric field at the interface of heterostructures, spin degeneracy of two-dimensional electrons is lifted at zero field to a greater or less extent. This effect is called the Rashba effect and has attracted much interest in connection with spintronics.

B. Valence band

The valence band structure in quantum wells and heterostructures is fairly complicated because of the interaction between the two different valence bands which are degenerate without the quantum potential. The degeneracy of the heavy hole band and the light hole bands is lifted by the quantum potential, but they interact with each other through the $\boldsymbol{k} \cdot \boldsymbol{p}$ perturbation. In quantum wells, modulation doped superlattices, and heterostructures, electrons are subjected to quantum potentials, as shown in Fig. 2.10. The Landau levels can be calculated by adding a potential $V(z)$ to the Hamiltonian H_v for the valence band

$$H_t = H_v + V(z). \tag{2.162}$$

By solving the Schrödinger equation with H_t, we can obtain the valence band structure. In systems with carriers such as heterostructures or modulation doped superlattices, the potential $V(z)$ is influenced by the electronic distribution determined from the solution. Therefore the calculation should be made self-consistently. The energy dispersion and the Landau levels in quantum wells and heterojunctions were investigated theoretically by Ando [23–25]. An example of the energy dispersion for a multiple quantum well is depicted in Fig. 2.11 (without magnetic fields) [23]. It is seen that the ground states of the heavy hole

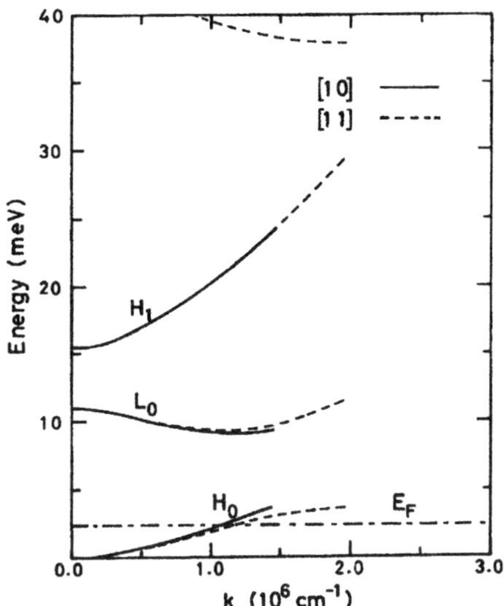

FIG. 2.11. Subband energy dispersion in a GaAs/AlGaAs multiple quantum well [23]. The solid and broken lines show dispersions along the [10] and [11] directions, respectively. The following parameters were assumed in the calculation: 2D hole concentration $p = 2.0 \times 10^{11}\mathrm{cm}^{-2}$, thickness of the GaAs layer $L_z = 11.2$ nm, thickness of the AlGaAs layer $L_b = 15$ nm, thickness of the spacer layer between the GaAs and AlGaAs layers $L_s = 1$ nm, height of the barrier potential $V_0 = 92$ meV.

band (H_0) and the light hole band (L_0) are split by the quantum potential. It should be noted that the energy of the lowest subband of the light hole band L_0 decreases with increasing k in the small k region; that is to say, the light holes have a negative effective mass. This phenomenon is caused by the repulsive interaction with the heavy hole band H_1. With further increasing k, the (L_0) band is repelled by the (H_0) band when they approach each other. In this way, both (H_0) and (L_0) show a complicated dispersion in this anti-crossing region. When the potential has an asymmetric shape as in a single heterostructure as shown in Fig. 2.9 a similar dispersion is obtained, but in this case, each subband shows a spin splitting due to the asymmetric potential.

When magnetic fields are applied perpendicular to the 2D layer, the energy states are quantized into Landau levels. Corresponding to the complicated dispersion curves and the interaction between the subbands, the Landau level structure is also quite complicated. Here we demonstrate a calculation by Ando [24, 25].

A quantum potential is regarded as a part of a periodic superlattice potential and written as

$$V(z) = \sum_{j \geq 0} V_j \cos \frac{2\pi j}{d} z. \tag{2.163}$$

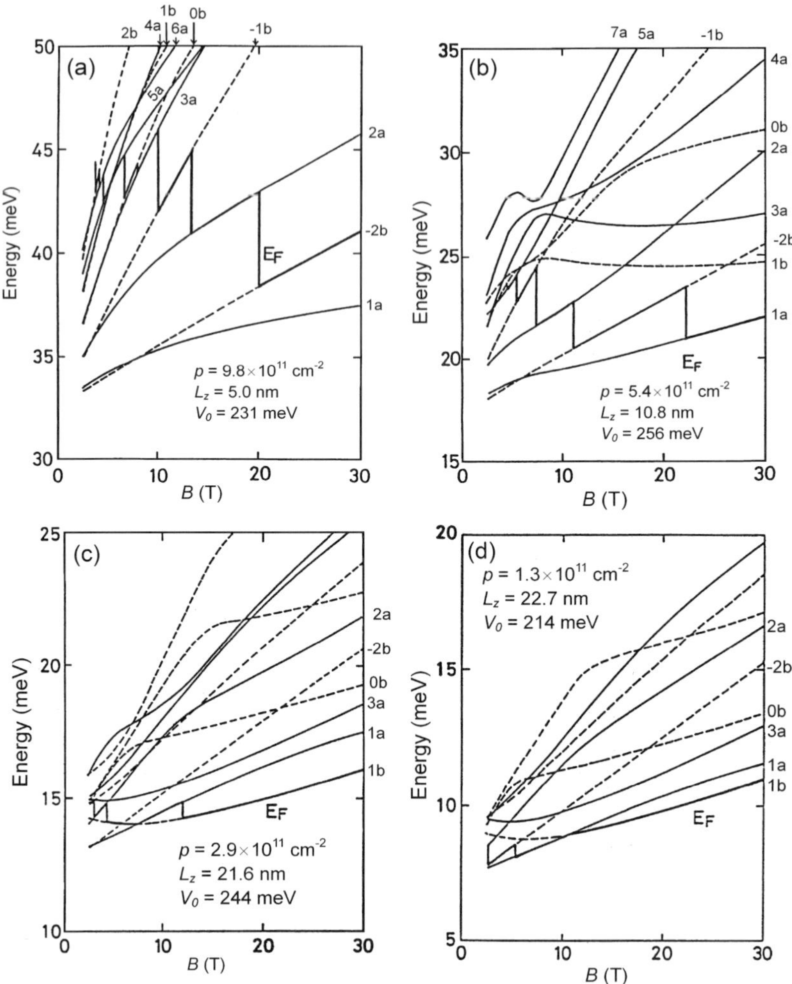

FIG. 2.12. Landau levels of the valence band in modulation doped quantum wells with different hole concentration p, well width L_z, and the barrier potential height V_0. (a) $p = 9.8 \times 10^{11}\text{cm}^{-2}$, $L_z = 5.0$ nm, $V_0 = 231$ meV. (b) $p = 5.4 \times 10^{11}\text{cm}^{-2}$, $L_z = 10.8$ nm, $V_0 = 256$ meV. (c) $p = 2.9 \times 10^{11}\text{cm}^{-2}$, $L_z = 21.6$ nm, $V_0 = 244$ meV. (d) $p = 1.3 \times 10^{11}\text{cm}^{-2}$, $L_z = 22.7$ nm, $V_0 = 214$ meV.

Here d is the period of the superlattice. Following the effective mass approximation, the envelope function is expanded by plane wave functions in the z-direction and harmonic oscillator wave functions in $x - y$ plane. Then the wave function is represented by two sets of functions as follows.

$$\Psi_{a,n} = (\phi_{\frac{3}{2}}, \phi_{\frac{1}{2}}, \phi_{-\frac{1}{2}}, \phi_{-\frac{3}{2}}) \sum_j \begin{pmatrix} A_{1j} u_{n-1} \cos \frac{2\pi j}{d} z \\ A_{2j}\ u_n \sin \frac{2\pi j}{d} z \\ A_{3j} u_{n+1} \cos \frac{2\pi j}{d} z \\ A_{4j} u_{n+2} \sin \frac{2\pi j}{d} z \end{pmatrix}, \tag{2.164}$$

$$\Psi_{b,n} = (\phi_{\frac{3}{2}}, \phi_{\frac{1}{2}}, \phi_{-\frac{1}{2}}, \phi_{-\frac{3}{2}}) \sum_j \begin{pmatrix} B_{1j} u_{n-1} \cos \frac{2\pi j}{d} z \\ B_{2j}\ u_n \sin \frac{2\pi j}{d} z \\ B_{3j} u_{n+1} \cos \frac{2\pi j}{d} z \\ B_{4j} u_{n+2} \sin \frac{2\pi j}{d} z \end{pmatrix}. \tag{2.165}$$

$\Psi_{a,n}$ and $\Psi_{b,n}$ correspond to the $\uparrow$ and $\downarrow$ spin states which are degenerate with each other in the absence of magnetic field. If we neglect the warping term in the $x - y$ plane, $\Psi_{a,n}$ and $\Psi_{b,n}$ are eigenfunctions of the Hamiltonian. In reality, there is a warping and each state has a coupling with other states whose n values are different by a multiple of 4. Therefore, the wave functions for both a and b sets are expressed by the following formula,

$$\Psi_{a,N} = \sum_{n=N,N\pm4,N\pm8,\cdots}^{n+2\geq0} \Psi_{a,n},$$

$$\Psi_{b,N} = \sum_{n=N,N\pm4,N\pm8,\cdots}^{n+2\geq0} \Psi_{b,n}. \tag{2.166}$$

$$(N = 0, 1, 2, 3)$$

Examples of the Landau levels for modulation doped quantum wells are shown in Fig. 2.12. It can be seen that the Landau level structure is quite complicated involving non-uniform spacing and non-linearity, due to the subband interaction.

2.5.3 *Superlattice*

A superlattice is a kind of periodic multiple quantum well with relatively thin barrier layers. The electron wave function in the well layers penetrates into the barrier layers and if the overlap between the penetrated wave function becomes substantial in the barrier layers, conduction becomes possible across the barrier layers. In other words, the subbands are broadened and a miniband with a small band width is formed in the direction perpendicular to the layers. The conduction through the minibands across the layers may exhibit a non-linear conductivity including a negative resistance. Such phenomena were first predicted by Esaki and Tsu [26]. When we apply magnetic fields in the perpendicular direction, Landau levels are formed in each miniband. When magnetic fields are applied parallel to the layers, the character of the energy levels depends on the relative dimension of the cyclotron radius in comparison to the superlattice period. When the magnetic field is weak so that the radius is large, Landau levels become just those for an anisotropic energy band. When the field is large so that the radius becomes comparable with the superlattice period, the Landau levels are broadened, as the energy depends on the position of the center coordinate of the cyclotron motion. Such a problem will be treated in Section 5.4.3.

2.5.4 *Quantum wire and quantum dot*

One-dimensional (1D) and zero-dimensional (0D) electron systems are realized in quantum wires (QWR) and quantum dots (QD). Many different techniques have been developed for fabricating QWR and QD. One of the motivations is to develop new high efficiency optical devices such as semiconductor lasers. The structures can be fabricated by microprocessing such as electron beam lithography and etching processes. Besides such techniques for fabricating small devices from bulk or film crystals into small sizes by processing (top-down technique), techniques for growing crystals insitu by self-assembling (bottom-up technique) have also been developed. Regarding QWR, for example, techniques such as T-shaped QWR [27], QWR grown on V-grooves [28–30], QWR grown at giant steps on a slightly tilted vicinal surfaces [31] have been reported.

Regarding QD, different techniques have been developed, such as an etching process to form a mesa-shaped island [32], gated vertical QD made from a double barrier heterostructure [33], a process of growing self-assembled QD called Stranski-Krastanow (SK) mode [34,35], or QD grown by MOCVD in tetrahedral-shaped recesses (TSRs) [36]. The SK mode utilizes an epitaxial technique with a strained thin film. When we grow an epitaxial thin film of a crystal on a substrate of a different crystal that has a smaller lattice constant, first a film with a uniform thickness is grown on the substrate (wetting layer). However, as the wetting layer thickness becomes thicker, many quantum dots are formed on the wetting layer as protruding parts due to a large strain arising from the lattice mismatch. In this way, a large number of self-assembled quantum dots with a size as small as a few tens of nanometers can be grown on the substrate. The position and the size of the dots are random, but recent advance of technology has enabled to fabricate quantum dots whose size distribution is reasonably small.

In quantum dots, electronic states are fully quantized, and form a zero-dimensional state. Here we consider the energy states of electrons confined in a dot under magnetic fields. In actual quantum dots, the confining potential can be approximated by a parabolic potential. It is a better approximation than a simple square well potential because of the surface potential. When the magnetic field is applied parallel to the z-axis, the quantum dot potential in the perpendicular direction is written as

$$V(x, y) = \frac{m^*}{2}\omega_0^2(x^2 + y^2). \tag{2.167}$$

The electronic energy levels under such a harmonic potential on a two-dimensional $x - y$ plane are equally spaced. The Hamiltonian in the presence of both magnetic field and the quantum dot potential is represented by using the symmetric gauge as

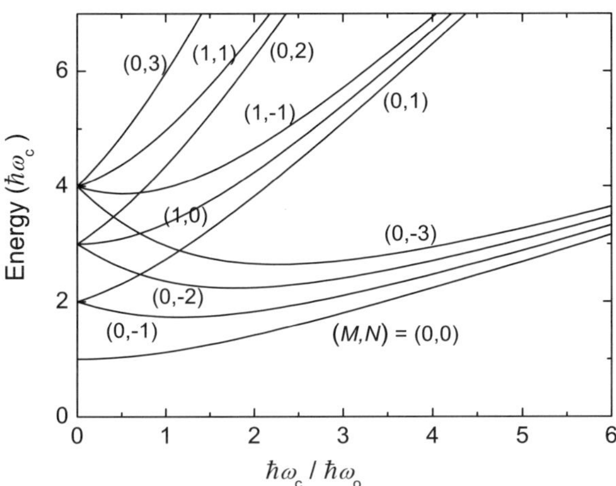

FIG. 2.13. Energies of the electronic states in quantum dots in the presence of magnetic fields for different (M, N)

.

$$\mathcal{H} = \frac{p^2}{2m^*} + \frac{1}{2}\hbar\omega_c L_z + \frac{m^*}{2}\left[\left(\frac{\omega_c}{2}\right)^2 + \omega_0^2\right](x^2 + y^2). \tag{2.168}$$

The Schrödinger equation with this Hamiltonian can be readily solved because both the magnetic field potential and quantum potential have harmonic forms. The solution is obtained as was done for (2.82).

$$\mathcal{E}(N, M) = (2N + |M| + 1)\hbar\left[\left(\frac{\omega_c}{2}\right)^2 + \omega_0^2\right]^{\frac{1}{2}} + \left(\frac{\hbar\omega_c}{2}\right)M. \tag{2.169}$$

The states have a finite angular momentum M, so that they show Zeeman splitting. Figure 2.13 shows the energy diagram of quantum dots in magnetic fields. This diagram is called a Darwin-Fock diagram [37, 38]. These levels have been investigated by cyclotron resonance, magneto-photoluminescence, and magneto-tunneling experiments.

2.6 Electronic states in magnetic fields and electric fields

2.6.1 *Crossed magnetic and electric fields*

The degeneracy of the Landau levels with respect to the orbit center is lifted when electric fields E are applied perpendicularly to the magnetic field. The problem of the crossed magnetic and electric fields is of importance when we analyze the transverse magneto-conductivity.

Let us assume that the magnetic fields $\boldsymbol{B} = [0, 0, B]$ and electric fields $\boldsymbol{E} = [E, 0, 0]$ are applied in the z and x directions, respectively. The Schrödinger equation is

$$\left[\frac{1}{2m}(\boldsymbol{p} + e\boldsymbol{A})^2 + e\boldsymbol{E}\cdot\boldsymbol{r}\right]\Psi = \mathcal{E}\Psi. \tag{2.170}$$

Putting

$$\Psi = e^{ik_y y + ik_z z}\varphi(x), \tag{2.171}$$

we get

$$\left[-\frac{\hbar^2}{2m}\frac{\partial^2}{\partial x^2} + \hbar\omega_c k_y x + \frac{m\omega_c^2}{2}x^2 + eEx\right]\varphi(x) = \left(\mathcal{E}_\perp - \frac{\hbar^2 k_y^2}{2m}\right)\varphi(x). \tag{2.172}$$

It is easy to see that the above equation can be treated in a similar manner as we did in Section 2.2, if we replace x with x' with

$$x = x' - k_y l^2 - \frac{eEl^2}{\hbar\omega_c} = x' + X. \tag{2.173}$$

Equation (2.173) shows that the center coordinate of the cyclotron motion is shifted by the electric field by $eEl^2/\hbar\omega_c$,

$$X = -k_y l^2 - \frac{eEl^2}{\hbar\omega_c}. \tag{2.174}$$

The Schrödinger equation is then

$$\left[-\frac{\hbar^2}{2m}\frac{\partial^2}{\partial x'^2} + \hbar\omega_c k_y x + \frac{m\omega_c^2}{2}x'^2\right]\varphi(x) = \left(\mathcal{E}_\perp + eEk_y l^2 + \frac{m}{2}\frac{E^2}{B^2}\right)\varphi(x). \tag{2.175}$$

The energy eigenvalue of this equation is

$$\mathcal{E}_\perp = \left(N + \frac{1}{2}\right)\hbar\omega_c - eEk_y l^2 - \frac{m}{2}\frac{E^2}{B^2} \tag{2.176}$$

$$= \left(N + \frac{1}{2}\right)\hbar\omega_c + eEX + \frac{m}{2}\frac{E^2}{B^2}. \tag{2.177}$$

The meaning of each term in the above formula is readily understood: the second term is the potential energy of electrons centered at X, and the third term is the kinetic energy of for the drift velocity

$$v_d = \frac{E}{B}. \tag{2.178}$$

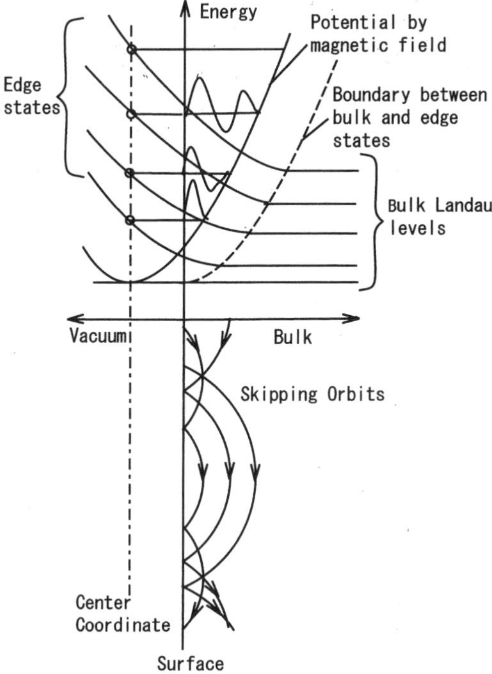

FIG. 2.14. Energy and orbits of electrons near the edge of the sample.

2.6.2 *Edge states*

When electrons are located near the edge of the samples, cyclotron motion cannot fulfill the entire cyclotron orbit if the distance between the center of the cyclotron orbit and the edge is smaller than the radius of the cyclotron motion. In such a case, electron orbit is bounced at the edge plane, and bounced electrons conduct a one-dimensional motion as a whole being repeatedly reflected at the edge plane. Such an orbit is called a skipping orbit. The energy states with a skipping orbit are different from those in the interior of the sample where the cyclotron orbit is completed. Let us consider the skipping electrons under magnetic fields near the sample edge, as shown in Fig. 2.14. Here let the magnetic fields be applied perpendicular to the plane. The potential of the magnetic field is parabolic, centered at the center coordinate X as shown in (2.59). As electrons cannot penetrate outside the sample, the edge surface is considered to be an infinite barrier. Then the electrons near the edge are regarded as localized electrons confined in the potential formed by the magnetic potential and the surface potential. If the center of the cyclotron orbit is more to the left of the broken parabolic line which is dependent on the energy, there is a barrier within the orbit, and electrons conduct a skipping orbit. Electrons are confined in the potential as shown by a thick solid line. Thus the energy of such a state is always higher than the interior. Such states are called the edge states. The energies of the edge states as a function of the center coordinate are shown in the figure. In actual semiconductor surface edges, the potential profile near the surface is more gentle than the abruptly rising barrier as shown in Fig. 2.14 due to the surface potential. As a result, the shape of the energy profile is slightly different from Fig. 2.14.

The edge states play important roles in quantum transport in high magnetic fields, such as quantum Hall effect or magneto-tunneling effect, as will be described in the next chapter. More general cases in a highly anisotropic energy band having open orbits, as in organic conductors, were treated by Osada and Miura [39].

2.7 High magnetic field effects on Bloch electrons

2.7.1 *Breakdown of effective mass theory*

The effective mass theory is very successful in obtaining the energy levels under magnetic fields. However, it is of importance to know that there is a validity limit when the magnetic fields become very high and hence the radius of the cyclotron motion becomes very small. As shown in (2.66), the cyclotron radius decreases inversely proportional to $\sqrt{B}$. Since the effective mass theory is based on the $\boldsymbol{k} \cdot \boldsymbol{p}$ perturbation theory, the matrix elements of the $\boldsymbol{k} \cdot \boldsymbol{p}$ term $(\hbar/m)k \cdot p_{nn'}$ should be small in comparison to the energy difference between the bands $\mathcal{E}_{nn'}$. Since

$$k \sim l, \qquad p_{nn'} \sim \frac{\hbar}{a}, \qquad a = \text{lattice constant}, \qquad \mathcal{E}_g \sim \frac{p_{max}^2}{m} \sim \frac{\hbar^2}{ma^2},$$

the effective mass theory is valid only if

$$\frac{\frac{\hbar}{m} k \cdot p_{nn'}}{\mathcal{E}_{nn'}} \ll 1. \tag{2.179}$$

This leads to

$$\left(\frac{\hbar}{m}\frac{1}{l}\frac{\hbar}{a}\right) / \left(\frac{\hbar^2}{ma^2}\right) = \frac{a}{l} \ll 1. \tag{2.180}$$

Namely, when B becomes extremely high and the magnetic length is l decreased such that

$$l = \sqrt{\frac{\hbar}{eB}} \rightarrow a, \tag{2.181}$$

the effective mass theory breaks down. The breakdown under the condition (2.181) is reasonable, because the effective mass theory is based upon the assumption that the effect of the periodic lattice potential can be averaged out to lead to the effective mass. When the magnetic length l becomes comparable with the atomic spacing of the crystal lattice in very high magnetic fields, such an averaging is not valid any more. The Landau levels are degenerate with respect to the center coordinate X, as we saw in Section 2.2. Actually, when the cyclotron radius is large, the energy is independent of the position of the center. However under the condition $l \approx a$, the potential that electrons feel would differ depending on the center. Thus the degeneracy of the Landau levels will be lifted. As a result, the Landau levels should have finite width in energy.

How high are the magnetic fields required to observe such a broadening? If we define α by

$$\alpha \equiv \left(\frac{a}{l}\right)^2, \tag{2.182}$$

$$\alpha = \frac{e}{\hbar}a^2 B = \frac{2\pi a^2 B}{h/e} = \frac{2\pi\Phi}{\Phi_0}, \tag{2.183}$$

where

$$\Phi_0 = \text{unit flux quantum} = \frac{h}{e} = 4.1325 \times 10^{-15}\ \text{Tm}^2.$$

The magnetic flux density corresponding to $\alpha = 1$ is $B_0 = e/\hbar a^2$. In other words, B_0 is the flux density at which one unit flux quantum penetrates a unit cell.

The broadening of the Landau levels in extremely high magnetic fields was first investigated by Harper based on the tight binding approximation [40]. Thus the broadening of the Landau levels under the condition (2.181) is called Harper broadening. Here let us consider the energy spectra following the theory of Harper. The wave function of Bloch electrons is expressed in the tight binding approximation as

$$\Psi_{\boldsymbol{k}}(\boldsymbol{r}) = \sum_j e^{i\boldsymbol{k}\cdot\boldsymbol{r}_j}\phi(\boldsymbol{r} - \boldsymbol{r}_j), \tag{2.184}$$

where $\phi(\boldsymbol{r} - \boldsymbol{r}_j)$ is an atomic wave function. The energy is obtained under such an approximation as follows;

$$\mathcal{E} = \mathcal{E}_a - A - C \sum_{j(\text{Nearest Neighbor})} e^{i\boldsymbol{k}\cdot\boldsymbol{r}_j}. \tag{2.185}$$

The coefficients A and C are

$$A = -\int \phi^*(\boldsymbol{r} - \boldsymbol{r}_i)V'\phi^*(\boldsymbol{r} - \boldsymbol{r}_i)d\boldsymbol{r}, \tag{2.186}$$

$$C = -\int \phi^*(\boldsymbol{r} - \boldsymbol{r}_i)V'\phi^*(\boldsymbol{r} - \boldsymbol{r}_j)d\boldsymbol{r}, \tag{2.187}$$

where V' is the potential and j runs over the nearest neighbors. For simplicity, we consider the case of a simple cubic lattice. Then the energy can be written as

$$\mathcal{E}(\boldsymbol{k}) = \mathcal{E}_a - A - C[e^{ik_x a} + e^{-ik_x a} + e^{ik_y a} + e^{-ik_y a} + e^{ik_z a} + e^{-ik_z a}]$$

$$= \mathcal{E}_a - A - C[\cos ak_x + \cos ak_y + \cos ak_z]. \tag{2.188}$$

When magnetic fields B are applied to the z direction, the energy states are obtained by a Landau Peierls substitution with $\boldsymbol{A} = [0, Bx, 0]$,

$$\mathcal{E}\left(\frac{\nabla}{i} + \frac{e}{\hbar}\boldsymbol{A}\right) u(\boldsymbol{r}) = Wu(\boldsymbol{r}). \tag{2.189}$$

The exponential terms in (2.188) $e^{ik_x a}u(\boldsymbol{r})$ etc. are transformed to $e^{\nabla_x a}u(\boldsymbol{r})$. The operators in the exponent are treated by taking the expanded form. For $e^{\pm\nabla_x a}$,

$$e^{\pm\nabla_x a}u(\boldsymbol{r}) = u(\boldsymbol{r}) \pm a\frac{\partial}{\partial x}u(\boldsymbol{r}) + \frac{1}{2}a^2\left(\frac{\partial}{\partial x}\right)^2 u(\boldsymbol{r}) + \ldots = u(x \pm a, y, z), \tag{2.190}$$

and for $e^{\pm\nabla_y a}$

$$e^{ik_y a}u(\boldsymbol{r}) \to e^{\nabla_y a}e^{ieBax/\hbar}u(\boldsymbol{r}) = u(x, y + a, z)e^{ieBax/\hbar}. \tag{2.191}$$

The Schrödinger equation is thus obtained as

$$\begin{aligned}\frac{1}{2}[u(x+a) + u(x-a) + e^{ieBax/\hbar}u(y+a) + e^{-ieBax/\hbar}u(y-a)&\\ + u(z+a) + u(z-a)] = Wu(\boldsymbol{r}).&\end{aligned} \tag{2.192}$$

From the inspection, we can assume that the wave function is free electron-like in the y and z directions, so that we can put

$$u(\boldsymbol{r}) = e^{ik_y}e^{ik_z}u(x). \tag{2.193}$$

Then (2.192) becomes

$$\begin{aligned}\frac{1}{2}[u(x+a) + u(x-a)] + \cos(eaBx/\hbar - k_y a)u(x)&\\ = (W - \cos k_z a)u(x).&\end{aligned} \tag{2.194}$$

Putting

$$\frac{eaB}{\hbar} = \frac{a}{l^2} \equiv G, \tag{2.195}$$

we obtain

$$\frac{1}{2}[u(x+a) + u(x-a)] + \cos(Gx - k_y a)u(x) = Wu(x). \tag{2.196}$$

The argument of the cos is rewritten

$$Gx - k_y a = G(x - k_y l^2) = G(x + X), \tag{2.197}$$

using the center coordinate X. If we take the center at X, we obtain

$$\frac{1}{2}[u(x+a) + u(x-a)] + (\cos \alpha n - W)u(x) = 0, \tag{2.198}$$

where $\alpha = Ga$. Equation (2.198) is called the Harper equation, and it gives a basis for considering the problem of the Bloch electron in extremely high magnetic fields.

The Harper equation (2.198) is rewritten as

$$u_{n+1} + u_{n-1} - 2u_n + 2(1 + \cos \alpha n - W)u_n = 0,$$
$$\frac{\partial^2 u_n}{\partial x^2} + \frac{2}{a^2}(1 + \cos \alpha n - W)u_n = 0. \tag{2.199}$$

Therefore, we obtain

$$\frac{\partial^2 u}{\partial x^2} + K(x)u = 0, \tag{2.200}$$

with

$$K(x) = \frac{2}{a^2}(1 - W + \cos Gx). \tag{2.201}$$

Equation (2.200) indicates that it is a potential problem including a new periodicity created by magnetic fields in addition to the lattice periodicity. When G is small,

$$K(x) \simeq \frac{2}{a^2}\left[2 - W - \frac{1}{2}(Gx)^2\right]. \tag{2.202}$$

As this potential contains just the x^2 dependence, solution gives the harmonic oscillator function, and ordinary Landau levels are obtained. However, as G becomes large with large B, Landau levels are broadened with finite width. This is the Harper broadening. The Harper broadening is interpreted as a result of the lifting of the degeneracy of the Landau levels.

Hofstadter further developed the idea of the broadening of the Landau levels and found various interesting properties of the Bloch electrons in extremely high magnetic fields [41]. Starting from the Harper equation, Hofstadter derived the following equation,

$$u_{n+1} + u_{n-1} - 2\cos(\alpha n - \nu)u(n) = \mathcal{E}u(n). \tag{2.203}$$

This can be rewritten as,

$$\begin{pmatrix} u(n+1) \\ u(n) \end{pmatrix} = \begin{pmatrix} \mathcal{E} - 2\cos(\alpha n - \nu) & -1 \\ 1 & 0 \end{pmatrix} \begin{pmatrix} u(n) \\ u(n-1) \end{pmatrix}$$
$$= A(n) \begin{pmatrix} u(n) \\ u(n-1) \end{pmatrix}. \tag{2.204}$$

Solving this equation, Hofstadter obtained the energy diagram as shown in Fig. 2.15. Since it looks like a butterfly, this beautiful diagram is often called the Hofstadter butterfly diagram. The abscissa indicates $\alpha = aG = (a/l)^2$, and is proportional to magnetic field. The point $\alpha = 1$ corresponds to a field where the cyclotron radius decreases to the size of the lattice constant. As α can be transformed as $Ba^2/(h/e) = \phi/\phi_0$, the magnetic flux penetrating the unit cell becomes equal to a unit flux h/e at $\alpha = 1$. So the right edge of the diagram

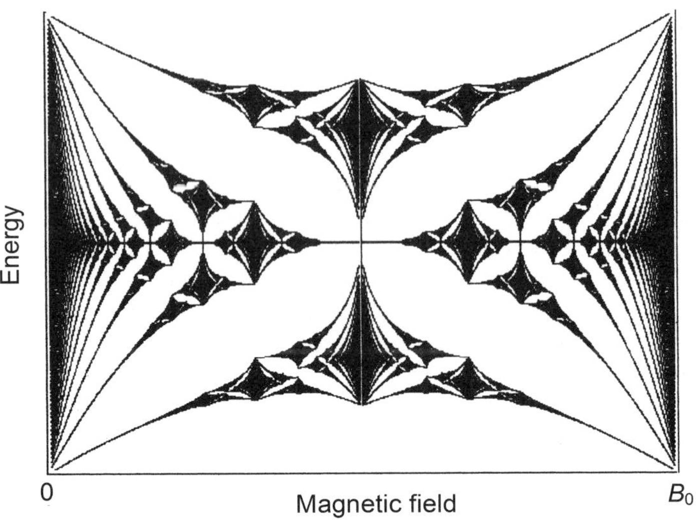

FIG. 2.15. Energy spectrum of Bloch electrons in extremely high magnetic fields (Hofstadter diagram). (First published in [41].)

shows a very high magnetic field as mentioned in the previous section. At $B = B_0$, the energies come back to the same as at $B = 0$ and it is periodic with respect to B_0. The diagram shows many interesting features. First, we should notice that the energy is that of Landau levels in the small α limit (low field limit), but it shows a broadening and complicated splitting in each level. Next, it is symmetric with respect to $\alpha = 1/2$. This means that the energy spectrum returns to the low field one again when α becomes 1. That is to say it is periodic as a function of magnetic field. When α is a rational number represented by two integers p and q with $\alpha = p/q$, the spectrum break up into q distinct bands. For example, we can easily see that for $\alpha = 1/2, 1/3, 1/4$, the band is represented by two, three and four vertical lines. The spectra as a whole exhibit fractal natures. These properties come from the new periodicity created by the very high magnetic fields.

The magnetic field B_0 for realizing $\phi/\phi_0 = 1$ ($\alpha = 1$) in real crystals is very high; for example if we assume $a = 0.15$ nm as in usual semiconductors, $B_0 = 1.83 \times 10^5$ T. This is a very difficult value to achieve even with the presently available ultrahigh magnetic fields. However, in some crystals with larger lattice constants, B_0 can be much smaller. In artificial lateral superlattices with $a = 100$–120 nm, fractal energy spectra expected from the Hofstadter's butterfly diagram were observed in the quantum Hall conductance [42]. Vogl and Strahberger calculated the optical absorption spectra in artificially modulated lateral superlattices on a

GaAs quantum well, and found that the fractal nature of the Hofstadter diagram should appear in the spectra [43].

2.7.2 *Magnetic breakdown*

Magnetic breakdown is another example of the effects of a high magnetic field on the motion of Bloch electrons. In magnetic fields, electrons move in a plane perpendicular to the field on the Fermi surface or equi-energy surface. The motion is determined by the equation of the acceleration (2.105). When the orbit in the $\boldsymbol{k}$ space is small in the Brillouin zone and well apart from any other orbits, such motions are independent each other. However, if there are other orbits nearby, electronic states corresponding to these orbits interact with each other in high magnetic fields where the magnetic quantizing energy (Landau level spacing) becomes comparable with the energy gap between the two states. As a result, new orbits are created for electrons to move around. Such an effect was first discovered in the de Haas-van Alphen effect in metals. In group II metals, the Fermi surface is touching the Brillouin zone edge. Figure 2.16 shows

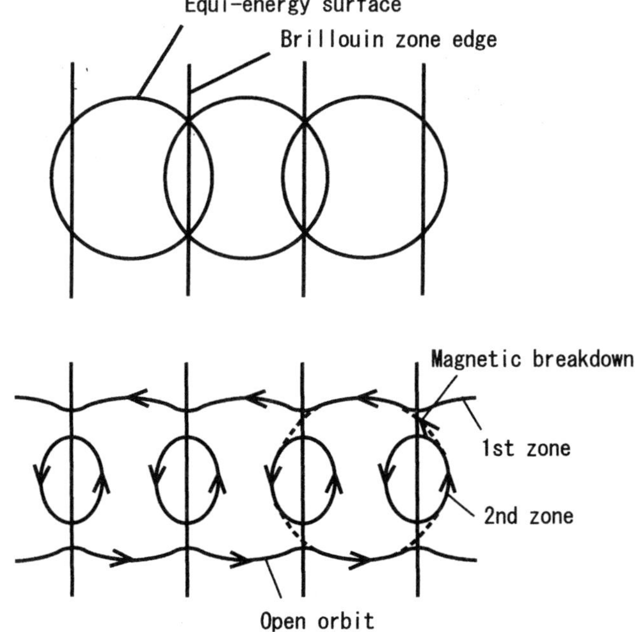

FIG. 2.16. Upper panel: Equi-energy surface of electrons in the plane perpendicular to the magnetic field without interaction when it extends over the Brillouin zone edge. Lower panel: Motion of electrons in the k-space without magnetic breakdown (solid line) and that when magnetic breakdown takes place (broken line).

an example of the electron motion in such a case in the extended zone. Circular orbits as shown in the upper panel of the figure are split into two kinds of orbits, open orbits and closed orbits near the zone edge in the second zone, as shown in the lower panel. The two kinds of orbits are formed by the periodic lattice potential. When magnetic fields are sufficiently high, electrons conduct motion across the open and the closed orbits as indicated by dotted line. Thus a new orbit that is nearly circular is formed as a result of a kind of tunneling effect in the $\boldsymbol{k}$ space. Such a phenomenon is called magnetic breakdown or magnetic breakthrough. The breakdown is observed as a sudden change of the de Haas van Alphen period and the magneto-resistance saturation which exists only for the a closed orbit. Many investigations have been made on the magnetic breakdown theoretically [44–53] and experimentally [50, 54–58]. Calculating the probability of the electron tunneling in the $\boldsymbol{k}$ space, it is considered to take place when the condition $\hbar\omega_c$ becomes comparable with the band gap at the zone edge [49].

Magnetic breakdown was first studied in metals, but recently a variety of phenomena associated with magnetic breakdown have been observed, for example, in organic conductors or in semiconductor devices with artificially fabricated quantum potentials. Another example is the phenomena which take place in energy bands with a camel's back structure as will be shown in the next chapters.

2.8 Bound electron states in magnetic fields

2.8.1 *Donor states*

The shallow impurity states in semiconductors, donors and acceptors are analogous to the electron in a hydrogen atom. The effective mass approximation which was useful for determining the electronic states in magnetic fields can be applied also for the donor states [7]. The energy of the donor states for an energy band with an energy dispersion,

$$\mathcal{E}(\boldsymbol{k}) = \frac{\hbar^2 k^2}{2m^*} \tag{2.205}$$

is represented as a solution of a Schrödinger equation

$$\left[\mathcal{E}\left(\frac{\nabla}{i}\right) + U(\boldsymbol{r})\right]\phi(\boldsymbol{r}) = \mathcal{E}\phi(\boldsymbol{r}), \tag{2.206}$$

where

$$U(\boldsymbol{r}) = -\frac{e^2}{4\pi\kappa\epsilon_0 r}$$

is a potential energy provided by a donor ion. The solution of (2.206) gives a hydrogen-like state. The energy of the ground state is represented by the effective Rydberg energy,

$$Ry^* = \frac{m^* e^4}{32\pi^2\hbar^2\kappa^2\epsilon_0^2}$$

$$= \frac{e^2}{4\pi\kappa\epsilon_0(2a_B^*)}, \tag{2.207}$$

with the effective Bohr radius

$$a_B^* = \frac{4\pi\hbar^2\kappa\epsilon_0}{m^*e^2}. \tag{2.208}$$

The states are represented by quantum numbers (n, l, m) just in the same manner as for a hydrogen atom. Thus the problem of the impurity states in magnetic fields as well as the exciton states that will be described in the next section, is reduced to a fundamental problem of a hydrogen atom in magnetic fields.

We now consider electronic states of a hydrogen atom in the presence of magnetic fields. The Coulomb interaction is expressed by

$$\mathcal{H}_{\text{Coulomb}} = -\frac{e^2}{4\pi\kappa\epsilon_0 r}. \tag{2.209}$$

In the presence of the Coulomb potential which has a spherical symmetry, it is more convenient to use the symmetric gauge rather than the Landau gauge for magnetic flux density B. The total Hamiltonian is

$$\mathcal{H} = \frac{\hbar^2}{2m^*}\left(\frac{\nabla}{i} + \frac{e\boldsymbol{A}}{\hbar}\right)^2 - \frac{e^2}{4\pi\kappa\epsilon_0 r} + gS\mu_B B. \tag{2.210}$$

Taking the symmetric gauge, and assuming that $\boldsymbol{B}$ is parallel to the z-axis, we obtain

$$\boldsymbol{B} \parallel z, \quad \boldsymbol{A} = B\left(-\frac{y}{2}, \frac{x}{2}, 0\right),$$

and

$$\mathcal{H} = \frac{\boldsymbol{p}^2}{2m^*} + \frac{\omega_c}{2}L_z + \frac{1}{8}m^*\omega_c^2(x^2+y^2) - \frac{e^2}{4\pi\kappa\epsilon_0 r} + gS\mu_B B. \tag{2.211}$$

It is not possible to obtain a solution of a Schrödinger equation derived from Eq. (2.211) analytically, because the flux density B or magnetic field H has a cylindrical symmetry, whilst the Coulomb energy has spherical symmetry. Therefore, various approximate solutions were derived in different range of B relative to the Coulomb interaction. The characteristic energy representing the intensity of magnetic field is the energy of the cyclotron motion, $\hbar\omega_c$, while that for the Coulomb interaction is the effective Rydberg energy Ry^* (2.207), that is the binding energy of the donor states. The relative strength of the magnetic field in comparison to the effect of the Coulomb interaction is often represented by the ratio of these two quantities,

$$\gamma = \frac{\hbar\omega_c}{2}/Ry^* = \frac{16\pi^2\hbar^3\kappa^2\epsilon_0^2}{m^{*2}e^3}B. \tag{2.212}$$

The dimensionless parameter γ is called the gamma parameter. When γ is much smaller than unity ($\gamma \ll 1$), the system is almost hydrogen atom-like and the magnetic field is regarded as a small perturbation. When γ is much larger than one ($\gamma \gg 1$), on the other hand, the system is almost Landau level-like, and the Coulomb interaction is regarded as a small perturbation. For $\gamma \approx 1$, the two perturbations should be treated with an equal weight so that the problem is more complicated.

It should be noted that γ is also expressed in terms of the the magnetic length l that is the characteristic length for the magnetic field, and the Bohr radius a_B^* that is the characteristic length for the Coulomb interaction.

$$\gamma = \left(\frac{a_B^*}{l}\right)^2 = \frac{eB}{\hbar}\frac{(4\pi)^2\hbar^4\kappa^2\epsilon_0^2}{m^{*2}e^4} = \frac{16\pi^2\hbar^3\kappa^2\epsilon_0^2}{m^{*2}e^3}B. \tag{2.213}$$

For a hydrogen atom, $\kappa = 1$ and $m^* = m$, so that the magnetic flux density B_c to achieve $\gamma = 1$ is

$$B_c = \frac{m^2e^3}{16\pi^2\hbar^3\epsilon_0^2} = 2.35 \times 10^5 \text{ T}. \tag{2.214}$$

The above B_c is too high to obtain even by the most advanced presently available technology. In the case of an actual donor state or excitons in semiconductors, the effective B_c^* is expressed as

$$B_c^* = \left(\frac{\kappa}{m^*/m}\right)^2 B_c. \tag{2.215}$$

In ordinary semiconductors, κ is larger than 1 and m^*/m is smaller than 1, and therefore, B_c^* is considerably smaller than B_c. For example, assuming typical values, $m^* \approx 0.1m$ and $\kappa \approx 10$, we obtain $B_c^* \approx 23.5$ T, which is well within the range available in modern magnet technology. We will see many examples of the energy states for $\gamma \gg 1$ in later chapters. In the following, we consider the magnetic energy levels in each range of γ.

A. Weak field limit ($\gamma \ll 1$)

When the magnetic field is weak so that $\gamma \ll 1$, we treat a term involving B as a perturbation acting on a hydrogen atom-like state. The unperturbed states are represented by quantum numbers (n, l, m). L_z can be commutated with $\mathcal{H}$, so that L_z is a good quantum number. For an s state, $< L_z > = 0$, so that apart from the spin term, the only term which is related to B in the Hamiltonian is $\frac{1}{8}m^*\omega_c^2(x^2+y^2)$. Taking this term as a perturbation, we can calculate a first order perturbation energy. For a $1s$ state, the correction of energy due to this term is

$$\Delta\mathcal{E}_{1s} =< 1s|\tfrac{1}{8}m^*\omega_c^2(x^2+y^2)|1s> = \frac{4\pi^2\kappa^2\epsilon_0^2\hbar^4}{e^2m^{*3}}B^2 = \sigma B^2 \quad , \tag{2.216}$$

$$\sigma = \frac{4\pi^2\kappa^2\epsilon_0^2\hbar^4}{e^2 m^{*3}}. \tag{2.217}$$

In a similar manner, the perturbation energy for an ns state is

$$\Delta\mathcal{E}_{ns} = \frac{n^2(5n^2+1)}{6}\sigma B^2. \tag{2.218}$$

As the energy is increased with increasing B, $\Delta\mathcal{E}_{1s}, \Delta\mathcal{E}_{2s}, \cdots$ are called the diamagnetic shift. That is to say the diamagnetic shifts of the s states are proportional to $\boldsymbol{B}^{\ 2}$.

For the p states, we have to add an orbital Zeeman energy arising from the term $\omega_c/2 \cdot L_z$ in Eq. (2.211). The $2p$ state splits into three states, $< L_z >= \pm 1,\ 0$. The energy of the p state in the weak field limit is predominated by the orbital Zeeman energy

$$\mathcal{E}_{L_z=\pm 1,\ 0} = L_z \frac{\hbar\omega_c}{2}. \tag{2.219}$$

Electronic transition from $1s$ state to $2p$ state is observed in far infrared absorption or photoconductivity. Stradling *et al.* studied the absorption arising from the $1s$ - $2p$ transition, and found that absorption comprises of many peaks [59–61]. The splitting is due to the so-called central cell correction or chemical shift of the $1s$ state. The energy of the $1s$ state is different depending on the species of each donor state. The extension of the $1s$ wave function is small and localized in the vicinity of the donor nuclei, so that the binding energy is sensitively modified by the specific potential shape of the donor atoms. On the other hand, the energy of the excited states as $2p$ states are almost independent of the donor atoms because the wave function is more extended and the effective mass approximation better holds. From the peak position of the $1s$ - $2p$ transition, we can identify the donor species. Thus the far infrared absorption spectra in magnetic fields can be utilized for identifying residual impurities in semiconductor crystals with an extremely high sensitivity.

B. High field limit ($\gamma \gg 1$)

When the effect of magnetic fields is much larger than that of the Coulomb interaction, the latter effect is considered to be a small perturbation to the magnetically formed energy states. To calculate the energy levels in such a situation, the adiabatic approximation is often used. This is a method which separates the wave function into a component in the $x-y$ plane and that in the z-direction. In the $x-y$ plane, only the magnetic field effect is taken into account and the Coulomb interaction is considered in the z-component. Namely, the total wave function is decomposed

$$\Psi_{NM\lambda} = \phi_{NM}(x,y) f_{NM\lambda}(z), \tag{2.220}$$

where $\phi_{NM}(x,y)$ is a wave function where the Coulomb interaction is zero. Based on such a wave function, Yafet, Keyes, and Adams calculated the ground state of an hydrogen atom in the high field limit by a variational method [62]. The outline of their theory (YKA theory) is as follows.

As the effective Coulomb potential for calculating the z-component $f_{NM\lambda}(z)$ the following expression is assumed.

$$V_{nm}(z) = -\int |\phi_{NM}|^2 \frac{e^2}{4\pi\kappa\epsilon_0} dxdy. \tag{2.221}$$

Using this potential, the Schrödinger equation for $f_{NM\lambda}$ is given by

$$\left[-\frac{d^2}{dz^2} + V_{NM}(z)\right] f_{NM\lambda}(z) = \mathcal{E}_{NM\lambda}(z). \tag{2.222}$$

The eigenstates of this equation are characterized by a quantum number λ which is associated with each set of N and M. The total energy is characterized by a set of quantum numbers (N, M, λ) and is given by

$$\mathcal{E}_{NM\lambda} = \left(N + \frac{1}{2}\right)\hbar\omega_c + \mathcal{E}^{(z)}_{NM\lambda}. \tag{2.223}$$

For $\gamma \to \infty$, the states tend to Landau level-like, so that their B dependence tends to the one which is proportional to B. This is in contrast to the B^2 dependence in the case of the weak field limit.

Yafet, Keyes, and Adams obtained the ground state energy using a variational function [62],

$$\varphi = (2^{\frac{3}{2}} a_\perp^2 a_\parallel \pi^{\frac{3}{2}})^{-\frac{1}{2}} \exp\left[-\frac{x^2+y^2}{(2a_\perp)^2}\right] \cdot \exp\left[-\frac{z^2}{(2a_\parallel)^2}\right], \tag{2.224}$$

where $a_\perp$ and $a_\parallel$ are variational parameters representing the extension of the wave function in the parallel and perpendicular direction to the magnetic field, respectively. Putting $\varepsilon = a_\perp/a_\parallel$, the variational energy can be obtained as

$$\mathcal{E} = \frac{1}{2a_\perp^2}\left(1 + \frac{\varepsilon^2}{2}\right) + \frac{\gamma^2 a_\perp^2}{2} - \frac{\sqrt{2}\varepsilon}{a_\perp\sqrt{\pi}\sqrt{1-\varepsilon^2}} \log\frac{1+\sqrt{1-\varepsilon^2}}{1-\sqrt{1-\varepsilon^2}}. \tag{2.225}$$

In Eq. (2.225), the energy is represented in a unit of the effective Rydberg energy. Minimizing the energy $\mathcal{E}$ with respect to $a_\perp$ and $a_\parallel$, the energy is obtained as

$$\mathcal{E}(B) = \gamma - \frac{1}{2a_\perp^2}\left(1 + \frac{\varepsilon^2}{2}\right) - \frac{\gamma^2 a_\perp^2}{2} + \frac{\sqrt{2}\varepsilon}{a_\perp\sqrt{\pi}\sqrt{1-\varepsilon^2}} \log\frac{1+\sqrt{1-\varepsilon^2}}{1-\sqrt{1-\varepsilon^2}}. \tag{2.226}$$

The magnetic field dependence of $a_\perp$ and $a_\parallel$ is shown in Fig. 2.17 (a), where the magnetic field is expressed in a unit of $\gamma = \hbar\omega_c/Ry$.

The binding energy of the 1s state $\mathcal{E}_B$ is given by

$$\mathcal{E}_B = \gamma - \mathcal{E}(B). \tag{2.227}$$

The magnetic field dependence of the binding energy $\mathcal{E}_B$ is shown in Fig. 2.17 (b). As the magnetic field is increased, $\mathcal{E}_B$ increases. This is interpreted as the

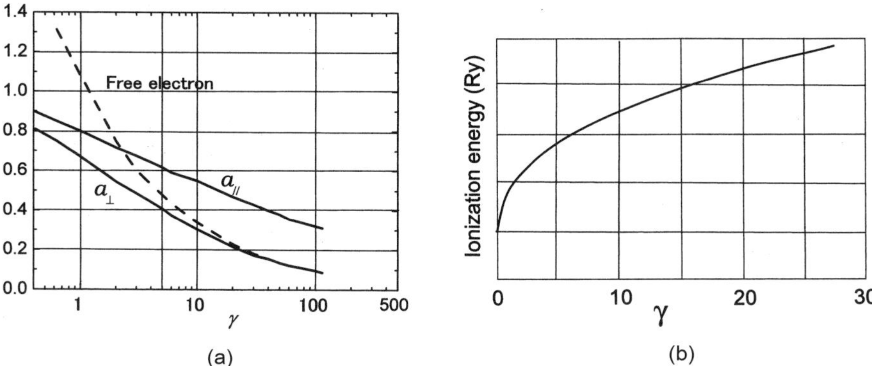

FIG. 2.17. (a) The magnetic field dependence of the wave function extension of the ground state in the direction perpendicular ($a_\perp$) and parallel ($a_\parallel$) to the magnetic field [62]. The wave function extension of free electron is also shown. (b) The magnetic field dependence of the binding energy of a donor state.

increase of the Coulomb interaction due to the shrinkage of the wave function by the magnetic field. Because of the increase of the binding energy in high magnetic fields, the carrier density in the conduction band decreases with increasing magnetic field at low temperatures. It is given as

$$n_c = \frac{nN_T}{N_T + N_I e^{\frac{E_B}{kT}}}, \tag{2.228}$$

where n_c is the number of carriers in the conduction band, N_I is the density of the donors, and

$$N_T = \frac{eB}{\pi^2 \hbar^2 c}(2m^* kT)^{\frac{1}{2}} \tag{2.229}$$

is the effective density of states in the conduction band.

The decrease of the carrier density in high magnetic fields is called the magnetic freeze out effect, in analogy to the carrier freeze out effect caused by decreasing temperature. The effect was studied first in n-type InSb [63] in which the magnetic field dependence of $\mathcal{E}$ is particularly large. It is found that as the magnetic field is increased at low temperatures, the resistance is increased by several orders of magnitude. The increase of the Hall coefficient implies that the carrier concentration is decreased by magnetic fields.

The huge magneto-resistance at low temperatures can also be interpreted in terms of the collapse of the impurity band. In semiconductors with a relatively large donor concentration and small effective mass like InSb, the wave functions of the donor states overlap with each other and the impurity band is formed. The electrical conduction is carried by the transfer of electrons among donor atoms through the impurity band. The impurity band is usually mingled with the conduction band for small effective mass semiconductors. Such semiconductors are called degenerate semiconductors. When high magnetic fields are applied to the impurity band, the wave function of the donor states is decreased so that their overlaps are diminished. Thus the break of the impurity band causes the metal-insulator transition. The problem of the donor levels in the high field limit has been studied by many authors [64–66].

C. Intermediate range ($\gamma \approx 1$)

The intermediate range ($\gamma \approx 1$) where the effects of magnetic field and the Coulomb interaction are comparable is the most difficult range to deal with. Often a variational method is employed to calculate the energy levels. In choosing variational functions, we have to keep in mind that for

$$B \rightarrow 0 \ (\gamma \rightarrow 0),$$

the wave function should have a hydrogen-atom-like exponential form,

$$\Psi \rightarrow \exp(-\alpha x),$$

whereas for $B \rightarrow \infty$ ($\gamma \rightarrow \infty$), it should have a Landau-level-like Gaussian form,

$$\Psi \rightarrow \exp(-\beta x^2).$$

Putting

$$\Psi = \sum_j c_j \phi_j,$$

the energy

$$\mathcal{E} = \frac{< \Psi|\mathcal{H}|\Psi >}{< \Psi|\Psi >}$$

is minimized. Various variational functions were introduced by many authors [67–73].

Here we show a result calculated by Makado and McGill by the variational method [69]. They employed a variational function,

$$\Psi_i = r^{|m|+q} \sin^{|m|}\theta \exp(im\varphi) \cos^q\theta \exp(-\alpha_i r^2 \sin^2\theta) \exp(-\beta_i r), \qquad (2.230)$$

with a spherical polar coordinates. Here α_i and β_i are the variational parameters, and $q = 0, 1$. Figure 2.18 shows the energies of several bound states in magnetic fields. We can see that some of the hydrogen-atom-like levels at zero field tend to be associated with the lowest Landau level, whilst some others tend to be associated with higher lying Landau levels.

It is an intriguing problem how each of the hydrogen-atom-like levels (n, l, m) at low fields is connected with the Landau-level-like levels $(N.M, \lambda)$ at high fields in between. This level conjunction problem has long been an unsolved fundamental problem of quantum physics. As mentioned above, the states characterized

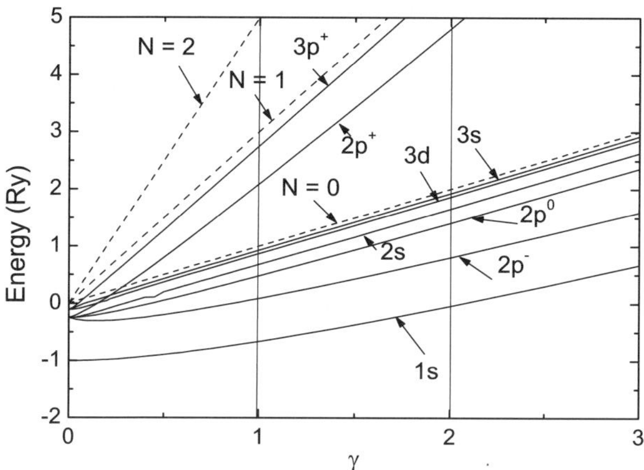

FIG. 2.18. Magnetic field dependence of the different bound levels in a hydrogen model [69].

by (n, l, m) originate from a potential with spherical symmetry whereas those characterized by (N, M, λ) have a cylindrical symmetry. The difference in the symmetry causes a difficulty in their conjunction.

Concerning the problem of the conjunction between the Rydberg series in the low field limit and the Landau-level-like series in the high field limit, Shinada *et al.* proposed a model of nodal plane conservation based on Kleiner's hypothesis on the nodal plane [74]. The wave functions of (n, l, m) and (N, M, λ) have certain nodal planes where the functions take zero. According to their model, the total number of the nodal planes is conserved in the conjunction. Also, states with the same number of nodal planes in the both limits are mixed together and their mutual crossing is solved.

On the other hand, Lee *et al.* proposed a non-crossing rule which insists that states with the same symmetry (for instance $m = 0$ even parity hydrogen states) do not cross with each other in a magnetic field [75]. It leads to a conclusion that all the $m = 0$ states at zero field lie below the lowest Landau level in magnetic fields.

Acceptor states in magnetic fields can be treated in a way similar to donor states [76]. However, the theory is more complicated because of the degeneracy of the valence band as we saw in Section 2.4.3.

2.8.2 *Exciton states*

A. *Three-dimensional excitons*

Exciton states are similar to impurity states in the sense that electrons are bound to a positive charge of holes, but as both electrons are holes are mobile, the motion of the center of gravity should be involved in the kinetic equations. The Hamiltonian of exciton states in a magnetic field is represented with the symmetric gauge by [77]

$$\begin{aligned}\mathcal{H} &= \frac{1}{2m_e^*}(\boldsymbol{p}_e + c\boldsymbol{A}(\boldsymbol{r}_e))^2 + \frac{1}{2m_h^*}(\boldsymbol{p}_e - e\boldsymbol{A}(\boldsymbol{r}_h))^2 - \frac{e^2}{4\pi\kappa\epsilon_0 r} \\ &= -\frac{\hbar^2}{2m_e^*}\nabla_e^2 - \frac{\hbar^2}{2m_e^*}\nabla_e^2 - \frac{e^2}{4\pi\kappa\epsilon_0 r} + \frac{ie\hbar}{m_e^*}\boldsymbol{A}(\boldsymbol{r}_e))^2 \cdot \nabla_e \\ &\quad -\frac{ie\hbar}{m_h^*}\boldsymbol{A}(\boldsymbol{r}_h))^2 \cdot \nabla_h + \frac{e^2}{2m_e^*}A(\boldsymbol{r}_e)^2 + \frac{e^2}{2m_h^*}A(\boldsymbol{r}_h)^2, \end{aligned} \tag{2.231}$$

where m_e^* and m_h^*, $\boldsymbol{r}_e$ and $\boldsymbol{r}_h$ are effective masses and position coordinate of electrons and holes, and suffices e and h stand for electrons and holes in ∇. The difference of (2.231) from (2.210) is that it contains terms for holes, because holes are also mobile as well as electrons. However, it can be rewritten in an easier form for handling, by transforming $\boldsymbol{r}_e$ and $\boldsymbol{r}_h$ to a relative coordinate

$$\boldsymbol{r} = \boldsymbol{r}_e - \boldsymbol{r}_h, \tag{2.232}$$

and the center of mass coordinate

$$\boldsymbol{\rho} = \frac{m_e^*\boldsymbol{r}_e + m_h^*\boldsymbol{r}_h}{m_e^* + m_h^*}. \tag{2.233}$$

By a canonical transformation

$$\Psi(\boldsymbol{\rho}, \boldsymbol{r}) = \exp\left(i\left[\boldsymbol{K} - \frac{e}{\hbar}\boldsymbol{A}(\boldsymbol{r})\right]\boldsymbol{\rho}\right)F(\boldsymbol{r}), \tag{2.234}$$

with the wave vector of the center-of-mass motion of excitons,

$$\boldsymbol{K} = \boldsymbol{k}_e - \boldsymbol{k}_h, \tag{2.235}$$

we obtain

$$\begin{aligned}\Bigg[-\frac{\hbar^2}{2\mu^*}\nabla^2 - \frac{e^2}{4\pi\epsilon r} + ie\hbar\left(\frac{1}{m_e^*} - \frac{1}{m_h^*}\right)\boldsymbol{A}(\boldsymbol{r})\cdot\nabla \\ + \frac{e^2}{2\mu^*}A(\boldsymbol{r})^2 - \frac{2e\hbar}{m_e^* + m_h^*}\boldsymbol{A}(\boldsymbol{r})\cdot\boldsymbol{K}\Bigg] F(\boldsymbol{r}) \\ = \left[\mathcal{E} - \frac{\hbar^2 K^2}{2(m_e^* + m_h^*)}\right] F(\boldsymbol{r}). \end{aligned} \tag{2.236}$$

where we used a reduced mass of excitons μ^* which is given by

$$\frac{1}{\mu^*} = \frac{1}{m_e^*} + \frac{1}{m_e^*}. \tag{2.237}$$

In (2.236), the first term represents the kinetic energy of the relative motion of electrons and holes, the second term the Coulomb potential energy between electrons and holes, the third term the Zeeman energy of the envelope function which is zero for s-states, the fourth term the diamagnetic energy. The fifth term can be converted to $-e(\boldsymbol{v} \times \boldsymbol{B}) \cdot \boldsymbol{r}$ ($\boldsymbol{v}$ is the velocity of the center of mass motion) which is equivalent to the Lorentz force acting on the motion of the exciton and considered to be an effective electric field. This term is in ordinary cases very small, because of the small wave vector of the incident light to create excitons. In the direct allowed transition, the absorption coefficient of exciton lines is proportional to $|F(0)|^2$, so that the only visible lines are usually those of s-states and d states (d_0).

In addition to the orbital terms expressed in (2.231), there is a spin Zeeman term

$$\mathcal{H}_z = g_e \mu_B \boldsymbol{s}_e B + g_h \mu_B \boldsymbol{s}_h B, \tag{2.238}$$

where g_e, g_h, $\boldsymbol{s}_e$, $\boldsymbol{s}_h$ are g-factors and the spins of electrons and holes. When taking account of the exchange interaction between the spins $\boldsymbol{s}_e$ and $\boldsymbol{s}_h$ the spin states split into singlet and triplet states. As the dominant terms are just the first and the second terms, the Hamiltonian for exciton states in a magnetic field is reduced to a problem similar to that for donor states. Similarly to (2.208) and (2.207), excitons in a three-dimensional space are characterized by the Bohr radius,

$$a_B^* = \frac{4\pi\hbar^2\kappa\epsilon_0}{\mu^* e^2}, \tag{2.239}$$

and the binding energy (Rydberg energy)

$$\mathcal{E}_n = -\frac{Ry^*}{n^2} \qquad (n = 1, 2, \cdots), \tag{2.240}$$

$$Ry^* = \frac{\mu^* e^4}{32\pi^2\hbar^2\kappa^2\epsilon_0^2} = \frac{\hbar^2}{2\mu^*}\frac{1}{a_B^{*2}}. \tag{2.241}$$

The only different point from the donor states is that m^* is replaced by μ^* here.

B. Two-dimensional excitons

In quantum wells and superlattices, excitons should have a nearly two-dimensional character. If excitons are located in a perfect two-dimensional space, the problem is connected with an interesting fundamental problem of a hydrogen atom in a two-dimensional space. The Schrödinger equation of a hydrogen atom is given by

$$\left[-\frac{\hbar^2}{2m}\Delta - \frac{e^2}{4\pi\epsilon_0 r}\right]\Phi = \mathcal{E}\Phi, \tag{2.242}$$

To obtain the energy and the Bohr radius for the $1s$ state, let us put the wave function

$$\Phi = C\exp\left(-\frac{r}{a}\right).$$

In a three-dimensional space, we can readily obtain

$$a = a_B = \frac{4\pi\hbar^2\epsilon_0}{me^2} \tag{2.243}$$

and the energy

$$\mathcal{E} = -\frac{\hbar^2}{2m}\frac{1}{a_B} = -Ry. \tag{2.244}$$

If we calculate a and $\mathcal{E}$ in a similar manner assuming a two-dimensional space, we get

$$a = \frac{a_B}{2}, \tag{2.245}$$

and

$$\mathcal{E} = -4Ry. \tag{2.246}$$

This indicates that in a two-dimensional space, electrons move in a smaller orbit around the proton in comparison to the three-dimensional case, so that the Coulomb binding energy is four times larger. The diamagnetic shift is proportional to $< x^2 + y^2 >$, so that it should be smaller in a 2D space than in a 3D space reflecting the smaller Bohr radius. The diamagnetic shift is given in (2.216) as

$$\Delta\mathcal{E}_{1s}(3D) = \sigma B^2.$$

In the 2D case, it can be obtained as

$$\Delta\mathcal{E}_{1s}(2D) = \frac{3}{16}\sigma B^2. \tag{2.247}$$

Akimoto and Hasegawa calculated the diamagnetic shift of the two-dimensional excitons including the excited states by the Wentzel-Kramers-Brillouin (WKB) method, and obtained the energy diagram as shown in Fig. 2.19 [78]. The motivation of their work was to explore the 2D excitons in layer-structured semiconductors such as GaSe. In these layered semiconductors, however, it was found that the anisotropy is not sufficiently large to observe the 2D character. It was after the development of quantum wells and superlattices that nearly 2D excitons are observed. In quantum wells, the dimensionality can be controlled by changing the well width L_w. Almost 2D in a very small L_w to nearly 3D in a large L_w. For most cases, however, it is difficult to achieve the 2D limit in a very thin well width, because the exciton wave function penetrates into the barrier layer due to the finite barrier height. Problems of nearly 2D excitons in quantum wells will be discussed in more detail in Section 5.2.3.

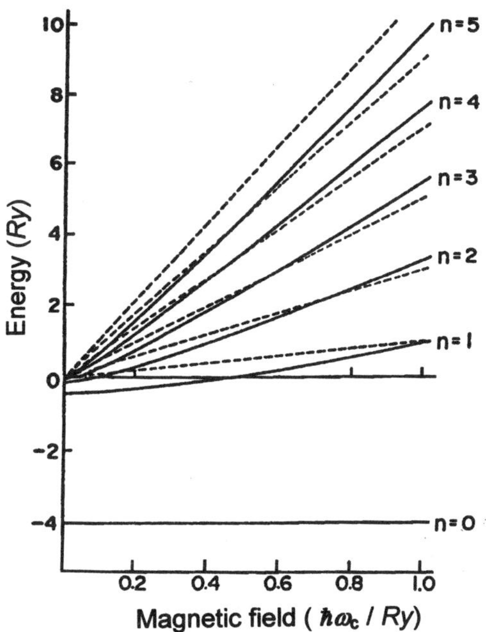

FIG. 2.19. Diamagnetic shift of 2D excitons calculated by Akimoto and Hasegawa [78].

3

MAGNETO-TRANSPORT PHENOMENA

3.1 Magneto-transport in high magnetic fields

3.1.1 *Magneto-conductivity and magneto-resistance*

The conductivity in the presence of magnetic field is expressed by the generalized Ohm's law as follows,

$$\boldsymbol{J} = \boldsymbol{\sigma} \cdot \boldsymbol{E}, \tag{3.1}$$

where $\boldsymbol{J}$ is the current density, $\boldsymbol{E}$ is the electric field vector and $\boldsymbol{\sigma}$ is the conductivity tensor. In a magnetic field, the conductivity tensor $\boldsymbol{\sigma}$ has non-diagonal components due to the Lorentz force. If the static magnetic field is applied in the z-direction, $\boldsymbol{\sigma}$ is represented as

$$\boldsymbol{J} = \begin{pmatrix} \sigma_{xx} & \sigma_{xy} & 0 \\ \sigma_{yx} & \sigma_{yy} & 0 \\ 0 & 0 & \sigma_{zz} \end{pmatrix} \cdot \boldsymbol{E}. \tag{3.2}$$

From Onsager's reciprocity relation,

$$\sigma_{yx} = -\sigma_{xy}. \tag{3.3}$$

Furthermore, if the system is isotropic,

$$\sigma_{yy} = \sigma_{xx}. \tag{3.4}$$

The resistivity tensor $\boldsymbol{\rho}$, the inverse tensor of the conductivity tensor, is defined by

$$\boldsymbol{E} = \boldsymbol{\rho} \cdot \boldsymbol{J}. \tag{3.5}$$

From Eqs. (3.1), (3.2), and (3.5), we obtain

$$\boldsymbol{\rho} = \begin{pmatrix} \rho_{xx} & \rho_{xy} & 0 \\ -\rho_{xy} & \rho_{yy} & 0 \\ 0 & 0 & \rho_{zz} \end{pmatrix} = \begin{pmatrix} \dfrac{\sigma_{yy}}{\sigma_{xx}\sigma_{yy} + \sigma_{xy}^2} & -\dfrac{\sigma_{xy}}{\sigma_{xx}\sigma_{yy} + \sigma_{xy}^2} & 0 \\ \dfrac{\sigma_{xy}}{\sigma_{xx}\sigma_{yy} + \sigma_{xy}^2} & \dfrac{\sigma_{xx}}{\sigma_{xx}\sigma_{yy} + \sigma_{xy}^2} & 0 \\ 0 & 0 & \dfrac{1}{\sigma_{zz}} \end{pmatrix}. \tag{3.6}$$

The expression of the resistivity tensor components in terms of those of the conductivity tensor is useful, because the former is the quantity which is measured in

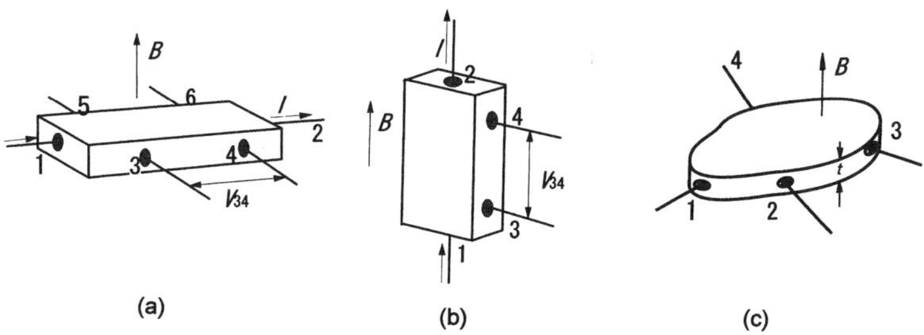

FIG. 3.1. Arrangement of contacts to a sample in the four probe method: (a) Transverse magneto-resistance and Hall effect, (b) Longitudinal magneto-resistance, (c) Transverse magneto-resistance and Hall effect of an arbitrary shape of the sample with a uniform thickness t.

experiments using a constant current condition, and the latter can be calculated theoretically as a response of electrons to the external fields $\boldsymbol{E}$ and $\boldsymbol{B}$.

In real experiments, the conductivity or the resistivity is measured by using the four probe method, as shown in Fig. 3.1. By separating the contacts to the sample for measuring the voltage (contacts 3 and 4) from those for supplying current (contacts 1 and 2), we can avoid the effect of contact resistance. The transverse magneto-resistance is measured in the arrangement, as shown in Fig. 3.1 (a). The current is perpendicular to the magnetic field. By supplying a constant current density J_x through contacts 1 and 2, we measure the electric field E_x as a potential drop across contacts 3 and 4. In this case, $J_y = J_z = 0$, so that

$$E_x = \rho_{xx} J_x. \tag{3.7}$$

Thus we can obtain a component of the resistivity tensor,

$$\rho_{xx} = \frac{\sigma_{yy}}{\sigma_{xx}\sigma_{yy} + \sigma_{xy}^2}. \tag{3.8}$$

$$\rho_{xy} = -\frac{\sigma_{xy}}{\sigma_{xx}\sigma_{yy} + \sigma_{xy}^2}. \tag{3.9}$$

The Hall voltage is obtained by measuring a voltage across contacts 3 and 5. The electric field E_y in this direction is

$$E_y = \rho_{yx} J_x = R_H B J_x. \tag{3.10}$$

Here

$$R_H = \frac{\rho_{yx}}{B} = \frac{\sigma_{xy}}{\sigma_{xx}\sigma_{yy} + \sigma_{xy}^2}\frac{1}{B} \tag{3.11}$$

is the Hall coefficient.

The longitudinal magneto-resistance is measured using a configuration of Fig. 3.1 (b), with magnetic field parallel to the current in the z-direction. In this case, the electric field measured across contacts 3 and 4 is

$$E_z = \rho_{zz} J_z, \tag{3.12}$$

$$\rho_{zz} = \frac{1}{\sigma_{zz}}. \tag{3.13}$$

When the sample has a well-defined rectangular shape as in (a) or (b), the electric field E and the current density J can be calculated from the actual voltage V_{34} which appears across contacts 3 and 4 and the current I supplied across contacts 1 and 2, by multiplying the geometrical factor of the sample. For example, $J = I/S$ and $E = V/l$ (S is the cross-sectional area, and l is the distance between contacts 3 and 4). When the sample has an arbitrary shape but a uniform thickness as in (c), it is not possible to know the accurate values of J and E. However, if we measure the resistance by two different combinations of the contacts, we can obtain approximate values of the resistivity ρ and the Hall coefficient R_H by

$$\rho = t\frac{\pi}{\ln 2}\left(\frac{r_{12,34} + r_{23,41}}{2}\right) f(r_{12,34}/r_{23,41}), \tag{3.14}$$

and

$$R_H = \frac{1}{B}(r_{13,24}(B) - r_{13,24}(0)), \tag{3.15}$$

where t is the thickness of the sample, $r_{12,34} = V_{34}/I_{12}$, *etc.*, and f is a function of $r_{12,34}/r_{23,41}$ varying between 1 and 0, which can be obtained numerically from a figure in the literature [79,80]. This method is called "Van der Pauw method".

In order to obtain the component of the conductivity tensor σ_{xx} from the transverse magneto-resistance more directly, the Corbino disk symmetry as shown in Fig. 3.2 is sometimes employed. The Corbino disk is a hollow-like disk with electrodes on the outer and inner side surfaces, and the current flows in the radial direction. We measure the voltage across the outer and inner surfaces. If we define the x axis radially and the y axis azimuthally, and supply a radial current (x-direction), then

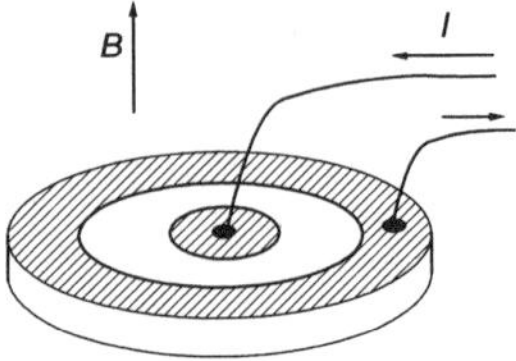

FIG. 3.2. Corbino disk.

$$E_y = -\rho_{xy}J_x + \rho_{yy}J_y = 0,$$

from the cylindrical symmetry. Therefore, we obtain

$$E_x = \rho_{xx}J_x + \rho_{xy}J_y \tag{3.16}$$

$$= \frac{1}{\sigma_{xx}}J_x. \tag{3.17}$$

Thus we can obtain σ_{xx} in a much simpler form than in Eqs. (3.7) and (3.8). Therefore, Corbino disks are conveniently used for measuring just σ_{xx}, although this is not a four probe method.

3.1.2 *Effect of scattering*

Let us consider the components of the conductivity tensor in more detail from the motion of electrons in the presence of both a static magnetic field and an electric field. The classical equation of motion of electrons is described as

$$m^*\frac{d\boldsymbol{v}}{dt} = -m^*\frac{\boldsymbol{v}}{\tau} - e(\boldsymbol{E} + \boldsymbol{v} \times \boldsymbol{B}). \tag{3.18}$$

If the magnetic field is in the z-direction and the electric field is in the x-direction, Eq. (3.18) is reduced to

$$m^*\frac{dv_x}{dt} = -m^*\frac{v_x}{\tau} - eE - ev_yB, \tag{3.19}$$

$$m^*\frac{dv_y}{dt} = -m^*\frac{v_y}{\tau} + ev_xB. \tag{3.20}$$

Here the first terms of the equations represent the scattering of electrons by various mechanisms with a relaxation time of τ. When the electron scattering is neglected, assuming $\tau \to \infty$, the solution of the equation is

$$x = A\cos(\omega_c t), \tag{3.21}$$

$$y = A\sin(\omega_c t) - \frac{E}{B}t, \tag{3.22}$$

where A is a constant and $\omega_c = eB/m^*$ is the cyclotron angular frequency. The motion is a cycloid motion as shown in Fig. 3.3 (a). The center of the circular motion of electrons moves towards the $-y$-direction, which is perpendicular to both the electric field and the magnetic field, with a velocity

$$v_d = \frac{E}{B}. \tag{3.23}$$

This velocity is called the drift velocity. Equation (3.22) shows that apart from the circular motion, electrons never move in the direction parallel to the electric field $\boldsymbol{E}$ in a long-term average.

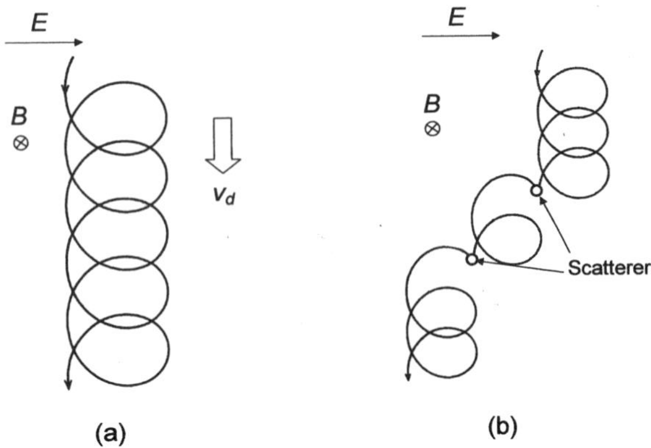

FIG. 3.3. Cycloid motion of an electron in the presence of a magnetic field B and a perpendicular electric field E. (a) without scattering, (b) with scattering.

Electrical conduction in the $\boldsymbol{E}$ direction occurs by the scattering of electrons. Namely, when the scattering takes place, the center of the cyclotron motion jumps to a new position. As such a jump takes place randomly at every scattering, the motion of the center coordinate is random. When the static electric field in the x-direction is present, however, the coordinate moves to the x-direction on average, as shown in Fig. 3.3 (b). This can be understood by obtaining the steady state solution of (3.19) and (3.20). If we assume the left sides of these equations to be zero taking the long time average, we obtain

$$\begin{aligned}\sigma_{xx} = \sigma_{yy} &= -\frac{nev_x}{E} \\ &= \frac{ne^2\tau}{m^*}\frac{1}{1+(\omega_c\tau)^2} \\ &= ne\mu\frac{1}{1+(\omega_c\tau)^2},\end{aligned} \tag{3.24}$$

$$\begin{aligned}\sigma_{xy} &= -\frac{nev_y}{E} \\ &= \frac{ne^2\tau}{m^*}\frac{\omega_c\tau}{1+(\omega_c\tau)^2} \\ &= ne\mu\frac{\omega_c\tau}{1+(\omega_c\tau)^2} = \omega_c\tau\sigma_{xx},\end{aligned} \tag{3.25}$$

where

$$\mu = \frac{e\tau}{m^*} \tag{3.26}$$

is the carrier mobility. In the long run, the electron moves along the electric field in average after repeated scatterings. Thus the conduction in the direction of the electric field occurs due to the scattering.

Equation (3.25) can be further transformed to

$$\sigma_{xy} = -\frac{ne}{B} - \frac{1}{\omega_c \tau}\sigma_{xx}. \tag{3.27}$$

In the limit of $\omega_c \tau \to \infty$ (3.27) leads to

$$\sigma_{xy} \to -\frac{ne}{B}. \tag{3.28}$$

3.1.3 *Quantum theory of transport phenomena in high magnetic fields*

As we saw in the classical treatment discussed in the previous section, change of the center coordinate by carrier scattering plays an important role in transport phenomena in magnetic fields. Components of the conductivity tensor can also be derived by quantum mechanical treatment. Quantum theory of the transport property of carriers in magnetic fields was developed by Kubo, Miyake, and Hashizume [2]. In the following, we survey the theory to derive the conductivity.

Suppose the magnetic field is in the z-direction, and there is a scattering potential $U(\boldsymbol{r})$. Using the quasi-momentum $\boldsymbol{P}$ defined by (2.21), the Hamiltonian is given by

$$\mathcal{H} = \mathcal{H}_0 + U(\boldsymbol{r}), \tag{3.29}$$

where

$$\mathcal{H}_0 = \frac{\boldsymbol{P}^2}{2m} = \mathcal{E}_0 \tag{3.30}$$

is the Hamiltonian representing the electron motion without the scattering potential. The magnetic field causes the cyclotron motion, but the conductivity is determined by the motion of the center coordinates (X, Y). So that we separate the the variables (x, y) into (X, Y) and (ξ, η) as in (2.28). Then the kinetic equations for P_x, P_y, ξ, and η can be expressed as,

$$\begin{cases} \dot{P}_x = \left(\dfrac{i}{\hbar}\right)[\mathcal{H}, P_x] = -eB\dfrac{\partial \mathcal{E}_0}{\partial P_y} - \dfrac{\partial U(\boldsymbol{r})}{\partial x}, \\ \dot{P}_y = \left(\dfrac{i}{\hbar}\right)[\mathcal{H}, P_y] = eB\dfrac{\partial \mathcal{E}_0}{\partial P_x} - \dfrac{\partial U(\boldsymbol{r})}{\partial y}, \end{cases} \tag{3.31}$$

$$\dot{\xi} = \frac{\partial \mathcal{E}_0}{\partial P_x} - \frac{1}{eB}\frac{\partial U}{\partial y}, \qquad \dot{\eta} = \frac{\partial \mathcal{E}_0}{\partial P_y} + \frac{1}{eB}\frac{\partial U}{\partial x}, \tag{3.32}$$

$$\dot{x} = \left(\frac{i}{\hbar}\right)[\mathcal{H}, x] = \frac{\partial \mathcal{E}_0}{\partial P_x}, \qquad \dot{y} = \left(\frac{i}{\hbar}\right)[\mathcal{H}, y] = \frac{\partial \mathcal{E}_0}{\partial P_y}. \tag{3.33}$$

From these relations, we can derive equations of motion for the center coordinates,

$$\dot{X} = \frac{1}{eB}\frac{\partial U}{\partial y},$$

$$\dot{Y} = -\frac{1}{eB}\frac{\partial U}{\partial x}. \tag{3.34}$$

We can see that the center coordinates are affected by the scattering due to the potential U and this leads to the current. The interesting point is that the motion of the center coordinates is caused by the gradient of the potential (electric field) in the perpendicular direction. This is because of the action of the Lorentz force caused by the magnetic field.

The components of the conductivity tensor can be obtained by starting from the general Kubo formula with respect to the current $\boldsymbol{j}_\mu$ in the μ direction,

$$\sigma_{\mu\nu}(\omega) = V^{-1}\int_0^\infty dt e^{-\omega t}\int_0^\beta d\lambda < \boldsymbol{j}_\nu(-i\hbar\lambda)\boldsymbol{j}_\mu(t) > \qquad (\mu,\nu = x,y,z) \tag{3.35}$$

giving the amplitude and phase of the AC current with a frequency ω. The tensor components for DC case ($\omega = 0$) can be obtained as follows.

$$\sigma_{xx} = \frac{e^2}{2kTV}\int_{-\infty}^{\infty} dt < \dot{\boldsymbol{X}}\dot{\boldsymbol{X}}(t) >,$$
$$\sigma_{xy} = \frac{e^2}{2kTV}\int_{-\infty}^{\infty} dt < \dot{\boldsymbol{X}}\dot{\boldsymbol{Y}}(t) >.$$

Here, β is $1/kT$ and

$$\boldsymbol{j}_\mu = \int \Psi(\boldsymbol{r})^\dagger j_\mu \Psi(\boldsymbol{r})d\boldsymbol{r}, \tag{3.36}$$

with the single electron expression of the current in the μ direction j_μ,

$$\dot{\boldsymbol{X}} = \int \Psi(\boldsymbol{r})^\dagger \dot{X}\Psi(\boldsymbol{r})d\boldsymbol{r},$$
$$\dot{\boldsymbol{Y}} = \int \Psi(\boldsymbol{r})^\dagger \dot{Y}\Psi(\boldsymbol{r})d\boldsymbol{r}. \tag{3.37}$$

In the above equations, $\Psi(\boldsymbol{r})$ is the many electron wave function in the second quantization. Taking account of the actual potential, σ_{xx} and σ_{xy} can be calculated. As we have seen in the classical treatment (3.28), in the strong magnetic field limit,

$$\sigma_{xy} \to -\frac{ne}{B}, \qquad R_H = \frac{1}{ne}. \tag{3.38}$$

On the other hand, σ_{xx} is calculated taking account of different scattering mechanisms. By using relations $\dot{X} = (i/\hbar)[U, X]$, *etc.*, it can be shown that

$$\sigma_{xx} = \frac{2\pi e^2}{\hbar V}\int_{-\infty}^{\infty} d\mathcal{E}\left(-\frac{\partial f}{\partial \mathcal{E}}\right)\sum_{NXp_z}\sum_{N'X'p'_z}(X-X')^2$$
$$< |< NXp_z|U|N'X'p'_z >|^2 >_s \delta(\mathcal{E}-\mathcal{E}_N(p_z))\delta(\mathcal{E}_-\mathcal{E}_{N'}(p'_z)) \tag{3.39}$$

Replacing the term associated with the scattering by τ^{-1}, we obtain

$$\sigma_{xx} \simeq \frac{n_{eff}e^2}{kT}\frac{1}{2}(X - X')^2\tau^{-1}. \tag{3.40}$$

Here $<>_s$ denotes the average over the scatters and the primes indicate the parameters after the scattering. The above formula shows that σ_{xx} is proportional to the scattering probability τ^{-1} and to the square of the shift of the center coordinate per scattering $\Delta X = X - X'$.

There are elastic scattering and inelastic scattering of carriers. An example of the former case is impurity scattering. A typical example of inelastic scattering is phonon scattering where electrons absorb or emit the energy of phonons at every collision.

In the case of elastic scattering, the feature of the conductivity is different depending on the extension of the potential (the force range) a relative to the extension of the electron wave function (the magnetic length) l. As $\Delta X = -l^2 q_y$ ($\boldsymbol{r}$ is the Fourier component of the scattering potential), the dominant contribution is given by different components of q_y. For a short-range potential with $a \ll l$, the largest contribution to the conductivity occurs for $|\Delta X| \sim l$. Therefore, $X - X'$ in (3.39) can be approximated by l. For a long-range potential with $a \gg l$, on the other hand, we can set $|\Delta X| \sim l^2/a$.

3.2 Shubnikov-de Haas effect

3.2.1 *Shubnikov-de Haas effect in bulk semiconductors and semimetals*

Equation (3.39) can be rewritten in the form

$$\sigma_{xx} \propto \sum_E \frac{\partial f(\mathcal{E})}{\partial \mathcal{E}}(X - X')^2 \mathcal{D}(N, k)\mathcal{D}(N', k'), \tag{3.41}$$

where $f(\mathcal{E})$ is the Fermi distribution function, $\mathcal{D}(N, k)$ and $\mathcal{D}(N', k')$ are the density of states in the initial and the final states in a scattering. If $kT \ll \mathcal{E}_F$, $\partial f/\partial \mathcal{E}$ shows a sharp peak at the Fermi energy $\mathcal{E}_F$. Therefore, σ_{xx} is proportional to the density of states at the Fermi level $\mathcal{E}_F$. As the density of states diverges when the energy is equal to the Landau level energy, the conductivity increases divergently every time the Fermi level $\mathcal{E}_F$ crosses the N-th Landau level $\mathcal{E}_N$, ie.

$$\mathcal{E}_F = \mathcal{E}_N. \tag{3.42}$$

More specifically, if the band structure is a simple parabolic dispersion in the three-dimensional case,

$$\sigma_{xx} \sim \int d\mathcal{E} \sum_{N,N'} \left(-\frac{\partial f}{\partial \mathcal{E}}\right) \frac{L_{NN'}(\mathcal{E})}{[\mathcal{E} - (N + \frac{1}{2})\hbar\omega_c]^{1/2}[\mathcal{E} - (N' + \frac{1}{2})\hbar\omega_c]^{1/2}}, \tag{3.43}$$

where $L_{NN'}(\mathcal{E})$ is a slowly varying function of $\mathcal{E}$. For $N = N'$, the integral diverges logarithmically $\mathcal{E} = (N + \frac{1}{2})\hbar\omega_c$. In the actual system, the Landau levels are broadened by the scattering, so that the divergence is avoided, but the conductivity shows a sharp change at this point. In two-dimensional systems, the density of states shows a delta-function-like divergence at the Landau levels. Therefore, the conductivity shows an oscillatory change with strong peaks every time $\mathcal{E}_F$ crosses the Landau levels. Here again, the divergence is suppressed by the broadening of the Landau levels due to the scattering. The oscillatory change of the magneto-resistance with sharp peaks at the condition (3.42) is called the Shubnikov-de Haas effect. The Shubnikov-de Haas effect is similar to the de Haas-van Alphen effect that appears in the magnetization. However, a difference is that the former occurs due to the dissipative phenomena involving carrier scattering whilst the latter is a phenomenon in a thermodynamical quantity.

The condition of the Shubnikov-de Hass effect can be written as

$$\mathcal{E}_F = \left(N + \frac{1}{2}\right)\hbar\omega_c. \tag{3.44}$$

As σ_{xx} exhibits sharp peaks, the transverse magneto-resistance ρ_{xx} shows maxima at the condition (3.44) from the relation (3.8). The longitudinal magneto-resistance ρ_{zz} also shows maxima in the same condition. When the energy band is a simple parabolic band, (3.44) can be transformed to

$$N + \frac{1}{2} = \frac{m^*\mathcal{E}_F}{\hbar e}\frac{1}{B}. \tag{3.45}$$

If $\mathcal{E}_F \cong$ constant as is the case mostly for $N \gg 1$, the oscillation is periodic with respect to $1/B$. In the case of a non-parabolic band, maxima appear whenever the Fermi level crosses the Landau level.

In semimetals such as Bi or graphite, the Hall conductivity σ_{xy} is almost zero due to the cancellation of the contribution of the same amount of electrons and holes. Therefore, from (3.8), $\rho_{xx} \cong 1/\sigma_{xx}$. The Shubnikov-de Haas peaks thus appear as minima in the transverse magneto-resistance [81]. As there are Landau levels of both electrons and holes, minima appear for electron levels and hole levels. Figure 3.4 shows experimental traces of the Shubnikov-de Haas effect in Bi and Bi-Sb alloys with different Sb concentrations x for magnetic fields applied parallel to the binary axis [82]. We can see spike-like drops at conditions (3.42). It is also clear that each peak has a longer tail in the high field side, reflecting the three-dimensional density of states of the Landau levels. In Bi, the overlap energy $\mathcal{E}_0$ between the electron band and the hole band is 38.5 meV. When we apply magnetic fields, $\mathcal{E}_0$ diminishes as the energy of the lowest Landau levels of electrons and holes increase and cross each other. At a magnetic field for such a crossing, semimetal to semiconductor transition takes place. The semimetal to semiconductor transition in Bi for $\boldsymbol{B} \parallel$ binary axis was observed in far-infrared transmission spectra at $B = 88$ T [83]. In Bi-Sb alloys, $\mathcal{E}_0$ decreases as the Sb concentration is increased, and such a transition can be observed at a lower field as the last minimum of the series of the Shubnikov-de Haas oscillation. For a sample of $x = 4.4\%$, there is a sharp rise in the magneto-resistance due to the semimetal to semiconductor transition.

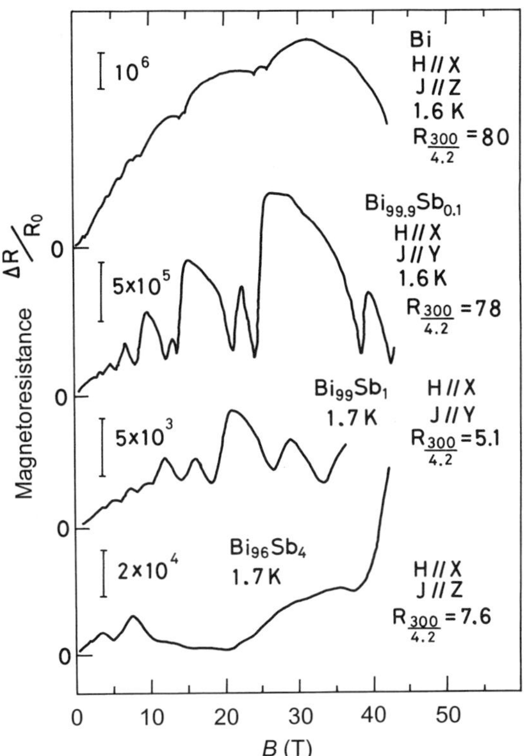

FIG. 3.4. Shubnikov-de Hass effect in Bi and Bi-Sb solid solutions with different Sb concentrations [82].

The Shubnikov-de Haas effect is also described by the semiclassical treatment, which we saw in Section 2.3. The Bohr-Sommerfeld quantization condition (2.106) for the Shubnikov-de Haas effect peak to occur is expressed as

$$A(\mathcal{E}_F, k_z) = (N + \gamma)\frac{2\pi}{l^2} = (N + \gamma)\frac{eB}{\hbar}, \tag{3.46}$$

where A is the extremal cross-section of the Fermi surface, defined by

$$\frac{\partial A}{\partial k_z} = 0, \tag{3.47}$$

on the Fermi surface. γ is the phase factor, usually 1/2. Thus the Shubnikov-de Haas effect is a useful means for obtaining information about the cross section of the Fermi surface.

For observing the Shubnikov-de Haas oscillation, a low temperature is essential, as a well-defined Fermi level is necessary. As temperature is elevated, the sharp edge of the electron distribution at the Fermi level is broadened by thermal excitation and the oscillation is obscured. The oscillatory part of the magneto-resistance has been studied theoretically by many authors. According to Roth and Argyres, a detailed calculation derives an expression of the magneto-resistance as follows [84]. The longitudinal magneto-resistance $\rho_{\parallel}$ is:

$$\rho_{\parallel} \cong \rho_0 \left[1 + \sum_{r=1}^{\infty} b_r \cos\left(\frac{2\pi\mathcal{E}_F}{\hbar\omega_c} r - \frac{\pi}{4}\right)\right], \tag{3.48}$$

where

$$b_r = \frac{(-1)^r}{r^{1/2}} \left(\frac{\hbar\omega_c}{2\mathcal{E}_F}\right)^{1/2} \frac{2\pi^2 rkT/\hbar\omega_c}{\sinh(2\pi^2 rkT/\hbar\omega_c)} \cos(\pi\nu r) \exp\left(\frac{-2\pi\Gamma r}{\hbar\omega_c}\right). \tag{3.49}$$

In (3.49), $\cos(\pi\nu r)$ is a result of the spin of the electron. The transverse magneto-resistance $\rho_{\perp}$ is:

$$\rho_{\perp} = \rho_0 \left[1 + \frac{5}{2} \sum_{r=1}^{\infty} b_r \cos\left(\frac{2\pi\mathcal{E}_F}{\hbar\omega_c} r - \frac{\pi}{4}\right) + R\right], \tag{3.50}$$

$$R = \frac{3}{4}\frac{\hbar\omega_c}{2\mathcal{E}_F} \left\{\sum_{r=1}^{\infty} b_r \left[\alpha_r \cos\left(\frac{2\pi\mathcal{E}_F}{\hbar\omega_c} r\right) + \beta_r \sin\left(\frac{2\pi\mathcal{E}_F}{\hbar\omega_c} r\right)\right] - \ln\left(1 - e^{4\pi\Gamma/\hbar\omega_c}\right)\right\}, \tag{3.51}$$

where b_r is given by (3.49) and

$$\alpha_r = 2r^{1/2} \sum_{s=1}^{\infty} \frac{1}{[s(r-s)]^{1/2}} e^{-4\pi s\Gamma/\hbar\omega_c}, \tag{3.52}$$

$$\beta_r = r^{1/2} \sum_{s=1}^{r-1} \frac{1}{[s(r-s)]^{1/2}}. \tag{3.53}$$

In (3.50) R is in practice much smaller than the first term. These expressions are a little complicated, but if we see the fundamental frequency ($r = 1$), it is clear that the resistance should oscillate with a frequency shown in (3.45) apart from a phase factor. In (3.49) and (3.52), Γ indicates the effect of damping arising from the broadening of the Landau levels. Γ/k is called Dingle temperature. It is a measure of the broadening and damping of the oscillation. In the above expressions of the Shubnikov-de Haas effect, if we take notice of only the temperature dependence of the fundamental frequency, we obtain the following simple relation for the amplitude of the oscillatory part of the magneto-resistance,

$$\Delta\rho \propto \frac{\chi}{\sinh\chi} \tag{3.54}$$

with $\chi = 2\pi^2 kT/\hbar\omega_c$. $\Delta\rho$ is a monotonically decreasing function of χ. Using this relation, the effective mass m^* can be obtained from the temperature dependence of the amplitude of the Shubnikov-de Haas oscillation.

3.2.2 *Shubnikov-de Haas effect in two-dimensional systems*

As the density of states of the Landau levels are delta-function-like, the Shubnikov-de Haas effect is more pronounced in two-dimensional systems than in three-dimensional cases. The Shunikov-de Haas effect in a 2D system was first reported by Fowler *et al.* for Si-MOS-FET [85, 86]. Since then, many studies have been made on the quantum transport phenomena in two-dimensional systems and these have led to the discovery of the quantum Hall effect. The oscillatory conductivity of the Shunikov-de Haas effect was theoretically derived by Ando as follows [14, 87, 88]:

$$\Delta\rho_{xx} \propto \frac{\chi}{\sinh\chi} \exp\left(-\frac{\pi}{\omega_c\tau}\right) \cos 2\pi(\nu - 1/2). \tag{3.55}$$

As a typical example of the Shubnikov-de Haas effect of Si-MOS-FET, we show the data by Kawaji and Wakabayashi in Fig. 3.5 [89]. Si has six valleys as shown in Fig. 2.8. When the growth direction is along the $< 100 >$ axis, the quantum subbands formed in the valleys in the $< 100 >$ and $< \bar{1}00 >$ direction (hatched valleys) have lower energies than the other valleys, because of the larger effective mass in the direction of the quantizing potential. With normal carrier concentration, electrons are populated only in the lowest subbands of these two valleys. Each Landau level has a four-fold degeneracy; two by the valleys, and two by spins. The spin degeneracy is lifted by magnetic field and the valley degeneracy is lifted by potentials. Accordingly, the Shubnikov-de Haas peak for each Landau level splits into four peaks. The larger splitting is due to the spin Zeeman effect and the smaller splitting is due to the valley splitting. In Si-MOS, the carrier concentration in the 2D channel can be controlled by the gate voltage V_G, so that instead of changing the magnetic field, we can sweep V_G at a constant magnetic field to observe the Shubnikov-de Haas effect.

In Fig. 3.5, each set of four split peaks belonging to the same Landau levels has nearly the same height. This feature was first explained by Ando *et al.* [90] analyzing the experimental data of Kobayashi and Komatsubara [91], as shown in Fig. 3.6. According to the quantum theory of the transverse conductivity, as discussed in Section 3.1.3, the peak height of the Shubnikov-de Haas oscillation can be derived from an expression like (3.39),

$$(\sigma_{xx})_{max} \sim \frac{e^2}{V}\mathcal{D}_{max}\frac{1}{\tau}(X - X')^2. \tag{3.56}$$

The change of the center coordinate of the N-th Landau level at the scattering is considered to be

$$X - X' \simeq 2r_c^{(N)}, \tag{3.57}$$

where $\mathcal{D}_{max}$ is the maximum of the density of the states and $r_c^{(N)}$ is the cyclotron orbit radius for the N-th Landau level,

$$r_c^{(N)} \simeq \left(N + \frac{1}{2}\right)^{1/2} l. \tag{3.58}$$

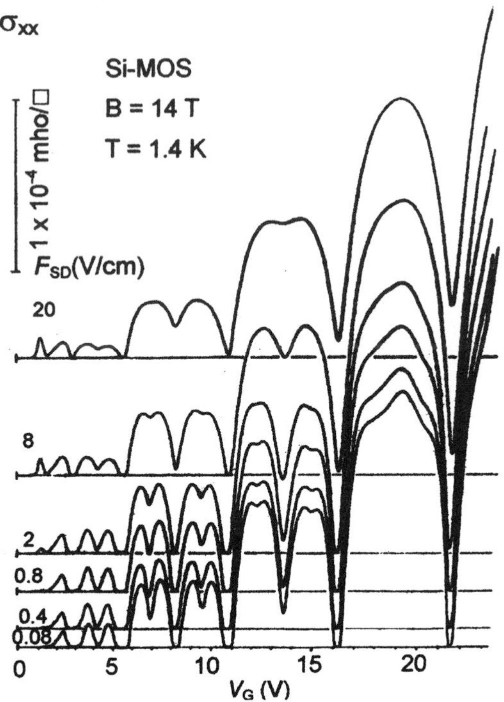

FIG. 3.5. Shubnikov-de Hass effect in an n-channel inversion layer of Si-MOS-FET [89]. The measurement was carried out for a Corbino disk sample. Transverse conductivity σ_{xx} is shown as a function of the gate voltage V_G which is almost proportional to the electron concentration. E_{SD} is the electric field across the source and drain electrodes.

On the other hand, $\mathcal{D}_{max}$ has a relation with the broadening of the 2D Landau levels. For the short-range scattering as in the case of impurity scattering, Ando showed by a self-consistent Born approximation that it is represented by the following formula [14, 92].

$$\rho(\mathcal{E}) = \frac{1}{2\pi l^2} \sum_N \frac{2}{\pi} \left[1 - \left(\frac{\mathcal{E} - \mathcal{E}_N}{\Gamma}\right)^2\right]^{1/2} \frac{1}{\Gamma}, \tag{3.59}$$

where Γ is the line width brought about by the impurity scattering, and represented by

$$\Gamma^2 \sim \frac{4N_i|V|^2}{2\pi l^2}, \tag{3.60}$$

with N_i being the density of scatters and V is the matrix element of the scattering. As the total density of states in one Landau level is expressed as (2.73),

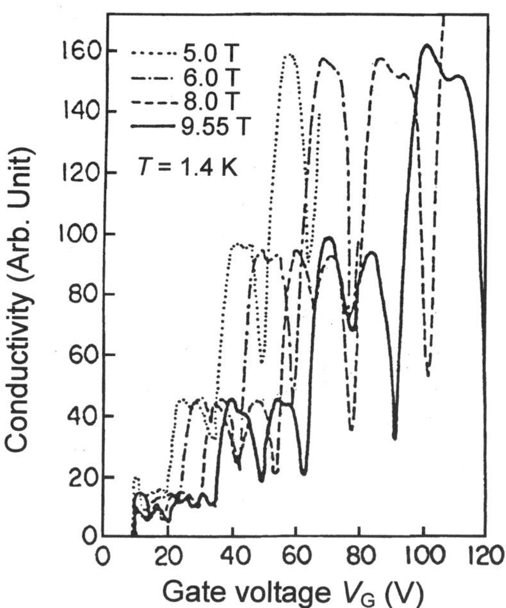

FIG. 3.6. Shubnikov-de Haas oscillation in Si-MOS at different magnetic fields [90].

$$D_{\max}\Gamma \cong \frac{1}{2\pi l^2}. \tag{3.61}$$

As we can set $\Gamma\tau \sim \hbar$ from the uncertainty principle, we can obtain

$$\begin{aligned}(\sigma_{xx})_{\max} &\sim \frac{e^2}{V}\frac{1}{2\pi l^2}\cdot\frac{1}{\Gamma}\cdot\frac{1}{\tau}(2l^2)\left(N+\frac{1}{2}\right)\\ &\sim \frac{4e^2}{Vh}\left(N+\frac{1}{2}\right).\end{aligned} \tag{3.62}$$

An interesting point is that the peak height does not depend on magnetic field nor on material parameters, which is actually observed in the experiment, as shown in Fig. 3.6. The simple relation above well explains the peak height of the Shubnikov-de Haas effect in Si-MOS-FET.

Another remarkable feature observed in Fig. 3.5 is that there are some ranges where σ_{xx} is completely zero. The zero σ_{xx} can be understood by considering the fact that the density of states becomes zero between the Landau levels. The fraction of the number of states of a Landau level filled with electrons,

$$\nu = \frac{n}{1/2\pi l^2} = \frac{n}{(eB/h)} \tag{3.63}$$

is called the filling factor. The zero conductivity occurs whenever the Fermi level crosses the gap between the Landau levels. Thus the condition of zero conductivity is that the filling factor ν of the Landau level becomes an integer number.

However, at first sight it looks strange that there are some finite ranges of the carrier density where the conductivity is zero, as the Fermi level should always lie on the finite density of states except for just one point even if the Landau level is discrete. The zero σ_{xx} was explained in terms of the localized states in the tail part on both sides of the Landau level in which the electrons are immobile. Immobile states in the Landau levels are made by the Anderson localization [93]. According to the scaling theory of Abrahams *et al.*, two-dimensional electron system are always localized at low enough temperatures [94]. However, in high magnetic fields, there are some mobile states in the center of the Landau levels. Therefore, there are extended states in the central part and localized states in the tail part in each Landau level in high magnetic fields. Such localized states are the origin of the zero conductivity. These states are also responsible for the quantum Hall effect, as will be discussed in the next section. Peaks of the Shubnikov-de Haas effect are designated by labels, $(N, \uparrow, +)$, $(N, \uparrow, -)$, $(N, \downarrow, +)$, and $(N, \downarrow, -)$, where N denotes the Landau level number, the arrows indicate the spin direction and $\pm$ denotes the valley. We can see that at low drain-source field E_{SD}, the lowest peak $(0, \uparrow, +)$ almost disappears. This means that the lowest Landau level is almost localized. As E_{SD} is increased, the peak becomes discernible because the electrons are heated up by the bias voltage.

3.3 Quantum Hall effect

3.3.1 *Integer quantum Hall effect*

In 1980, von Klitzing discovered a remarkable phenomenon in the Hall effect of Si-MOS-FET [95, 96]. Figure 3.7 shows the data of the Hall effect in Si-MOS-FET observed by von Klitzing, Dorda and Pepper [95]. In the region where the voltage along the current U_{pp} is zero in the vicinity of the integer filling of the N-th Landau level, the Hall voltage U_H shows a plateau, and the Hall resistance at the plateau is represented by

$$\rho_{xy} = \frac{h}{Ne^2}. \tag{3.64}$$

The characteristic features of the quantum Hall effect can seen in plots of σ_{xx} and σ_{xy}, as shown in Fig. 3.8 [97]. In the vicinity of the zero σ_{xx} at an integer ν, there are some plateau regions of the Hall voltage in some ranges of magnetic field or the gate voltage. Moreover, at the plateau, the Hall resistance takes a constant value which can be determined only by fundamental physical parameters, and is not dependent on any material parameters. When the filling factor (3.63) is an integer N,

$$N = \frac{n}{(eB/h)}. \tag{3.65}$$

If we assume that the above relation holds throughout the plateau region where $\sigma_{xx} = \sigma_{yy} = 0$, then we obtain Eq. (3.64) from (3.27). In the data of the Hall voltage as shown in Fig 3.7, von Klitzing *et al.* confirmed that the Hall resistance

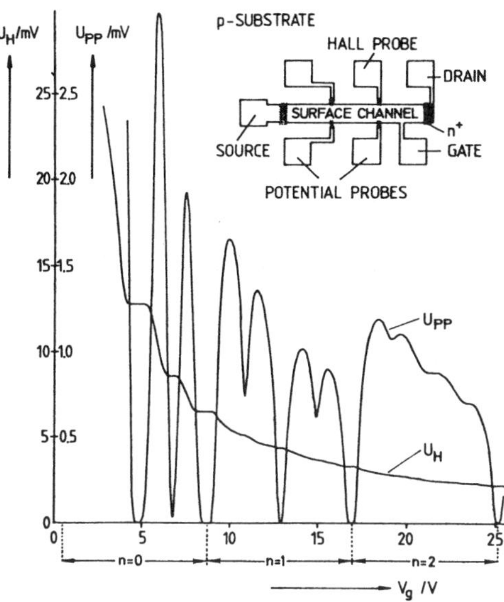

FIG. 3.7. Quantum Hall effect observed in Si-MOS [95]. U_H and U_{pp} are the Hall voltage and the voltage along the current, respectively.

actually is constant expressed by (3.64) within an accuracy of ppm in the plateau region. The phenomenon is called the quantum Hall effect. The value of ρ_{xy} for $N = 1$,

$$\frac{h}{e^2} = 25812.807449\ \Omega \quad \text{(recommended value of 2002 CODATA)}. \tag{3.66}$$

is called the "von Klitzing constant R_K"and now adopted as an international standard of the electrical resistance. A remarkable feature of the quantum Hall effect is that the plateau is almost completely flat and that the above value, which consists of only fundamental physical constants, can be determined extremely accurately. The quantity (3.66) has a similar form to the fine structure constant,

$$\alpha = \frac{e^2}{h}\frac{\mu_0 c}{2}, \tag{3.67}$$

so that the quantum Hall effect can also be employed for determining α very accurately.

The quantum Hall effect was first found in Si-MOS but it has been observed in many other two-dimensional systems. As the sample quality was improved, very pronounced plateaus are observed. A typical example of the data of the quantum Hall effect in GaAs/AlGaAs heterostructures is shown in Fig. 3.9 [98]. We can see that the plateau regions are very wide and ρ_{xy} rises from one plateau to the next very steeply.

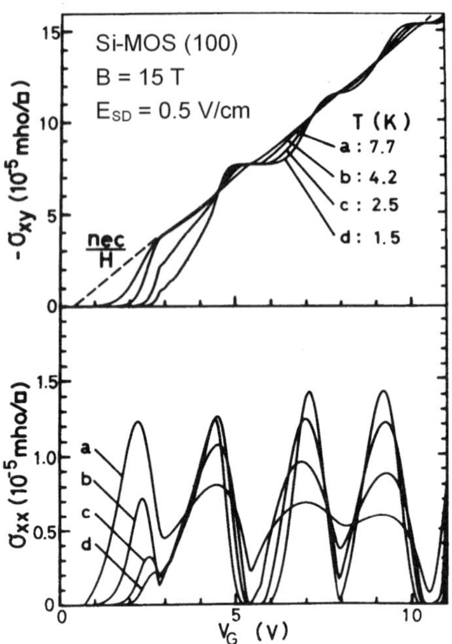

FIG. 3.8. Quantum Hall effect observed for the four $N = 0$ Landau levels in σ_{xx} and σ_{xy} in Si-MOS-FET at different temperatures [97]. By controlling the gate voltage V_G, the electron density n and the filling factor ν can be varied. In the region where $\sigma_{xx} = 0$, plateaus in σ_{xy} are observed. A straight broken line in the upper frame is a theoretical line expected for the classical Hall effect (ne/B).

The quantum Hall effect is now understood as a phenomenon associated with the localized states in the Landau level. As shown in Fig. 3.10 (a), there are localized states in the tail part of the both sides of the Landau levels (the hatched area in (a)). As shown in (b), the diagonal conductivity σ_{xx} is zero when the Fermi level ($\mathcal{E}_F$) is in this region. In the central part of the levels there are extended states and the conductivity σ_{xx} is finite when the Fermi level $\mathcal{E}_F$ is in this region. The Hall conductivity $-\sigma_{xy}$ is an integer 0 or 1 in the unit of e^2/h when $\mathcal{E}_F$ is in the localized states, and changes in between when $\mathcal{E}_F$ is in the extended states. For simplicity, the case for the first Landau level is shown in Fig. 3.10. But if we convert the range of the abscissa from [0, 1] to [$N-1$, N] and that of the ordinate from [0, e^2/h] to [$(N-1)e^2/h$, Ne^2/h], a similar shape can be obtained for the N-th Landau level.

The quantum Hall effect has been explained by several theoretical models. Laughlin showed elegantly that the effect is explained by considering the gauge invariance [99]. Below we introduce the argument of Laughlin on the quantum Hall effect. The arrangement of the sample, magnetic field, and the current for measuring the Hall effect is shown in Fig. 3.11 (a). If there is a periodic boundary

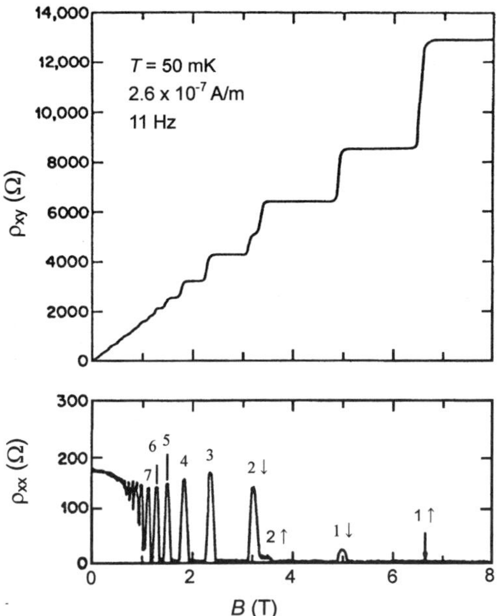

FIG. 3.9. Quantum Hall effect observed in a GaAs/AlGaAs heterostructure [98].

condition we can connect both edges of the sample and think of a ring with a radius r and the length w as shown in Fig. 3.11 (b). The magnetic field is applied in the perpendicular direction to the ring surface, and the bias voltage is applied to the axial direction of the ring (x-direction). We introduce a fictitious magnetic flux ϕ inside the ring. The vector potential in the azimuthal direction due to the flux is

$$A\theta = L\phi, \tag{3.68}$$

where L is the circumference of the ring, $L = 2\pi r$. As discussed in Section 2.2, if all the states are localized, the effect of magnetic flux on the electron energy can be cancelled by converting the electron wave function

$$\Psi \to \exp\left(ieAx/\hbar\right)\Psi. \tag{3.69}$$

For extended states, on the other hand, from the boundary condition,

$$A = N \cdot \frac{h}{eL} \tag{3.70}$$

is necessary. When there is an electric field along the x-axis, the electron energy is changed by the introduction of ϕ. As discussed in Section 2.5.1, in the crossed

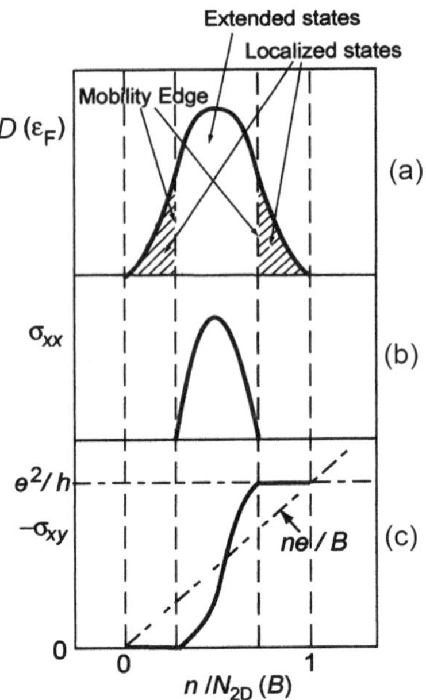

FIG. 3.10. Density of states and the conductivity as a function of the position of the Fermi level. The abscissa is represented in scale of the filling factor of the Landau level $\nu = n/N_{\rm 2D} = 2\pi l^2 n$. (a) Density of states $\mathcal{D}(\mathcal{E})$ of the first Landau level. (b) Conductivity σ_{xx} as a function of $\mathcal{E}_F$. (c) Conductivity $-\sigma_{xy}$ as a function of $\mathcal{E}_F$.

electric and magnetic field as shown in Fig. 3.11 (b), the energy of the electrons is

$$\mathcal{E}_{k,N} = \left(N + \frac{1}{2}\right)\hbar\omega_c + \frac{m^*}{2}\left(\frac{E}{B}\right)^2 - eEX, \tag{3.71}$$

and the momentum in the y-direction is

$$\hbar k_y = -eBX - m^*\frac{E}{B}. \tag{3.72}$$

The vector potential in the presence of a static field perpendicular to the sample surface is

$$\boldsymbol{A} = (0, Bx, 0). \tag{3.73}$$

Addition of a flux $\Delta\phi$ causes a change in $\boldsymbol{A}$ by ΔA in the y-direction,

$$\boldsymbol{A} \to (0, Bx + \Delta A, 0) = \left(0, Bx + \frac{\Delta\phi}{L}, 0\right), \tag{3.74}$$

so that it gives rise to a shift of X

$$X \to X - \frac{\Delta A}{B}. \tag{3.75}$$

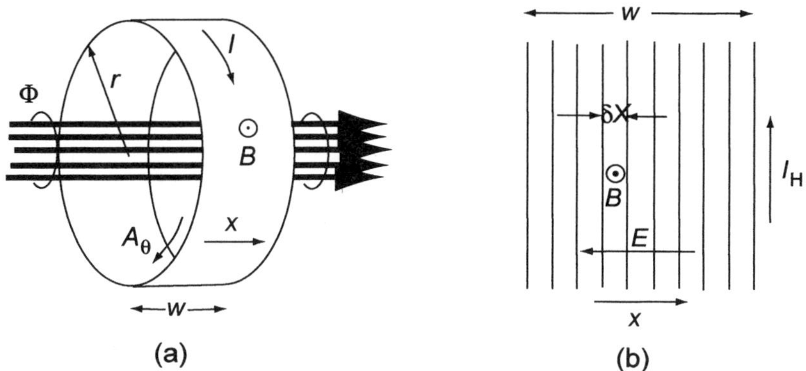

FIG. 3.11. (a) Ring after rolling a sample and connecting both edges. The bias electric field is applied parallel to the axis of the ring (x-direction). The Hall current is induced in the azimuthal direction of the ring. The external magnetic field is perpendicular to the surface of the ring. The radius of the ring is r. Fictitious magnetic flux Φ is introduced in the ring. (b) View from the top of the sample in the normal direction along the external magnetic field.

The shift of the center coordinate (3.75) gives rise to the change of energy

$$\Delta\mathcal{E} = eE\frac{\Delta A}{B}. \tag{3.76}$$

When we increase a flux by a unit flux $\Delta\phi = h/e$,

$$\Delta A = \frac{\Delta\phi}{L} = \frac{h}{Le}, \tag{3.77}$$

and

$$\Delta X = \frac{\Delta A}{B} = \frac{h}{LeB}. \tag{3.78}$$

ΔX is equal to the spacing between the center coordinates of neighboring states, as shown in (2.71) in Section 1.1. Therefore, the addition of the unit flux corresponds to the shift of the electrons to the neighboring states. If all the extended states are occupied, the energy change by this shift is equal to that for the move of integer number (n) of electrons from one electrode to the other over a potential V, and is written as

$$\Delta U = neV. \tag{3.79}$$

As the Hall current I is considered to be a derivative of the energy with respect to the flux,

$$I = \frac{\Delta U}{\Delta\phi} = n\frac{e^2}{h}V. \tag{3.80}$$

Therefore, we can obtain the quantized Hall conductivity,

$$\sigma_{xy} = \frac{I}{V} = n \cdot \frac{e^2}{h}. \tag{3.81}$$

Since the energy of the localized states does not change by a change of A, the above argument holds even if there are localized states. In this way, Laughlin's explanation showed that the quantum Hall effect is a result of a very fundamental quantum phenomenon described by the gauge invariance under the condition that $\sigma_{xx} = 0$ holds due to the Anderson localization.

The quantum Hall effect was also explained by an analysis based on the Kubo formula by Aoki and Ando [100]. They showed that if the states at the Fermi level $\mathcal{E}_F$ are localized σ_{xy} is constant, and is $-Ne^2/h$ when $\mathcal{E}_F$ is in the localized states between the N-th and N+1-th Landau level.

In the quantum Hall regime, the current path is quite peculiar because the Hall angle is almost 90°, and the current is pushed towards the edges of the current. Near the edges of the sample, the energy of the Landau levels is increased due to the potential near the edge, and the electrons display a skipping motion, as shown in Fig. 2.14. If there are N Landau levels below the Fermi level, there are N states which conduct current. The current paths by these edges states are called edge channels. The quantum Hall effect is also explained by the edge current which flows through the edge channels. Figure 3.12 (a) shows energies of Landau levels with a finite width as a function of the center-coordinate of the cyclotron motion X, and Fig. 3.12 (b) shows the edge currents in a Hall bar sample with six contacts in the quantum Hall regime. Electrons in the edge channels at the same edge conduct a motion in the same direction, and the velocity in the j-th channel is given by

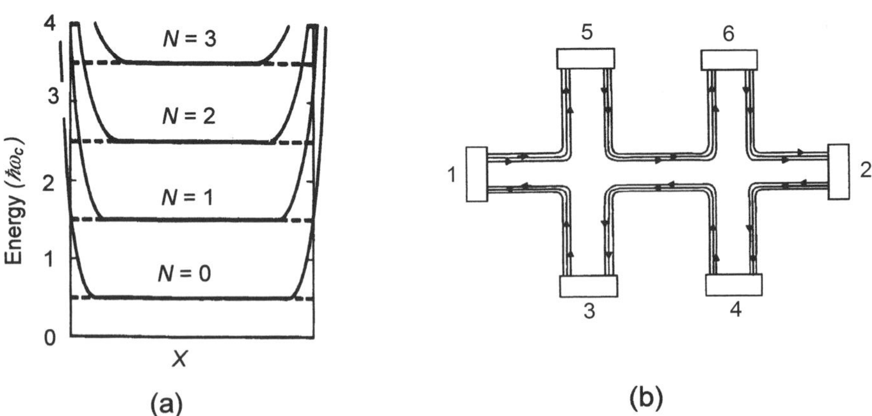

FIG. 3.12. (a) Energy of edge states in a sample with a finite width as a function of the center-coordinate of the cyclotron motion. (b) Edge currents in a Hall bar sample in the quantum Hall effect regime. A case when there are two channels is depicted.

$$v_j(k) = \frac{1}{\hbar}\frac{\partial \mathcal{E}_j(k)}{\partial k}, \tag{3.82}$$

where $\mathcal{E}_j(k)$ is the energy of the edge states at the center coordinate $X = -l^2 k$ (see Eq. (2.58)). The total current through the edge channels is then

$$J = \frac{-e}{\hbar}\frac{1}{2\pi}\sum_{j=1}^{q}\int_{k_j^{min}}^{k_j^{max}} dk \frac{\partial \mathcal{E}_j(k)}{\partial k} = -\frac{e}{h}\sum_{j=1}^{q}\left[\mathcal{E}_j(k_j^{max}) - \mathcal{E}_j(k_j^{min})\right], \tag{3.83}$$

where k_j^{max} and k_j^{min} are the maximum and minimum wave vector component of the edge states occupied by electrons, and q is the number of edge channels below the Fermi level. In the presence of the current J, there is a difference in the chemical potential between the left and right electrodes in Fig. 3.12 (a), and this difference gives the Hall voltage V_H. The Hall voltage is considered to be the average of the energy differences of the edge states in both the left and the right edges, so that

$$-eV_H = \frac{1}{q}\sum_{j=1}^{q}\left[\mathcal{E}_j(k_j^{max}) - \mathcal{E}_j(k_j^{min})\right] = -\frac{1}{q}\frac{Jh}{e}, \tag{3.84}$$

From (3.83) and (3.84), we obtain the Hall resistivity,

$$\rho_H = \rho_{xy} = \frac{V_H}{J} = \frac{h}{e^2}\frac{1}{q}. \tag{3.85}$$

Thus the Hall effect is quantized with an integer q. More detailed treatment including the mixing of the bulk states and the edge states as well as the many electrodes, the quantum Hall effect in terms of the edge states is explained by using the Landauer formula [101].

In the quantum Hall regime, the diagonal conductivity σ_{xx} is zero, but increasing the temperature, a certain number of electrons are thermally excited to higher Landau levels and from the fully occupied Landau level. Thus from the temperature dependence with a form $\sigma_{xx} = \sigma_0 \exp(-\Delta/2k_BT)$, we can obtain the energy gap Δ. For an even quantum number, Δ is $\hbar\omega_c$, and for an odd quantum number, Δ should be the spin splitting $g^*\mu_B B$. In GaAs, the effective g-factor is very small (–0.44), so that for the odd quantum number, Δ should be very small. In reality, however, such an excitation energy should be larger than expected from the g-factor of a single electron [102, 103]. This is due to the exchange interaction between different spin states. Due to the exchange interaction, when spins are excited from the ground state, several spins are collectively reversed to form a smoothly changing spin state rather than a reversal of just a single spin. Such an excitation is called Skyrmion. The Skyrmion excitation was observed not only in transport measurements but also in nuclear magnetic resonance (NMR) experiments [104] and magneto-optical experiments [105].

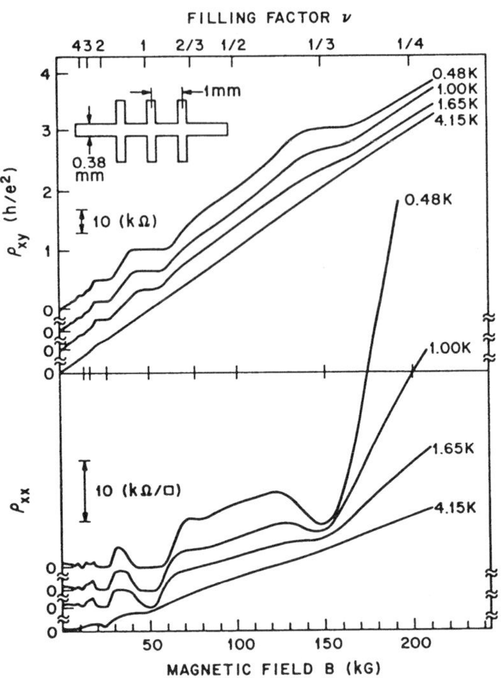

FIG. 3.13. The data of the fractional quantum Hall effect first observed by Tsui, Störmer, and Gossard in a GaAs/AlGaAs heterostructure [106]. Plateaus are observed in ρ_{xy} and minima in ρ_{xx} at fractional filling factors.

3.3.2 *Fractional quantum Hall effect*

As shown in the previous section, the quantum Hall effect discovered by von Klitzing is characterized by a vanishing σ_{xx} and a plateau of σ_{xy} at an integer filling factor. In 1982, a similar phenomenon was discovered at a fractional filling factor in a GaAs/AlGaAs heterostructure by Tsui, Störmer and Gossard [106]. Figure 3.13 shows the data of ρ_{xx} and ρ_{xy} in GaAs/AlGaAs reported by Tsui *et al.* [106]. The plateaus were observed at filling factors $\nu = 1/3$ or $2/3$, and simultaneously σ_{xx} shows sharp drops. Later it was found that the phenomenon takes place generally at a fractional filling with odd numbers of denominator such as $1/3, 2/3, 2/5, 3/5, 3/7, 4/7, \cdots$ in two-dimensional electron systems. This effect is now called the "fractional quantum Hall effect" (FQHE), as distinguished from the "integer quantum Hall effect" which occurs at an integer filling. The FQHE is generally observed in high quality samples at low temperatures (usually below $T < 1$ K). As the quality of the samples of GaAs/AlGaAs was improved, many more structures corresponding to different fractions were observed as an extremely sharp drop of σ_{xx} [107–109]. An example of such structures of σ_{xx} is shown in Fig. 3.14 [108]. The dip of σ_{xx} shows an activation behavior as

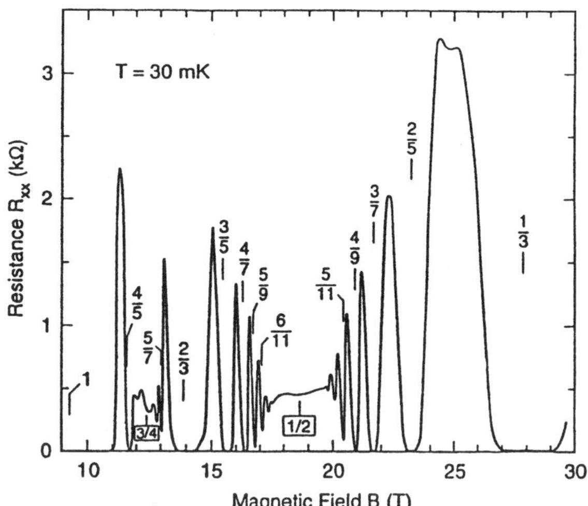

FIG. 3.14. The fractional quantum Hall effect observed in a high mobility GaAs/AlGaAs system at $T = 30$ mK by Du *et al.* [108]

$$\rho_{xx} \propto \exp\left(-\Delta/kT\right), \tag{3.86}$$

indicating the existence of some energy gap.

An explanation of the origin of the FQHE was also given by Laughlin [110] in terms of electron-electron interaction. As described in Section 2.1, electron wave function of the ground state ($N = 0$) in a magnetic field is written as

$$\Psi_{0m} = |0, m>= \frac{1}{(2\pi l^2 2^m m!)^{\frac{1}{2}}} \left(\frac{z}{l}\right)^m \exp\left(-\frac{|z|^2}{4l^2}\right), \tag{3.87}$$

in the symmetric gauge, where we put

$$z = x + iy. \tag{3.88}$$

If we assume that the sample is a disk with a radius of R, the maximum value M of m is determined by

$$(2M + 1)^{\frac{1}{2}} l = R, \tag{3.89}$$

so that

$$M \simeq \frac{1}{2}\left(\frac{R}{l}\right)^2. \tag{3.90}$$

Setting l a unit of length ($l = 1$), the wave function of a single electron in the ground state is generally written as

$$\Psi(z) = C \exp\left(-\frac{1}{4}|z|^2\right) f(z). \tag{3.91}$$

From (3.87), $f(z)$ is a polynominal of M-th order at most, so that (3.91) can be written as

$$\Psi(z) = C\exp\left(-\frac{1}{4}|z|^2\right)\prod_{m=1}^{M}(z - z_m). \tag{3.92}$$

z_m can be considered as a kind of vortex, since the phase is changed 2π by a rotation around it. Since the maximum value of m is M, the density of the vortices n_{vor} is

$$n_{\text{vor}} = \frac{M}{\pi R^2} = \frac{1}{2\pi l^2} = \frac{B}{\phi_0} = \frac{eB}{h}. \tag{3.93}$$

The density of electrons n_e, on the other hand, is

$$n_e = \frac{n_{\text{total}}}{\pi R^2}. \tag{3.94}$$

The filling factor is

$$\nu = \frac{n_{\text{total}}}{M} = \frac{n_e}{n_{\text{vor}}} = \frac{B}{\phi_0} = \frac{eB}{h}. \tag{3.95}$$

Laughlin considered that the ground state of the many electron wave function for a filling factor of $\nu = 1/m$ should be written as

$$\Psi_m = \left[\prod_{i>j}(z_i - z_j)^m\right]\exp\left(-\frac{1}{4}\sum_k |z_k|^2\right). \tag{3.96}$$

As the sign of the wave function should be reversed by commuting the two particles i and j, m in (3.96) should be an odd integer. The above wave function was derived by a superb intuition and analogy, but it was shown that it really represents the lowest energy of an interacting many electron system in a magnetic field. The state is a new type of quantum liquid. The many electron system represented by a wave function (3.96) is called the Laughlin state. Thus when the filling factor is $1/m$ with an odd integer, the fractional quantum Hall state is realized. For $\nu = 1 - 1/m$ such as 2/3, 4/5, $\cdots$, holes in the Landau level forms the Laughlin states, so that the FQHE occurs.

When ν deviates slightly from $1/m$ or $1 - 1/m$, the state consists of the Laughlin state with $1/m$ and some quasi-particles (quasi-electrons or quasi-holes) excited from the Laughlin state. As the number of quasi-particles is increased, the quasi-particles form a new Laughlin state, and the system forms a Laughlin state of electrons plus a Laughlin state of quasi-particles when

$$\nu = \frac{1}{m \pm \frac{1}{m_1}}. \tag{3.97}$$

When ν deviates further from (3.97), the quasi-particles excited from the state forms a new Laughlin state. In this way, many daughter states appear as sub-structures of a hierarchy represented by

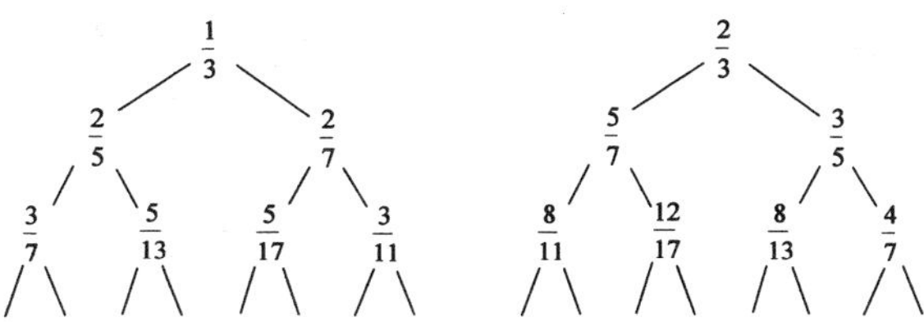

FIG. 3.15. Hierarchy of the fractional quantum Hall effect starting from the $\nu = 1/3$ (left) and $\nu = 2/3$ (right).

$$\nu = \cfrac{1}{m + \cfrac{\alpha_1}{m_1 + \cfrac{\alpha_2}{m_2 + \cfrac{\alpha_3}{m_3 + \cdots}}}}, \tag{3.98}$$

where m_i (i =1, 2, $\cdots$) is an even integer and α_i is 0 or ± 1. The hierarchy of structures originating from $\nu = 1/3$ and $2/3$ is shown in Fig. 3.15. Thus not only the $1/m$ or $1 - 1/m$ peaks, but also other observed FQHE peaks were successfully explained by the Laughlin theory [111–113].

Later, it was found that the FQHE can also be interpreted in terms of a different model, called "composite Fermion". In very high mobility samples, many sharp minima are observed for ρ_{xx}, as shown in Fig. 3.14. When we carefully inspect the data such as Fig.3.14, we find that the graph shows almost a mirror symmetry with respect to $\nu = 1/2$. Moreover, the strength of the peaks is not in the order of the hierarchy as shown in Fig. 3.15. Based on these observations Jain proposed a composite Fermion model [114, 115]. According to this model, a heavy Fermion is composed of an electron and associated with even number of unit fluxes $\Phi_0 = h/e$. In the composite Fermion model, when a total magnetic flux density B is applied to an electron system of the concentration of n, we consider that an even number $2p$ quantum fluxes are accompanying each electron to form a composite Fermion, and that the remaining flux is an effective field for composite Fermions as shown in Fig. 3.16. Defining the number of unit flux q in a flux density B, we obtain

$$B = q\Phi_0. \tag{3.99}$$

If we define the number of the unit flux q_0 and the flux density B_0 corresponding to a filling factor $\nu = 1/2N$ ($N = 1, 2, \cdots$),

$$q_0 = 2Nn. \tag{3.100}$$

$$B_0 = q_0 h/e. \tag{3.101}$$

The FQHE is regarded in this model as the Shubnikov-de Haas effect or the integer quantum Hall effect of composite Fermions for an effective field $B^* = |B - B_0|$. As discussed in Section 3.2, peaks of the Shubnikov-de Haas effect or

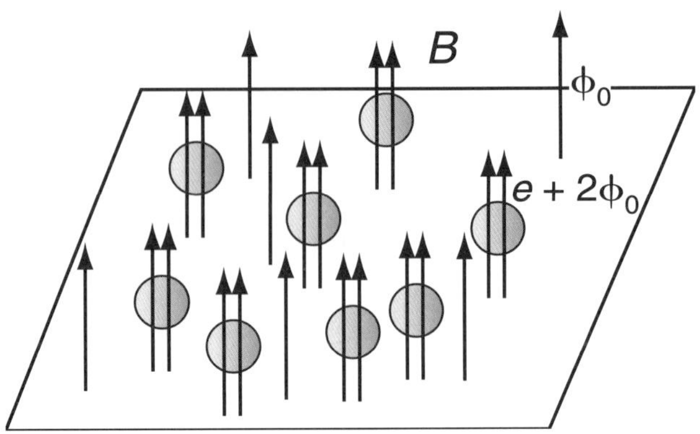

FIG. 3.16. Composite Fermion model. The case for $p = 1$ (two flux quanta attached to one electron) is shown.

the quantum Hall plateaus occur when a certain number of the Landau levels are completely filled with electrons. For composite Fermions, this condition for the p-th peak is given by

$$pN_{2D}(B^*) = p\frac{e}{h}B^* = n, \tag{3.102}$$

and then

$$p|q - 2Nn| = n. \tag{3.103}$$

Thus we obtain a filling factor for the p-th peak in the Shubnikov-de Haas effect of composite Fermions as

$$\nu = \frac{n}{q} = \frac{p}{2pN \pm 1}. \tag{3.104}$$

The above argument implies that the condition for the integer quantum Hall effect of composite Fermions is that for the FQHE of electrons. For instance, the FQHE of electrons at $\nu = \frac{1}{3}, \frac{2}{5}$ corresponds to the integer quantum Hall states of $\nu = 1, 2$ ($N = 1$). Related to the composite Fermion model, Willet *et al.* found an anomalous change of the velocity and the attenuation of the surface acoustic wave at $\nu = 1/2$ [116, 117]. Similarities of electrons for B to composite Fermions for B^* are also found. Kang *et al.* observed the geometrical resonance in the transport of antidot superlattices at $\nu = 1/2$ which is similar to the resonance at $B = 0$ [118]. The mean free path l of composite Fermions is found to be $\sim$2 μm, in comparison to $\sim$50 μm for electrons. From the temperature dependence of the amplitude of Shubnikov-de Haas effect oscillation, the effective mass of composite Fermions in GaAs has been determined [88, 108]. It was found that the composite Fermion mass is more than ten times larger than that of electrons. Figure 3.17 shows the effective mass of composite Fermions m^*_{cf} as a function of the filling factor ν, obtained from such an analysis [108]. It shows a large

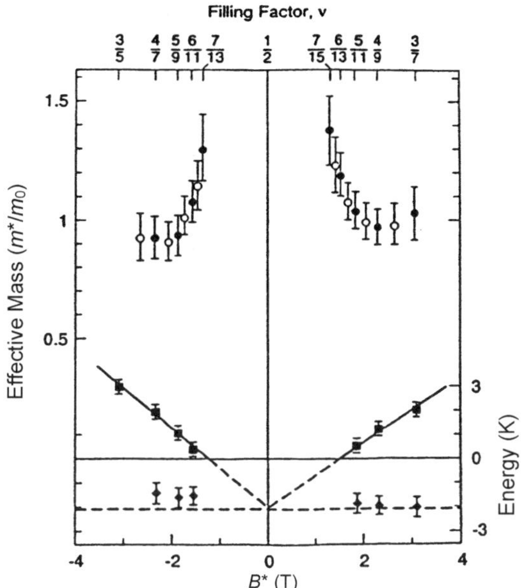

FIG. 3.17. Upper graph: Effective mass of composite Fermions determined from the temperature dependence of the Shubnikov-de Haas effect oscillation. Lower graph: Energy gap determined from the temperature dependence of the activation energy [108]. Negative values of the energy represent the broadening factor. The horizontal axis stands for the effective magnetic field for composite Fermions $B^* = B - B_0$.

enhancement at ν =1/2. From the linear dependence of the energy gap on B^*, m^*_{cf} is also obtained by assuming that it scales $\hbar\omega_c = \hbar eB^*/m^*_{cf}$. The evaluated values of m^*_{cf} is 1.0 m (for $B^* > 0$) and 0.82 m (for $B^* < 0$) which are consistent with the range of masses obtained from the Shubnikov-de Haas effect analysis. In recent experiments with very high quality samples, the minima of σ_{xx} emanating from $\nu = 1/6$ and 1/8 ($N = 3$ and 4 series in (3.102)) as well as $\nu = 1/2$ and 1/4 [109].

The fractional quantum Hall state is considered to be an incompressible liquid consisting of the interacting electrons which appears at low temperatures and it is different from the normal electron gas. When the density of electrons is small enough, the potential energy due to the Coulomb interaction between electrons becomes more important than the kinetic energy and electrons are considered to form a kind of solid crystal called a Wigner crystal [119, 120]. The Wigner crystal is considered to be more easily realized for two-dimensional (2D) systems than for the 3D case and is actually observed in electrons trapped on liquid He surface [121, 122]. Low density electrons in 2D systems (low ν number) are expected to form a Wigner crystal at high magnetic fields when the temperature is sufficiently low [123, 124]. The Wigner crystal itself should carry an infinite current as a superconductor, but it should be pinned by imperfections of the host crystal, so that the Wigner crystal should exhibit an insulating behavior. Many investigations have been made to explore Wigner crystals in 2D systems at high magnetic fields and low temperatures, and there is some evidence that Wigner crystals have been observed. Jiang *et al.* found that ρ_{xx} vanishes at $\nu = 1/5$, and shows an activated behavior with elevating temperature indicating that the Laughlin quantum liquid state forms the ground state. However, they observed a diverging increase of ρ_{xx} at ν slightly higher than 1/5 and at lower ν with decreasing temperature [124].

3.3.3 *Measurement of quantum Hall effect in pulsed magnetic fields*

Integer and fractional quantum Hall effects can be measured in pulsed magnetic fields. As the resistance change is very large including both a large resistance and almost zero resistance in the quantum Hall region, transient phenomena arising from the stray capacitance of the coaxial cables for the transmission of the signal sometimes distort the wave form of the signal. Therefore we have to be careful of the wave-form distortion. In order to reduce the effect of the stray capacitance of the cables, Takamasu, Dodo and Miura used an active shielding circuit by a voltage follower-type preamplifier for each contact of the sample, and succeeded

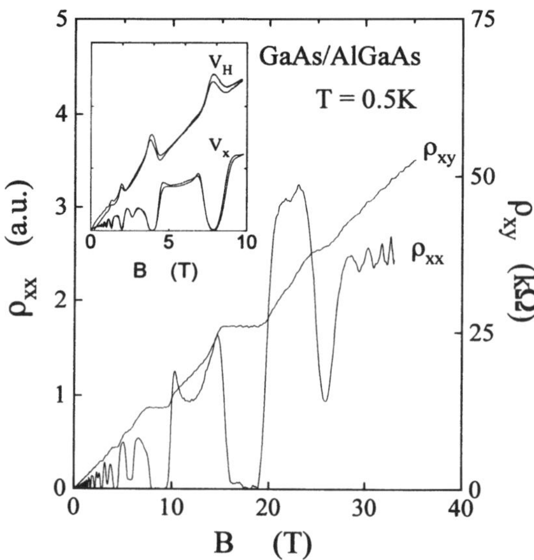

FIG. 3.18. Quantum Hall effect of GaAs/AlGaAs heterostructure observed in pulsed high magnetic fields [125]. The inset shows an example of data without using the active shielding circuit.

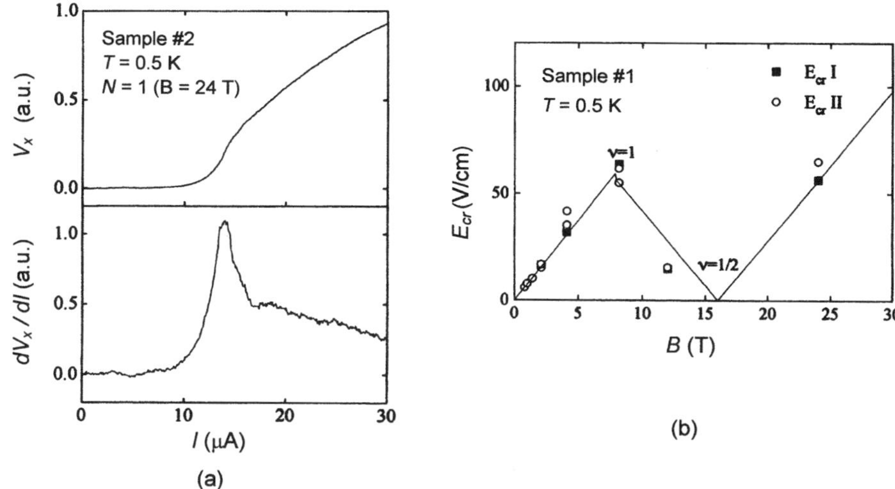

FIG. 3.19. Breakdown phenomena of the quantum Hall effect observed in pulsed high magnetic fields [125]. (a) Upper panel: Voltage V between the contacts for ρ_{xx} as a function of the current I. Lower panel: Derivative of V with respect to I to see the non-linear $V-I$ characteristics more clearly and define the critical electric field E_{cr}. $B = 24$ T (corresponding to $\nu = 1$). (b) Critical field E (V/cm) as a function of B for a different sample (Sample 1). E_{cr} I is determined from the peak in dV_x/dI and E_{cr} II is from the plateau width. Both results show a reasonably good agreement.

in measuring the quantum Hall effect avoiding such spurious signals [125].

Figure 3.18 shows the data of the quantum Hall effect in a GaAs/AlGaAs heterostucture measured in pulsed high magnetic fields. The data without the active shielding show a large distortion of the graphs of both ρ_{xx} and ρ_{xy}, whereas the main frame data show fine curves of the two quantities without such a distortion. The sharp minimum of ρ_{xx} at 26 T corresponds to a fraction $\nu = 2/3$, and thus we can study the fractional quantum Hall effect in pulsed fields.

One of the interesting aspects of the quantum Hall effect is the breakdown of the quantum Hall states. At sufficiently high bias current, the quantum Hall state breaks down and ρ_{xy} starts to deviate from the quantized value (3.64). At the same time, ρ_{xx} starts to increase from a minimum value. The breakdown phenomenon of the quantum Hall effect has been studied by many researchers, and explained by different models:

(1) an abrupt increase of scattering rate such as Cherenkov type emission of phonons by electrons moving faster than sound velocity [126–128];

(2) inter-Landau level transitions in the presence of a static potential [129];

(3) electron heating by applied electric field via the energy balance of the electron system [130, 131].

A good correspondence between the quantum Hall effect of electrons and composite Fermions suggests that there should be a breakdown of the FQHE at high electric field as of the integer quantum Hall effect. Takamasu, Dodo and Miura observed a breakdown of the FQHE at $\nu = \frac{1}{3}, \frac{2}{3}$, as well as at integer ν [125]. Figure 3.19 shows their experimental data of the breakdown. A sharp rise of the voltage along the current direction above the critical current is seen. The critical electric field E_{cr} corresponding to the critical current has a magnetic field dependence. For the IQHE, the E_{cr} increases linearly with increasing B. For the FQHE peaks at $\nu = 2/3$ and $1/3$, it should be noted that the E_{cr} fits well with the lines drawn from B_0 ($\nu = 1/2$) suggesting that it is determined by $B^* = |B - B_0|$.

3.4 Magneto-tunneling phenomena

3.4.1 *Magneto-tunneling in tunnel diodes*

Since the first discovery in a heavily doped $p - n$ Ge junction (Esaki diode) by Esaki [132], tunneling phenomena in semiconductors have been investigated in many different quantum heterostructures by many researchers. The simplest example is tunneling through a single quantum barrier, as shown in Fig. 3.20. We can fabricate such a tunneling diode, for example, from a GaAs/AlGaAs heterostructure. Assuming that the electronic states are those of free electrons in the regions I and III, the electron wave functions are represented as

$$\begin{aligned} \Psi_I &= A\exp(ikx) + B\exp(-ikx), \\ \Psi_{II} &= C\exp(-\alpha d), \\ \Psi_{III} &= D\exp(ikx). \end{aligned} \tag{3.105}$$

The transmission probability is then

$$TT^* = |D/A|^2 = \left[1 + \frac{1}{4}\frac{V_0^2}{\mathcal{E}(V_0 - \mathcal{E})}\sinh^2(\alpha d)\right]^{-1}, \tag{3.106}$$

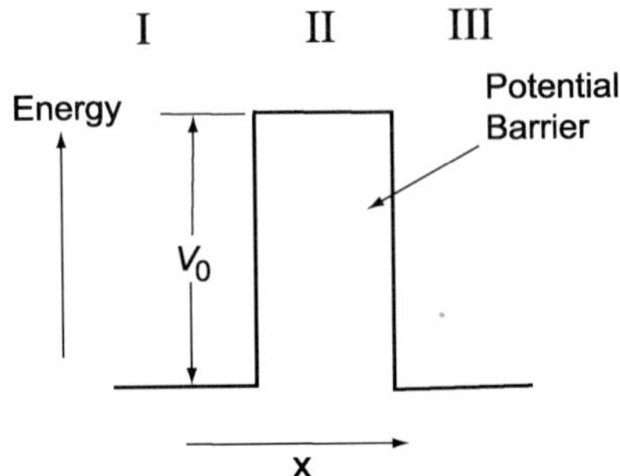

FIG. 3.20. Energy profile of a single barrier tunneling device. A potential barrier layer II exists in between the regions I and III.

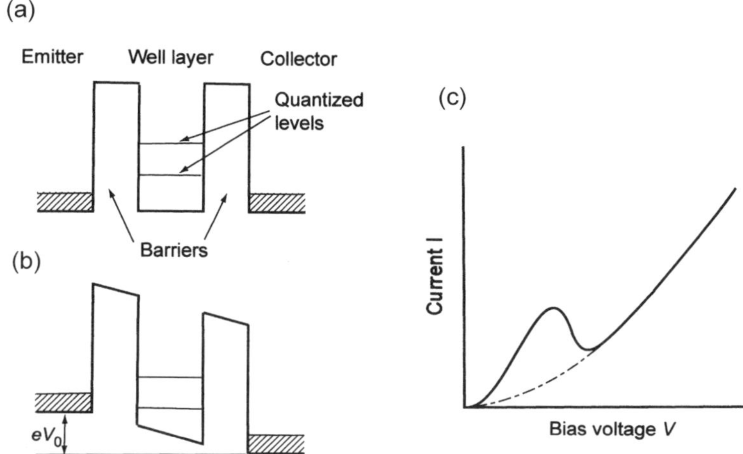

FIG. 3.21. (a) Structure of a double barrier resonant tunneling diode. (b) Energy diagram when the bias voltage V is applied between the emitter and collector electrodes. (c) Schematic voltage-current characteristics of a double barrier diode. The inset shows the energy profile when the emitter-collector voltage is applied.

where T denotes the probability amplitude of the tunneling. The tunneling current from the emitter electrode in the left to the collector electrode in the right through the barrier layer is represented as

$$\begin{aligned} J_{L\to R} &= \sum_k v_z f_L(1-f_R)TT^* \\ &= \frac{2e}{(2\pi)^3\hbar}\int\int d^2k_{\parallel}dk_z\left(\frac{\partial \mathcal{E}_z}{\partial k_z}\right) f_L(1-f_R)TT^*, \end{aligned} \tag{3.107}$$

where f_L and f_R denote the electron distribution function in the region I and region III.

A double barrier resonant tunneling diode has a structure as shown in Fig. 3.21 (a). In such a structure, a quantum well layer is sandwiched between the barrier layers, and heavily doped emitter and collector layers are put on both sides for contacts. In such a structure, quantum levels are formed in the quantum well, which are lying in higher energies than in the emitter or collector layers. However, by applying a bias voltage V_0 across the diode, the relative positions of the energy levels in each region can be varied as shown in Fig. 3.21 (b). Thus the levels in the quantum well can be tuned to those in the emitter. Electrons in the emitter layer can tunnel into the well layer when they can find an energy level with the same energy in the well layer. The tunneling process taking place between energy levels which are equalized is called resonant tunneling. Once the electrons tunnel into the well layer, they can tunnel out to the collector layer if there are empty states in the emitter layer. Therefore, the tunneling current flows under the condition that the emitter states have the same energy as those in the quantum well.

In terms of the wavelength of the electron wave packet, electrons in the energy state confined in the well have a wavelength corresponding to the standing wave between the barriers. Electrons in the emitter layer which have energy equal to those in the standing wave in the well have the same wavelength. Therefore the above resonant tunneling condition can be expressed in such a way that the tunneling occurs when the electrons which have the same wavelength as the standing wave in the well can pass through the structure just like the Fabry-Pérot interference filter. The tunneling probability in double barrier tunneling diodes can be calculated as in the single barrier structure, and it can be shown that the tunneling probability is 1 under the resonant condition.

The voltage-current characteristics of a double barrier tunneling diode are shown in Fig. 3.21 (c). At zero bias, the current is zero because the energy states in the emitter are lower than in the well. As the bias voltage is increased, the energy gradient increases. The current takes an extremum when the emitter levels reaches the confined energy level in the well, The current increases as the bias is further increased because more levels become available for the tunneling. The non-linear $V - I$ characteristics can be employed for switching circuits *etc.*

3.4.2 *Magneto-tunneling for $B \perp$ barrier layers*

When we apply magnetic fields to such double barrier tunneling diode, many interesting phenomena take place depending on the direction of the magnetic field relative to the layers. First let us consider the case where the field is perpendicular to the layers (B $\perp$layers). For this field direction, oscillatory conductivity is observed due to the Landau level formation. The behavior is different depending on the nature of the electron system in the emitter states; whether it is three-dimensional or two-dimensional. First let us consider the three-dimensional emitter states. In the emitter, electron states are quantized as shown in the left side of Fig. 3.22. Their energy is represented as

$$\mathcal{E}_e = \left(N_e + \frac{1}{2}\right)\hbar\omega_c(e) + \frac{\hbar^2 k_z^2}{2m^*}. \tag{3.108}$$

The energy in the well layer is fully quantized and given by

$$\mathcal{E}_w = \left(N_w + \frac{1}{2}\right)\hbar\omega_c(w) + E_0 - eV, \tag{3.109}$$

where E_0 is the energy difference of the lowest energy between the emitter states and the layer states without the field. The electron tunneling from the emitter states to the states in the well layer (well states) is allowed only when their energies and the Landau quantum numbers coincide with each other. Namely, the condition of the tunneling is

$$\mathcal{E}_e = \mathcal{E}_w,$$

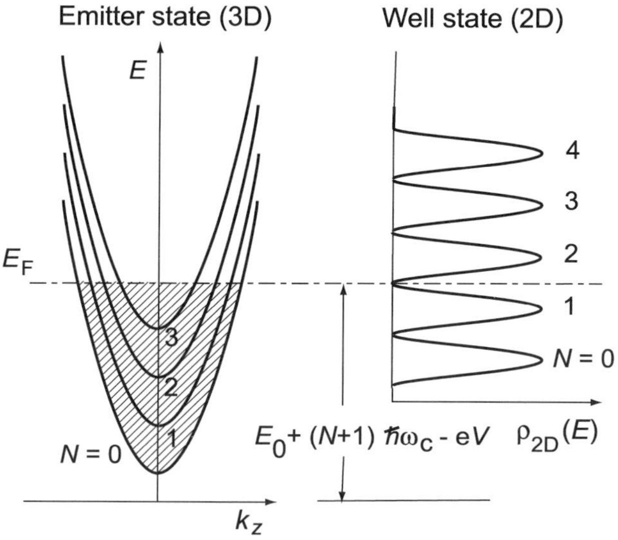

FIG. 3.22. Quantized energy levels in the emitter (left) and the well layer (right) when the magnetic field B is applied perpendicular to the layers and the bias voltage of V is applied.

$$N_e = N_w. \tag{3.110}$$

Thus when we tune the energy by the bias V, the conductivity shows a step-wise change. The steps appears at a condition

$$\mathcal{E}_F = \mathcal{E}_0 + (N+1)\hbar\omega_c(w) - eV. \tag{3.111}$$

Such oscillatory magneto-tunneling current was first observed by Mendez, Esaki and Wang [133].

When the emitter states are quantized by the triangular potential cause by the band bending due to doping, the emitter states are also two-dimensional. Then the tunneling occurs between the 2D states. In such an occasion, the allowed transition is between the states with the same Landau quantum number; $N_e = N_w$ as above. However, the resonant tunneling also occurs as weaker forbidden transition for $N_e \neq N_w$ as well. Therefore, when V or B is varied, a current peak is observed every time the Landau levels coincide with each other. Figure 3.23 shows a typical example of the oscillatory current as a function of magnetic field in a GaAs/AlGaAs double barrier resonant tunneling diode [134].

Applying a large bias voltage above the negative differential resistance region (see Fig. 3.21), we usually observe subsidiary structures due to the longitudinal optical (LO) phonon-assisted tunneling [135–138]. When a LO phonon is emitted at the tunneling from the emitter state to the quantum well state, the conservation rule (3.110) is violated and tunneling from the N-th Landau level of the

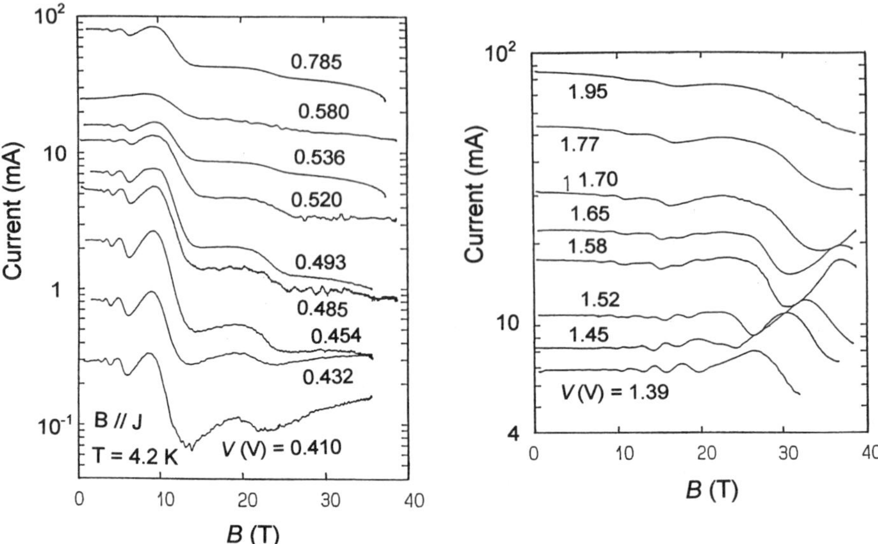

FIG. 3.23. Oscillatory magneto-tunneling current in a GaAs/AlGaAs double barrier resonant tunneling diode as a function of magnetic field applied perpendicular to the layers [134].

emitter to the N'-th Landau level of the quantum well becomes possible. The condition for the resonance transition is then

$$\mathcal{E}_{e0} = \mathcal{E}_{w0} + (N - N')\hbar\omega_c + \hbar\omega_o, \tag{3.112}$$

where ω_o is the angular frequency of the LO phonon and $\mathcal{E}_{e0}$ and $\mathcal{E}_{w0}$ are the bottoms of the emitter states and quantum states, respectively. These peaks were actually observed in the experiments. In addition, a weak peak is observed arising from the nonresonant tunneling transition involving elastic scattering from the N-th Landau level in the emitter to the $N+1$-th Landau level in the well [137].

3.4.3 *Magneto-tunneling for $B \parallel$ barrier layers*

When the magnetic field is applied parallel to the layers, the cyclotron orbits are interfered by the barrier potentials and suffer the reflection by the barrier walls. The energy of the Landau levels involving the skipping orbits becomes a function of the center coordinate of the cyclotron orbits forming the edge states as described in Chapter 2. Thus the Landau levels in the emitter layer $\mathcal{E}_L$, the well layer $\mathcal{E}_W$, and the collector layer $\mathcal{E}_R$ are as shown in Fig. 3.24. Here only the ground state in each region is shown as a function of the center coordinate X. In the emitter and the collector regions, the levels are flat independent of X when X is sufficiently far away from the barrier. As X approaches the barrier, however, the levels tend to the edges states and the energy is increased. The level $\mathcal{E}_W$

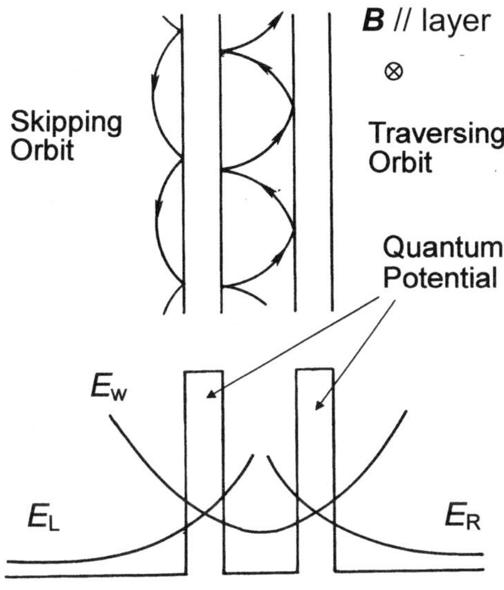

FIG. 3.24. Electron orbit (upper figure) and the energy levels (lower figure) in a double barrier tunneling diode when the magnetic field B is applied parallel to the layers. The horizontal axis stands for the center coordinate X of the cyclotron orbits. The ground state levels originating from the emitter E_L, the well layer E_W and the collector E_R are shown together.

(traversing orbit) depends on X everywhere if the cyclotron orbit is larger than the well width. Figure 3.25 shows the calculated results of the Landau levels ($\mathcal{E}_L$, $\mathcal{E}_W$, and $\mathcal{E}_R$) in a GaAs/AlGaAs double barrier tunneling diode in the presence of bias voltage between the emitter and the collector when the magnetic field B is applied parallel to the layers [134]. These levels are obtained by numerically solving the Schrödinger equation

$$\left[-\frac{\hbar^2}{2m^*}\frac{\partial^2}{\partial x^2}+\frac{e^2B^2}{2m^*}(x-X)^2+U(x)+eEx\right]\Phi(x)=\mathcal{E}\Phi(x), \tag{3.113}$$

where $U(x)$ is the quantum potential of the barrier layer and E is the electric field due to the bias voltage.

The tunneling of electrons from the emitter states to the well states takes place from one of the edge states originating from the emitter states to those in the well states through their crossing points. Therefore, in order for the tunneling to occur, the crossing points should be below the Fermi level. When we increase either bias voltage or magnetic field, the crossing points are shifted towards the higher energy side. When the crossing points cross the Fermi level, a

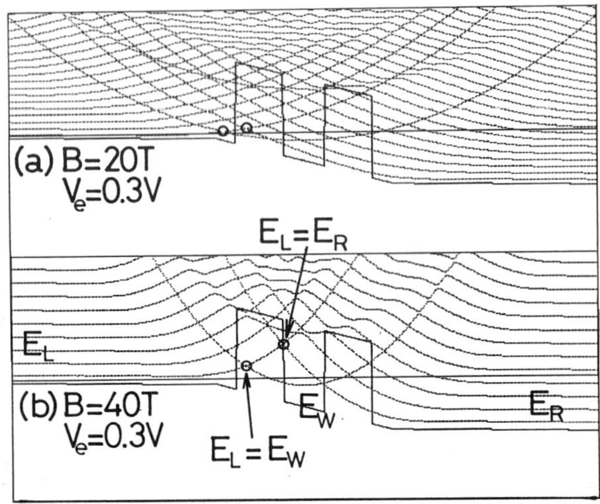

FIG. 3.25. Calculated Landau levels in a GaAs/AlGaAs double barrier tunneling diode when the magnetic field B is applied parallel to the layers, and the bias voltage V_e is applied between the emitter and the collector [134]. (a) $B = 20$ T, $V_e = 0.3$ V. (b) $B = 40$ T, $V_e = 0.3$ V.

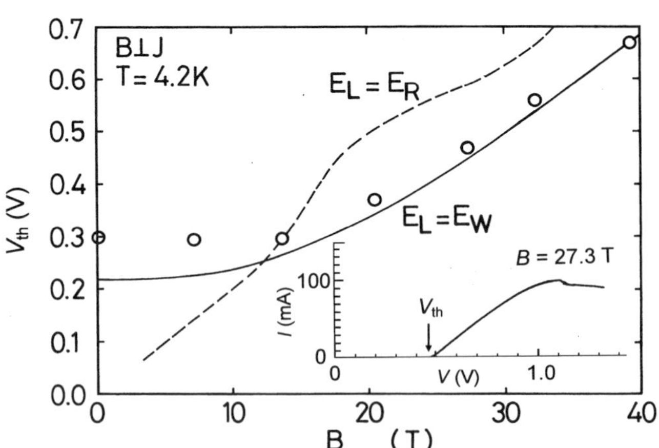

FIG. 3.26. The field dependence of V_{th} of a double barrier tunneling diode of AlGaAs/GaAs/AlGaAs [134]. The well width is 15 nm. The magnetic field is applied parallel to the layers. The solid and broken lines are the calculated results assuming the tunneling from the emitter states to the well states and to the collector states directly, respectively. The inset shows the $V - I$ characteristics.

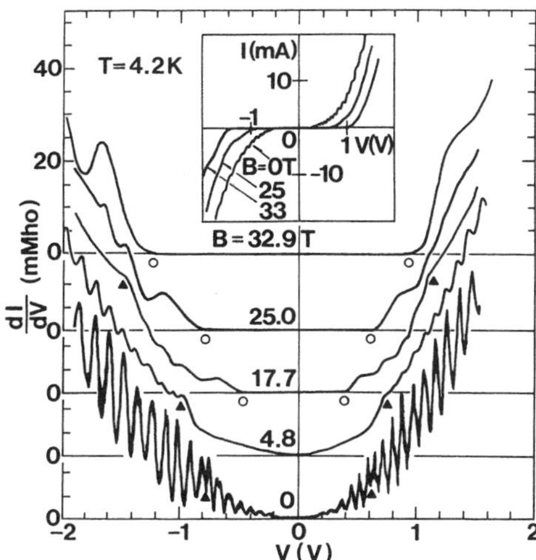

FIG. 3.27. Oscillatory magneto-tunneling current in a AlGaAs/GaAs/AlGaAs double barrier tunneling diode (well width = 50 nm) as a function of magnetic field applied parallel to the layers [139].

discontinuous change occurs in the transmission probability which gives rise to an oscillatory conductivity change.

Usually, only a few emitter states are occupied by electrons in high magnetic fields, so that it readily reaches the quantum limit. If there are not so many levels below the Fermi level, no conspicuous oscillation is observed. On the other hand, the threshold voltage V_{th} in the $V - I$ curve as defined in the inset of Fig. 3.26 shows a large magnetic field dependence [134]. As the field is increased, V_{th} increases. There are two possibilities: (1) tunneling from the emitter states to the well states first and then to the collector states (the solid line), (2) direct tunneling from the emitter states to the collector states (the broken line). The calculated curves were obtained from the cross of the Fermi level with the crossing points of the Landau levels as mentioned above. As is clear from Fig. 3.26 (b), the experimental data are in good agreement with the former assumption [134]

In contrast to a narrow quantum well, for samples with a wide quantum well, the level spacing in the layer is small, so that many peaks are observed. Figure 3.27 shows a trace of the oscillatory conductivity in an AlGaAs/GaAs/AlGaAs double barrier tunneling diode with a wide well layer (50 nm) for a parallel magnetic field. The peak positions are in good agreement with the calculation. Osada, Miura and Eaves found that the well states are classified into two different regimes depending on the B and V. They are skipping orbit, which has a reflection at one of the barrier walls, and traverse orbit, which has reflections at the both barriers [139].

Another interesting feature for the geometry of $B \parallel$ layers is that we can obtain the dispersion of the energy bands. The tunneling from the emitter states to the well states involves a shift of the center coordinate of the cyclotron motion X. In Chapter 2, we learnt that the center coordinate of the cyclotron motion X is related to k_y with a relation

$$X = -\frac{\hbar}{eB}k_y \tag{3.114}$$

The tunneling from the emitter states to the well states can be regarded as the shift of the center coordinate $\Delta X = s$ at the transition of electrons between the two regions, where s is the distance between the center of the emitter states and the well states. If the tunneling is considered to occur at the crossing points as shown in Fig. 3.24, the selection rule is $\Delta k_y = 0$. If we consider, however, that the tunneling is between the two states at the center of each region, we can consider that the emitter state with $k_y = 0$ would tunnel to the well state

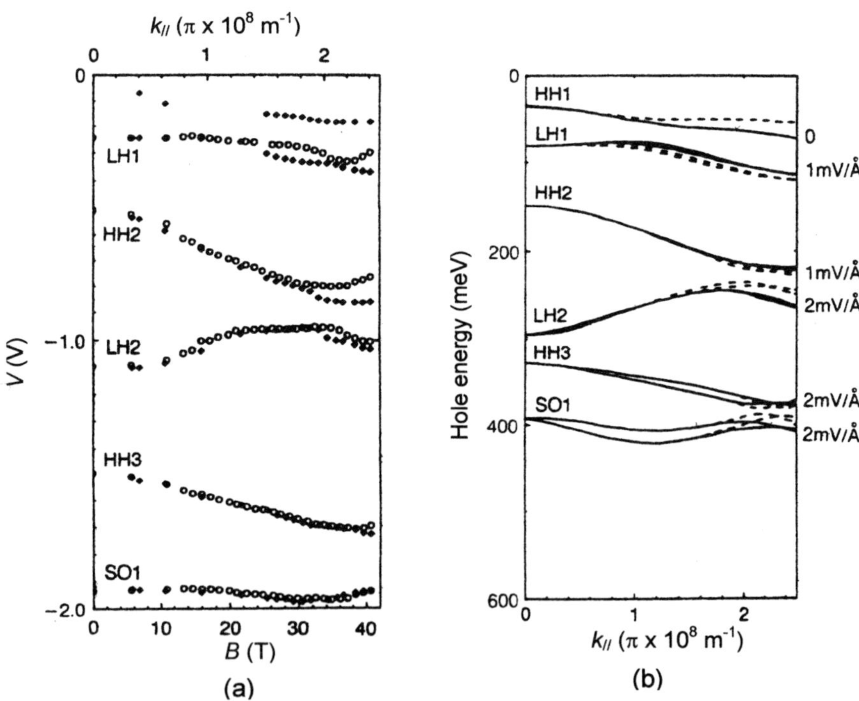

FIG. 3.28. (a) Peak positions in the $V-I$ characteristics of the p-type AlAs/GaAs/AlAs double barrier tunneling diode sample with a well width of 4.2 nm on a $V-B$ plane [141]. $B \parallel < 001 >$ ($k_\parallel$ along $< 010 >$). (b) Theoretically calculated energy dispersion of AlAs/GaAs/AlAs quantum well (valence band). Solid curves are for $k_\parallel$ along $< 010 >$ and broken curves are for $k_\parallel$ along $< 011 >$.

with $k_y = \Delta k_y = -seB/\hbar$. If we measure the tunneling current by varying magnetic field B and the bias voltage V, the resonant tunneling peaks appear at the positions where the energies of the states in the two regions with the above relation of k_y coincide. In other words, we can sweep k_y by magnetic field and the energy by the bias voltage.

Hayden *et al.* measured the magneto-tunneling for the p-type samples of an AlAs/GaAs/AlAs tunneling diode and obtained the energy dispersion of the valence band in the quantum well of GaAs [140–142]. In the $V-I$ characteristics of the p-type sample with a well width of 4.2 nm, several peaks were observed corresponding to the hole tunneling to the light hole band (LH), heavy hole band (HH) and the split off band (SO) in the well at different magnetic fields. A plot of the peak position as a function of magnetic field is shown in Fig. 3.28 (a). The vertical axis and the horizontal axis correspond to just the energy and k_y, respectively, with some numerical factors. Thus we can obtain the energy dispersion of the band directly, which is difficult to obtain by any other methods. Hayden *et al.* obtained a detailed information of the level crossing between the heavy hole and the light hole bands [140]. By changing the crystal orientation of the layers, the dispersion for different $\boldsymbol{k}$ vector directions can be obtained. Comparing with (b), we can see that the valence band dispersion thus obtained is qualitatively in good agreement with theoretical calculation.

3.4.4 *Chaotic behavior for tilted magnetic fields*

When the magnetic field is applied with a tilted angle θ from the normal to the layers, a complicated oscillation is observed in the magneto-tunneling current, because of the energy levels in the quantum well. Figure 3.29 represents the energy profile of a double barrier tunneling diode [143]. Considering the electron

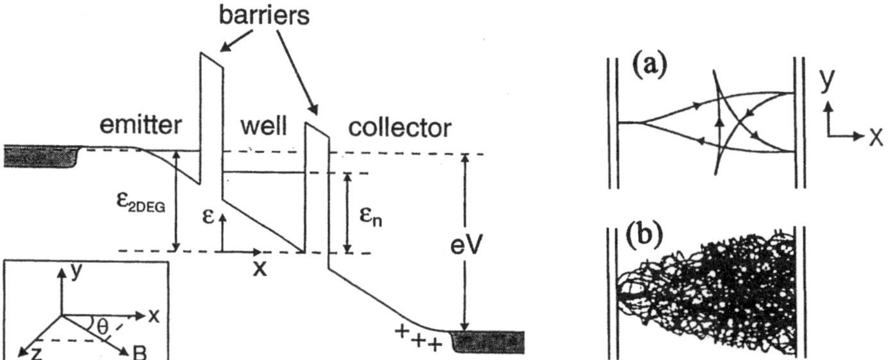

FIG. 3.29. Left: Energy profiles of a double barrier tunneling diode and the direction of magnetic field [143]. Right: (a) Projection in $x-y$ plane of closed orbit for $\theta = 20°$. (b) Chaotic orbit when the initial velocity is changed by 0.1% from the case of (a).

orbits in the quantum well, we can see that the Landau level structure must be very complicated. In a semiclassical picture, when electrons are injected from the emitter into the quantum well, electrons are bent by the Lorentz force and bounced at the barrier layers and conduct a back and forth motion. The electron motion is periodic only when the initial velocity, the magnetic field, and its angle satisfy some special conditions. Figure 3.29 (a) shows one of the examples of such periodic orbits. Projection in $x-y$ plane is shown for a tilt angle $\theta = 20°$, $B =$ 11.4 T and the electric field $E = 2.1 \times 10^{6}$ V/cm. Although the electron motion is bent by the Lorentz force before the reflection, the track is periodic. When θ is changed just by 0.1%, however, the orbit becomes completely different, as the electron never comes back to the same orbit at multiple reflections. That is to say, the electron motion is chaotic. The situation resembles the stadium billiard problem [144].

Quantum mechanically, the Hamiltonian is expressed as

$$\mathcal{H} = \frac{1}{2m^*}\left[p_x^2 + (p_y + exB\sin\theta - exB\cos\theta)^2 + p_z^2\right] + eE\left(\frac{1}{2}d - x\right), \quad (3.115)$$

with a vector potential

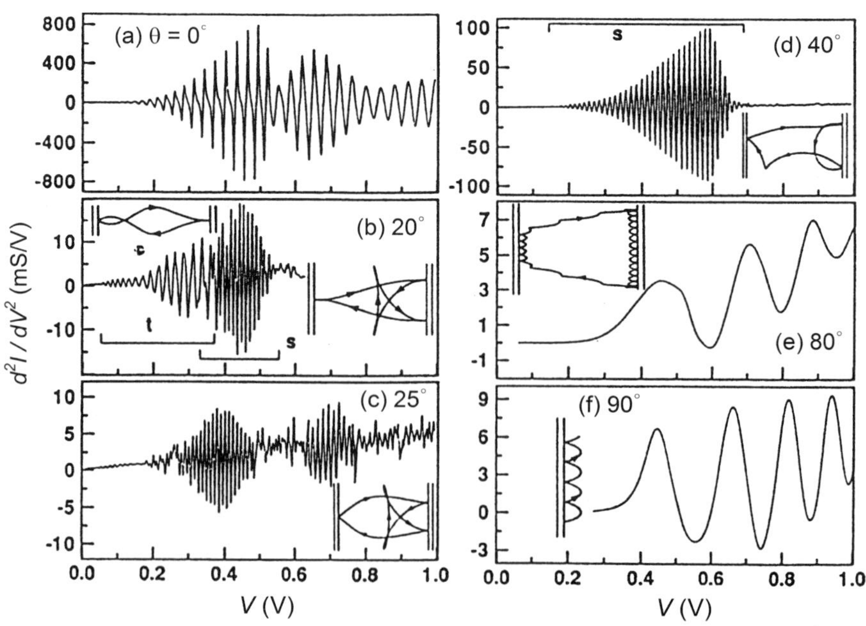

FIG. 3.30. Second derivative of the magneto-tunneling current in GaAs/AlGaAs tunneling diode for tilted magnetic field with different angles [143]. Well width is 120 nm. $B = 11.4$ T. Insets show periodic orbits associated with resonant structure.

$$\boldsymbol{A} = (0, exB\sin\theta - exB\cos\theta, 0), \tag{3.116}$$

where d is the well width. The Schrödinger equation with this Hamiltonian cannot be solved analytically as the motion perpendicular and parallel to the magnetic field cannot be separated in the presence of the quantum potential of the barrier layers. The equation can only be solved numerically, and it results in non-uniform spacing. Thus the system is really characterized as a chaos. A similar quantum mechanical chaotic system has also been found in the high Rydberg number states of hydrogen atoms in a magnetic field, for which the Coulomb potential and magnetic field potential are not separable [145]. The chaotic system in double barrier tunneling diodes was extensively studied by the groups of Eaves [143, 146, 147] and Boebinger [148].

For a tilted magnetic field, the oscillatory tunneling current shows a drastic change depending on the bias voltage, the magnetic field, and the tilted angle. A typical example is shown in Fig. 3.30 [143]. For $\theta = 0°$, the oscillation is rather periodic. For $\theta = 20°$, the oscillation is more complicated and classified into two series t and s as shown as insets. When θ is increased to $50°$, the oscillation becomes more irregular. The corresponding orbit is such as shown in the figure. For larger angles the oscillation period becomes larger and similar to the one shown in Fig. 3.27 reflecting the skipping orbit.

When we calculate the Poincaré section for such orbits, we can show that the motion is really chaotic for a tilted angle [146]. Actually, the system has various chaotic characteristics. An interesting feature has been found in higher magnetic fields, where the separation of the complicated Landau levels becomes sufficiently large. Figure 3.31 shows the experimentally observed oscillatory behavior in a double barrier tunnel diode with a well width of 22 nm when a magnetic field of 37 T was applied with a tilted angle of 40° [147]. Although the level scheme should be very complicated and populated densely in the well, the peaks appear quite regularly. By comparison with the calculation, it was found that the peaks appear only when the electron orbit yields to a periodic motion. The wave function corresponding to such a motion is called "scarred wave function" where the probability amplitude of the wave function is concentrated along nearly a single line forming some shape looking like a scar. Figure 3.31 shows the probability amplitude and the electron orbits of scarred and non-scarred wave functions. The scarred wave function is characterized by a number of nodes of the probability amplitude ν. The current peaks are observed in the order of ν as shown in Fig. 3.31 (b). The reason why the current peak appears only for the scarred wave function is obvious. From the emitter to the well, the resonant tunneling would occur more efficiently for the scarred case as the electron states are concentrated in a small region, as is clear from the comparison between Fig. 3.31 (c) and (d).

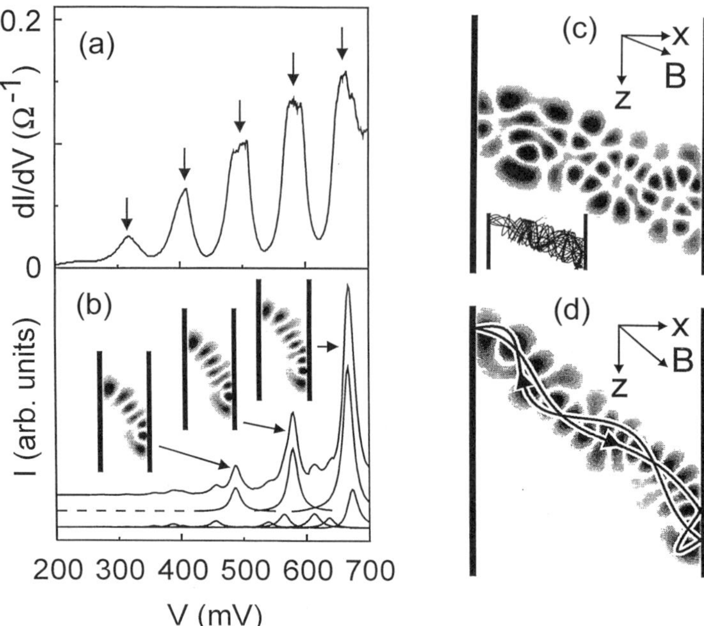

FIG. 3.31. Left: Magneto-tunneling current for a magnetic field of 37 T applied with a tilted angle of 40° from the normal to the layer. Width of the quantum well is 22 nm. (a) Experimental trace of dI/dV. (b) Calculated tunneling current. The curve in the middle shows the current due to the tunneling to the states with scarred wave function. The bottom curve shows the current due to the other states. The top curve shows the current due to all the transitions. The peaks correspond to $\nu =$ 7, 8, 9. Right: Comparison between calculated scarred wave function and unscarred wave function. Width of the quantum well is 120 nm. $B = 11.4$ T. $V = 100$ mV. The vertical lines on both sides indicate the barrier layers. (c) An unscarred function $\theta=$ 20 °. The energy is $\mathcal{E}_w = 141$ meV. (d) An unscarred wave function ($\nu = 17$). $\theta=$ 40 °. $\mathcal{E}_w = 97$ meV. The solid lines represent the classical orbits. (First published in [147].)

3.5 Magnetophonon resonance

3.5.1 *Magnetophonon resonance in bulk semiconductors*

In semiconductor crystals, there is a significant interaction between conduction electrons and longitudinal optical (LO) phonons. LO phonons have a nearly flat energy dispersion near the Γ-point of the Brillouin zone. LO phonons with small wave vector $\boldsymbol{q}$ have a strong interaction with conduction electrons or holes through the electrostatic potential produced by a macroscopic polarization due to the lattice displacement. In polar semiconductors, the strength of the electron-LO phonon interaction is characterized by a non-dimensional coupling constant called the Fröhlich constant [149],

$$\alpha = \frac{e^2}{\hbar}\left(\frac{m^*}{2\hbar\omega_o}\right)^{\frac{1}{2}}\left(\frac{1}{\epsilon_\infty} - \frac{1}{\epsilon_0}\right). \tag{3.117}$$

Here $\hbar\omega_o$is the LO phonon energy and ϵ_∞ and ϵ_0 are optical and static dielectric constants. Using α, the perturbation Hamiltonian of the interaction is given by

$$\mathcal{H}'_{e-p} = \sum_{\boldsymbol{q}} \frac{i\hbar\omega_o}{q}\left(\frac{\hbar}{2m^*\omega_o}\right)\left(\frac{4\pi\alpha}{V}\right)^{1/2}\left(a_{\boldsymbol{q}}e^{i\boldsymbol{q}\cdot\boldsymbol{r}} + a^\dagger_{-\boldsymbol{q}}e^{-i\boldsymbol{q}\cdot\boldsymbol{r}}\right), \tag{3.118}$$

where $\boldsymbol{q}$ is the wave vector of the LO phonon, and $a^\dagger_{-\boldsymbol{q}}$ and $a_{\boldsymbol{q}}$ are the creation and annihilation operators of the LO phonon with $\boldsymbol{q}$. The interaction almost dominates the carrier scattering in II-VI or III-V semiconductors at relatively high temperatures. It also has the effect of increasing the effective mass of electrons, as electrons move around together with phonon cloud. Such a quasi-particle comprising an electron and phonon cloud is called a polaron. The mass of polarons is obtained from $\mathcal{H}'_{e-p}$ as

$$m_p^* \sim m^*\frac{1}{1-\frac{\alpha}{6}} \sim m^*(1+\frac{\alpha}{6}). \tag{3.119}$$

In other words, the polaron mass is heavier than the bare mass of electrons by a fraction $\frac{\alpha}{6}$.

The unique character of LO phonon scattering is that it is an inelastic scattering with a constant energy change $\hbar\omega_o$. When the spacing of the Landau levels $n\hbar\omega_c$ (n is an integer) becomes equal to $\hbar\omega_o$, electrons are scattered resonantly by absorbing or emitting an LO phonon and this resonance induces a large effect in electronic transport properties as shown below. Gurevich and Firsov were the first to predict a large increase of the transverse magneto-resistance under the condition

$$n\hbar\omega_c = \hbar\omega_o. \tag{3.120}$$

Such a resonance in the Landau levels involving LO phonons is called magnetophonon resonance. From the quantum transport theory as described in Section 3.1.3, Gurevich and Firsov derived an expression for the conductivity tensor component when the LO phonon scattering is dominant as [150]

$$\sigma_{xx} \propto \int d\mathcal{E}\sum_{NN'}\frac{G_{NN'}(\mathcal{E})f(\mathcal{E})[1-f(\mathcal{E}+\hbar\omega_o)]}{\sqrt{\mathcal{E}-\hbar\omega_c(N+\frac{1}{2})}\sqrt{\mathcal{E}+\hbar\omega_o-\hbar\omega_c(N'+\frac{1}{2})}}, \tag{3.121}$$

where $G_{NN'}(\mathcal{E})$ is a slowly varying function of $\mathcal{E}$. The integral diverges when the two square roots in the denominator simultaneously become zero. From this condition, the relation (3.120) is easily derived.

The magnetophonon resonance was actually observed in n-InSb by Firsov *et al.* [151,152]. They found the maxima at the resonance in the transverse magneto-resistance and minima in the longitudinal magneto-resistance. The condition of the magnetophonon resonance (Eq. (3.120)) can be rewritten as

$$\frac{1}{B} = n \cdot \frac{e}{m^* \omega_o}. \tag{3.122}$$

This equation indicates that the oscillation is periodic as a function of the inverse of magnetic field, which is similar to the Shubnikov-de Haas effect . However, a large difference in the oscillation between the Shubnikov-de Haas effect and the magnetophonon resonance is the temperature dependence of the oscillation amplitude. The Shubnikov-de Haas oscillation amplitude is increased as the temperature is decreased. In the case of magnetophonon resonance, the oscillation amplitude is decreased at very low temperatures, while it also decreases at very high temperatures because of the broadening of the Landau levels. The reason for the decrease of the oscillation at low temperatures is due to the decrease of the number of LO phonons or electrons which have the same energy as the LO phonons. These are conditions needed for the absorption or emission of LO phonons by electrons. In usual semiconductors, the oscillation amplitude of the magnetophonon resonance becomes maximum at about 100–200 K [153]. Many investigations have been made on the magnetophonon resonance. It is a powerful means to study the effective mass of electrons, phonon modes and electron-phonon interaction. Excellent reviews of the work in early days are found in articles by Harper, Stradling and Hodby [154] and by Nicholas [155].

As an example of the experimental data for transverse magnetophonon resonance, we show in Fig. 3.32, data for the transverse magneto-resistance in n-type GaAs [156]. As σ_{xx} becomes the maximum at the magnetophonon resonance condition (3.120), the transverse magneto-resistance ρ_{xx} also takes maximum at these points from the relation (3.8). As the magneto-resistance is accompanied by a large background which is nearly proportional to magnetic field B, the oscillatory part is more easily seen by subtracting a linear function of B from the magneto-resistance. We can see that the amplitude becomes maximum at about 100 K. It is also clear that the amplitude is decreased exponentially as the field is decreased. Stradling and Wood derived an empirical formula [153]

$$\Delta\rho \propto \exp\left(-\gamma\frac{\omega_o}{\omega_c}\right)\cos\left(2\pi\frac{\omega_0}{\omega_c}\right). \tag{3.123}$$

Here, γ is an attenuation coefficient. This relation was verified by Barker theoretically [157].

From the magnetophonon resonance, we can obtain the carrier effective mass using a relation (3.120), if the LO phonon frequency is known. This is an alternative way to determine the effective mass to cyclotron resonance. On the contrary if the effective mass is known, the LO phonon energy involved in the scattering can be obtained. Moreover, information about the electron phonon interaction or carrier scattering mechanisms is obtained from the attenuation coefficient γ in (3.123). To determine the carrier effective mass, we have to take account of the fact that the magnetophonon effective mass m^*_{MPR} obtained from (3.120) is generally larger than the carrier effective mass. The polaron effect makes the mass heavier than the bare electron effective mass by a factor $\frac{\alpha}{6}$ as shown in (3.119).

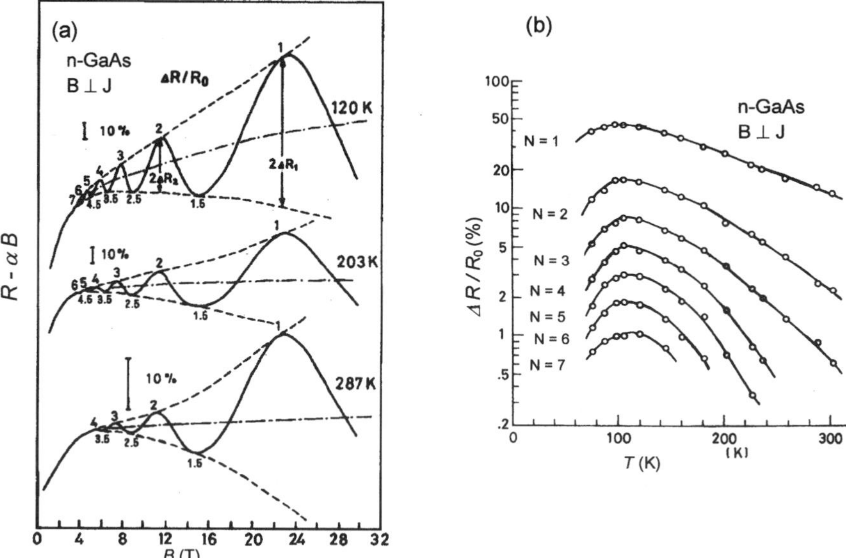

FIG. 3.32. Magnetophonon resonance oscillation in bulk n-type GaAs. In order to see the oscillation clearly, a linear function of B is subtracted from the data.

In the magnetophonon resonance, the resonant polaron effect takes place as will be discussed in Section 4.5. Therefore, the magnetophonon mass m^*_{MPR} is even larger than m^*_p. According to Parmer, the following equation holds [158],

$$m^*_{MPR} = \left(1 + \frac{0.83\alpha}{3}\right) m^*_p = \left(1 + \frac{0.83\alpha}{3}\right)\left(1 + \frac{\alpha}{6}\right). \tag{3.124}$$

This relation is experimentally confirmed by the group of Stradling [153,159,160]. The numerical analysis of the magnetophonon mass is also made by Nakayama [161]. In addition to the polaron effect, the band non-parabolicity effect is significant in the magnetophonon mass, as it is measured in the high energy range. Therefore, the correction of both the polaron effect and the band non-parabolicity is indispensable to obtain the bare band edge mass from the magnetophonon resonance.

The magnetophonon resonance is a powerful technique for obtaining information about electron effective mass, especially, for investigating the temperature dependence in a wide temperature range, as the phenomena are observed even in a relatively high temperature range [159,162,163]. As the energy band gap $\mathcal{E}_g$ changes with temperature, the effective mass should show temperature dependence. Such temperature dependence of the effective mass on temperature was theoretically discussed by Ehrenreich [164], Lang [165], and Ravich [166]. The temperature dependence of $\mathcal{E}_g$ consists of two parts; a dilational change and a change due to the electron-phonon interaction. The temperature dependence of $\mathcal{E}_g$ at a constant pressure is expressed by

$$\left(\frac{\partial \mathcal{E}_g}{\partial T}\right)_p = -3\alpha B\left(\frac{\partial \mathcal{E}_g}{\partial p}\right)_T + \left(\frac{\partial \mathcal{E}_g}{\partial T}\right)_V, \tag{3.125}$$

where α is the linear expansion coefficient and B is the bulk modulus. The first term arises from the electron-phonon interaction and the second term from the dilation. Lang and Ravich found that not only the former but also the latter effect should contribute to the temperature dependence of the effective mass in narrow gap semiconductors. Stradling and Wood obtained the temperature dependence of the effective mass of InSb, InAs and GaAs from the magnetophonon resonance in the temperature range 40–300 K [159]. They found that the dilational change of the band gap alone explains the temperature dependence of the effective mass very well in the case of InSb, but for InAs and GaAs, its contribution was nearly a factor two smaller than the experimental values, whereas the consideration of the total gap change gives too large a change. The temperature dependence of the carrier effective mass is still a controversial problem in semiconductor physics.

Besides the determination of effective mass, magnetophonon resonance is also useful for obtaining information about LO phonons involved in the resonance. As an example of the coupling of different phonons, we show a magnetophonon resonance of Te, where anisotropic phonon coupling was observed. Te has a peculiar crystal structure with helical chains and the band structure is very complicated and anisotropic. As there are three atoms in a unit cell, there are altogether six optical modes of phonons. Near the zone center, there are three infrared active modes, A_2, E_l, and E_h, which have net electric polarizations. An interesting question is which phonon would be coupled with electrons in the magnetophonon resonance. Figure 3.33 shows the transverse magnetophonon resonance oscillation in p-type Te [167]. While the magnetic field $\boldsymbol{B}$ was applied perpendicular to the c-axis, the direction of the current $\boldsymbol{J}$ was either parallel (upper trace, $\boldsymbol{J} \parallel$ c) or perpendicular (lower trace, $\boldsymbol{J}\perp$ c) to the c-axis. In both cases, $\boldsymbol{J} \perp \boldsymbol{B}$ (transverse magneto-resistance). It is found that there is a large difference in the oscillation period between the two traces for different current directions, in spite of the fact that the Landau levels are common. This is due to the coupling of different phonons. From the discussion mentioned in Section 3.1.3, it can be readily shown that for $\boldsymbol{J} \parallel$ c, the LO phonons which have the polarization vector parallel to the c-axis (A_2 phonons) are predominantly responsible in magnetophonon resonance, whereas for $\boldsymbol{J} \perp$ c, those with the polarization vector perpendicular to the c-axis (E phonons) are responsible. Among E-modes, the E_l modes are more strongly coupled, as their population is larger due to the smaller energy.

The valence band structure of Te is known to have a camel's back structure along the c-axis, as shown in Fig. 3.34. The energy dispersion along the k_z-direction has double maxima, while it is an ordinary parabolic shape in the perpendicular direction. For the magnetic field perpendicular to the k_z-direction

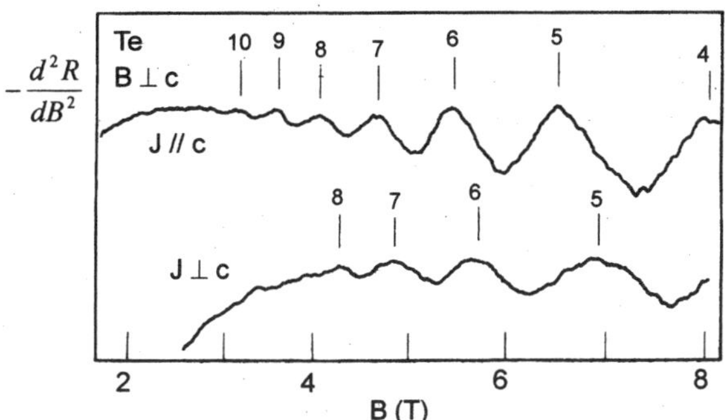

FIG. 3.33. Transverse magnetophonon resonance oscillation in p-type Te [167]. The upper and lower graphs stand for the oscillatory part of the transverse magneto-resistance for the current parallel and perpendicular to the c-axis, respectively, while the magnetic field was applied commonly perpendicular to the c-axis.

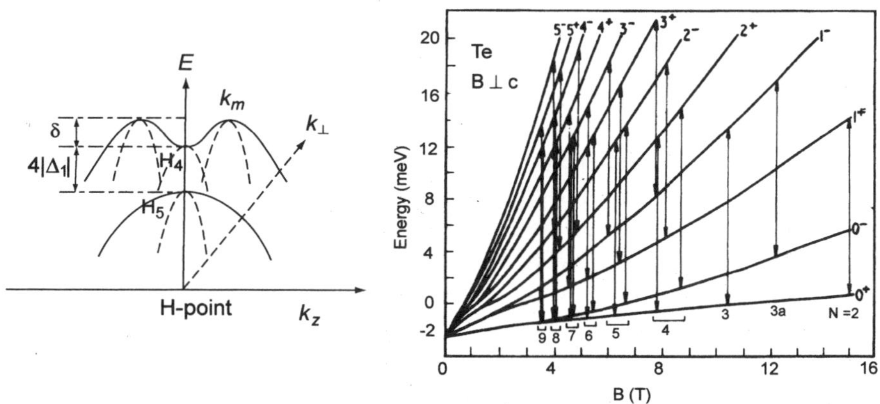

FIG. 3.34. Left: The dispersion of the two uppermost valence bands in p-type Te. The valence band top comprises two maxima at $k_z = k_m$ and $k_z = -k_m$ forming the camel's back structure. Right: Landau level energies as a function of magnetic field and possible transitions for $B \perp c$ [167]. Taking the energy of holes, the sign of the energy (vertical axis) is reversed from the left panel.

($\boldsymbol{B} \perp$ c-axis), the Landau level structure is very complicated, as shown in the right figure of Fig. 3.34 [167,168]. The Landau levels show non-linear field dependence involving a splitting at the energy corresponding to δ (see in the figure), due to the magnetic breakthrough. The spacing between the Landau levels is non-uniform. Such a Landau level strucure was confirmed by experiments of cy-

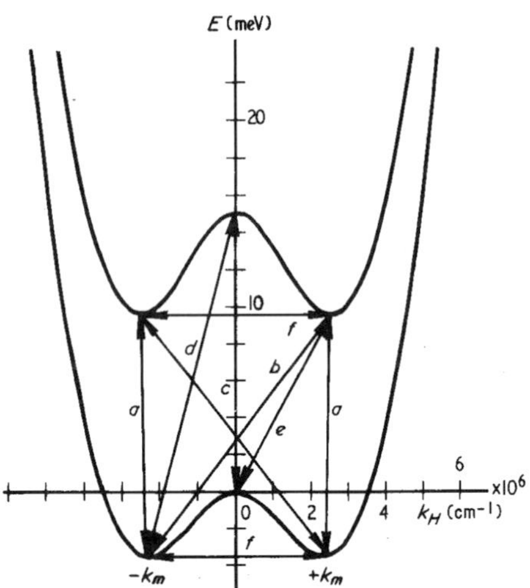

FIG. 3.35. Energy dispersion in the valence band of Te along the field direction when a magnetic field is applied parallel to the c-axis, and possible magnetophonon transition $(a - f)$ [167]. Due to the inversion asymmetry, there is a difference in energy $G\mu_0 B$ between $+k_m$ and $-k_m$ (not shown in the figure).

clotron resonance [169,170] and magneto-abosrption [171]. The peaks of the magnetophonon resonance actually comprise several different transitions as shown by vertical arrows in the figure. Taking account of all these involved transition energies, an excellent agreement was obtained between the experimental data and the calculation for both $\boldsymbol{J} \parallel$ c and $\boldsymbol{J} \perp$ c [167]. Note that such Landau levels formed in the energy band with a camel's back structure are similar to those in the conduction band of GaP, as we will study in detail in Section 4.10.1.

For $\boldsymbol{B} \parallel$ c, an interesting situation arises because of the peculiar energy dispersion of the Landau levels along the field direction (k_H) as shown in Fig. 3.35. If we include the effect of the inversion asymmetry, there occurs a splitting of the two valleys by $G\mu_0 B$ (G is a constant). When the magnetic field is parallel to the c-axis, we can expect to observe various transitions as shown in the figure $(a - f)$. Below $T = 30$ K, resistivity minima were observed at 2.85 T and 4.52 T [167]. These minima were identified as the transition of type f caused by acoustic phonons (LA phonon and TA phonon) which have the wave vector equal to the separation of the two valleys $2\boldsymbol{k}_m$.

In contrast to the magnetophonon resonance in the transverse magnetoresistance, the case of longitudinal magnetophonon resonance is more complicated. The longitudinal magneto-conductivity is represented as

$$\sigma_{zz} = \frac{e^2\hbar^2}{2\pi m^{*2}} \sum_N \int dk_z \frac{\partial f(\mathcal{E})}{\partial \mathcal{E}} k_z^2 \tau(N, k_z). \tag{3.126}$$

From this equation, it is clear that though the resonance occurs by the divergence of the density of states of Landau levels at $k_z = 0$, the electron wave vector component which contributes to the current is also k_z. Therefore the divergence at the resonance should be very weak, and the resonance should be explained by some second order effect. Gurevich and Firsov explained the longitudinal magnetophonon resonance by taking account of both the inelastic scattering by LO phonons and the elastic scattering by acoustic phonons simultaneously. On the other hand, Peterson introduced a concept of pseudoresonance to account for the longitudinal magnetophonon resonance showing by numerical calculation that the minima occurs at the resonance position as kinks where the derivative of the resistance has discontinuity [172]. The magnetophonon resonance in the longitudinal magneto-resistance in the Ohmic condition is not totally understood yet. Experimentally, it is known that the longitudinal magneto-resistance takes a minimum near the resonance condition, but the field position deviates slightly from the exact resonance. Experimentally, the longitudinal magnetophonon resonance is usually observed as resistance minima, but their field positions are slightly lower than the resonance condition (3.120) [173]. In n-GaAs and n-InAs, the minima appear at about 2% lower fields, but in n-InP, n-CdTe and c-CdSe, the deviation is much larger.

Another remarkable feature of the longitudinal magnetophonon resonance is the two-phonon process. By absorbing or emitting two LO phonons simultaneously the resonant conductivity change occurs at field positions satisfying the following condition:

$$2\hbar\omega_o = N\hbar\omega_c. \tag{3.127}$$

Apparently, the peaks for odd N are observed in between the normal magnetophonon resonance. They are very weak but are observed at relatively high temperatures, as expected from the larger phonon population [153, 174, 175].

3.5.2 *Magnetophonon resonance in two-dimensional electron systems and superlattices*

The magnetophonon resonance is observed not only in bulk crystals but also in heterostructures and two-dimensional systems. The magnetophonon resonance of two-dimensional electron systems was first observed by Tsui *et al.* in GaAs heterostructures [176]. Kido *et al.* extended the measurement to higher magnetic fields to observe the entire magnetophonon resonance series down to $N = 1$ peak. Figure 3.36 shows the data of the magnetophonon resonance in GaAs/AlGaAs heterostructures by Kido *et al.* [177]. As shown in the inset, the resonance field shifts to higher fields as the angle between the magnetic field and the normal of the 2D plane θ increases. The resonance field $B_N(\theta)$ for the angle θ is expressed as

$$B_N(\theta)\cos\theta = B_N(0). \tag{3.128}$$

This relation indicates that only the magnetic field component normal to the 2D plane is effective to the quantization of the energy levels, which is characteristic of the 2D electron system.

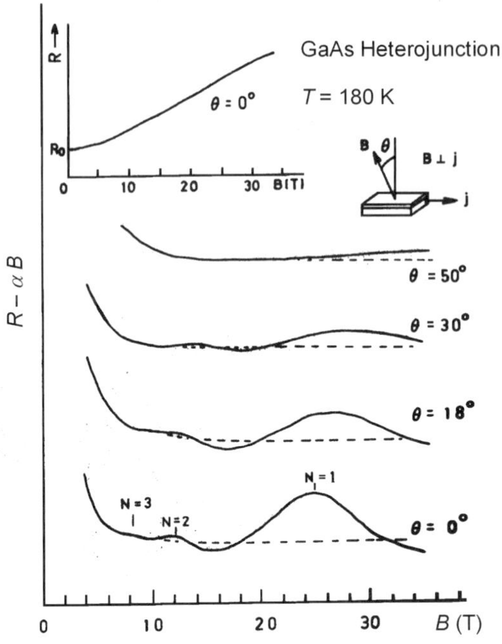

FIG. 3.36. Magnetophonon resonance oscillation in a GaAs/AlGaAs heterostructure for different magnetic field orientations [177]. The background B-linear magneto-resistance was subtracted from the raw data to extract the oscillatory part. The inset shows the raw data of the magneto-resistance for $\theta = 0$.

In the two-dimensional case, the conduction band edge is already higher than the bulk band edge by the subband energy due to the quantization. Therefore, the increase of the magnetophonon effective mass is more significant than in the bulk case, and the magnetophonon peaks in 2D systems appear in higher fields.

Two-dimensional carriers in GaAs/AlGaAs heterostructures are localized in the interface, where carriers can interact with bulk LO phonons of both GaAs and AlGaAs, as well as interface phonons. It is an interesting question which phonons are responsible for the magnetophonon resonance effect. So far, it has been found that in the case of GaAs heterostructures, LO phonons in GaAs rather than AlGaAs are responsible for bringing about the magnetophonon resonance. Brummell, Nicholas and Hopkins compared the effective masses obtained from cyclotron resonance and magnetophonon resonance. After the necessary corrections, they concluded that the bulk LO phonon energy involved in magnetophonon resonance is smaller than that in cyclotron resonance [178].

Magnetophonon resonance is also observed in the perpendicular transport in superlattices. In short period superlattices of GaAs, a mini-band with a narrow band width is formed in the growth direction (perpendicular to the superlattice layers, z-direction). When a current is supplied to the z direction, a characteristic

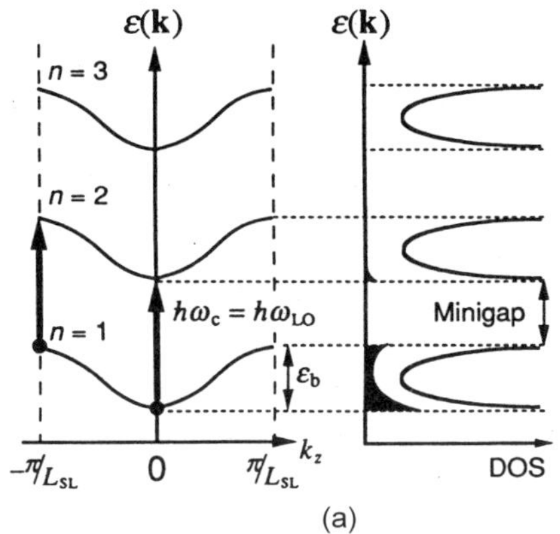

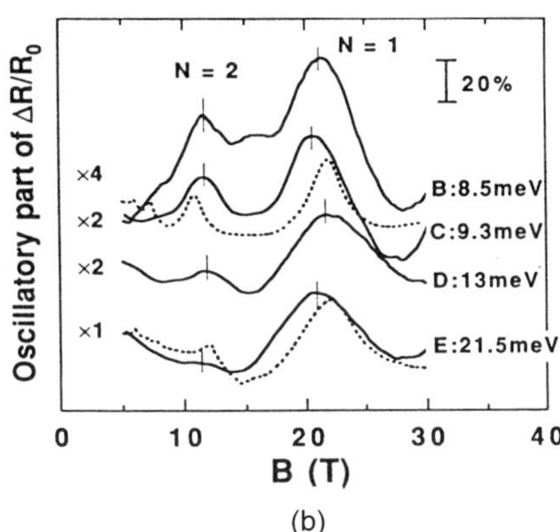

FIG. 3.37. (a) Dispersion relation and the density of states (DOS) of a short period superlattice in magnetic fields applied perpendicular to the superlattice layers [179]. $\mathcal{E}_b$ is the mini-band width. (b) magnetophonon resonance spectra for short period superlattice of GaAs with different mini-band width $\mathcal{E}_b$ whose values are indicated on the right side of each graph [179].

property of the mini-band conduction is observed. When the magnetic field is in the current direction, the mini-band is quantized to the Landau levels, and the energy is represented as

$$\mathcal{E}(k_z) = \mathcal{E}_z(k_z) + \left(N + \frac{1}{2}\right)\hbar\omega_c, \quad N = 0, 1, 2, \cdots, \tag{3.129}$$

where $\mathcal{E}_z(k_z)$ is the subband energy. Figure 3.37 (a) shows the energy diagram of the Landau levels. The resonant absorption of LO phonons occurs only when both the initial states and the final states can find the density of states in the mini-bands. The condition is

$$\hbar\omega_o - \mathcal{E}_b < n\hbar\omega_c < \hbar\omega_o + \mathcal{E}_b, \tag{3.130}$$

where $\mathcal{E}_b$ is the mini-band width. Noguchi *et al.* found in short period superlattices of GaAs with different $\mathcal{E}_b$ that the magnetophonon spectra remarkably depend on $\mathcal{E}_b$ [179]. They controlled $\mathcal{E}_b$ by varying the superlattice period. Figure 3.37 (b) shows the magnetophonon resonance spectra of GaAs/AlGaAs superlattices with different $\mathcal{E}_b$. As $\mathcal{E}_b$ is decreased, the magnetophonon resonance amplitude is increased and a plateau is observed in the minimum region between the magnetophonon resonance peaks. The increase of the amplitude is explained as a more pronounced quantization of the energy levels as the system approaches a two-dimensional electron system. The plateau between the $N = 1$ and $N = 2$ peaks appears in the region

$$(\hbar\omega_o + \mathcal{E}_b)/2 < \hbar\omega_c < \hbar\omega_o - \mathcal{E}_b, \tag{3.131}$$

because there are final states in the phonon absorption process.

3.5.3 *Hot electron magenetophonon resonance*

When the temperature is lowered, the magnetophonon resonance disappears as mentioned above due to the decrease of the number of LO phonons to be absorbed or the electrons which have a sufficiently high energy to emit LO phonons. However, when we supply a sufficiently high electric field, the magnetophonon resonance appears even at low temperatures. This is because electrons accelerated by the high electric field populate in energy states higher than those in the thermal equilibrium determined by the lattice temperature. These electrons are called hot electrons, and we can define the electron temperature T_e that is higher than the lattice temperature T_l. When the magnetophonon resonance condition is satisfied, the scattering of electrons by LO phonons is resonantly increased and it is effective to cool the hot electron temperature. Since the electron mobility depends on temperature in general, such resonant cooling of T_e gives rise to the magnetophonon resonance oscillation [154, 155]. The hot electron magnetophonon resonance is caused by the energy relaxation whereas magnetophonon resonance under the Ohmic condition is caused by the momentum relaxation as mentioned in the previous subsection. As the origin of the oscillation is the change of the electron temperature, the hot electron magnetophonon resonance can be observed similarly in both transverse and longitudinal magneto-resistance, without the mass shift.

Figure 3.38 shows experimental traces of the hot electron magnetophonon resonance in n-type InP observed at $T = 20$ K at different electric field E [154]. When E is low (0.2 V/cm), peaks "X" whose spacing is irregular are observed. These peaks originate from shallow impurity states but the details are not known. When E is sufficiently high (12 V/cm), a series of well-defined peaks is observed. It was found that in the hot electron magnetophonon resonance, the peaks are usually due to the electron transition from the Landau levels to the shallow impurity states (donor or acceptor states). Electrons or holes excited up to higher excited Landau levels (N) emit LO phonons and fall into the donor states (in the case of n-type samples) or acceptor states (in the case of p-type samples), The resonance condition of such a process is written as

$$N\hbar\omega_c = \hbar\omega_o + \mathcal{E}_d(B), \tag{3.132}$$

where $\mathcal{E}_d(B)$ is the depth of the impurity states in the magnetic field B. Hot electron magnetophonon resonance is actually observed in many semiconductors. Another feature of the hot electron magnetophonon resonance in InP is the peaks denoted as "2T, 3T, $\cdots$" in Fig 3.38 ($E = 2.4$ V/cm). These peak positions are different from the series expressed by (3.132) with a larger period implying a participation of smaller excitation energy. Stradling *et al.* explained the peak series as the process of simultaneous emission of two TA (transverse acoustic) phonons at the X-point [180]. Namely, if electrons in the higher Landau levels make transition to a lower Landau level by emitting two TA phonons with an energy $\hbar\omega_{\mathrm{TA}}$ and wave vector $\boldsymbol{q}$ and $-\boldsymbol{q}$ simultaneously, the momentum conservation rule is not violated. The resonance condition is

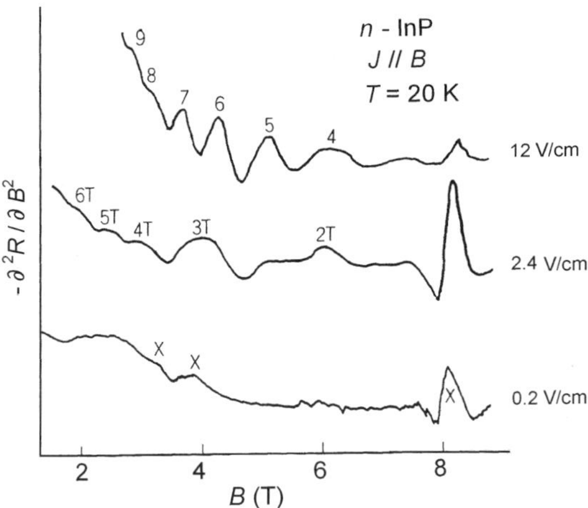

FIG. 3.38. Experimental traces of hot electron magnetophonon resonance in n-type InP at various electric field shown on the right of each graph. (First published in [154].)

$$2\hbar\omega_{\mathrm{TA}} = N\hbar\omega_c. \tag{3.133}$$

The transition probability should be very small, but because of the flat dispersion of the TA mode near the X-point and the small energy, it becomes visible at an intermediate electric field range.

When the electric field becomes very high, a reversal of the peak polarity is observed. Very high electric fields can be applied by applying a voltage between the electrodes across a thin film sample. Eaves, Guimaraes, and Portal measured the magneto-resistance in short channel $\mathrm{n^+nn^+}$ devices of a single crystal thin film of GaAs and InP with a thickness of a few μm [181]. In such a shape of the sample, the Hall voltage does not show up as it is shortened, so that the magneto-resistance is proportional to σ_{xx}. Therefore the magnetophonon resonance peaks appear as the magneto-resistance minima. Eaves *et al.* found that the resonance peaks are converted from minima to maxima as the electric field is increased higher than a few tens of kV/cm^{-1}. The data for GaAs are shown in Fig. 3.39 (a). At low current level (low electric field), conductivity maxima appear at the resonance positions as a normal magnetophonon resonance. However, as the current is increased, a hollow like structure appears on top of each resonance maximum and develops to eventually become an large minimum.

The sign reversal of the peaks was explained by the non-vertical transition as shown in Fig. 3.39 (b) based on a model of QUILLS (quasi-elastic inter-Landau level scattering) [181,182]. The magnetophonon resonance transition probability of electrons in the N-th Landau level to the ground state is proportional to the square of a matrix element

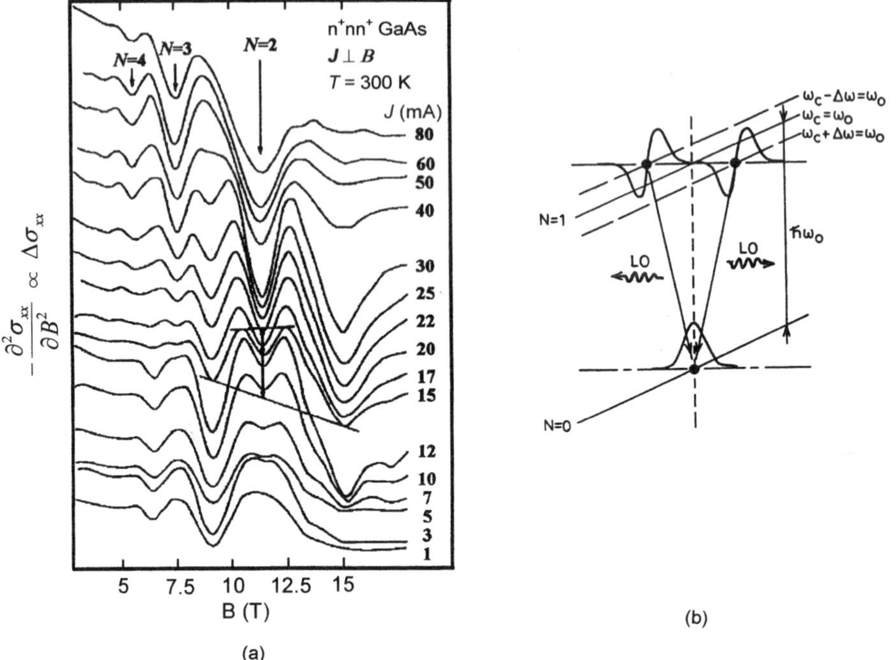

FIG. 3.39. (a) Examples of the experimental traces for magnetophonon resonance in n-type GaAs at various current level as shown on the right [181]. The channel length is 1 μm. $T = 300$ K. (b) Energy levels and wave function of the Landau levels with high electric field and the electron transition between $N = 1$ and 0 levels emitting an LO phonon [183] .

$$M = \langle 0, X, k_z | \exp(i\boldsymbol{q} \cdot \boldsymbol{r}) | N, X', k_z' \rangle, \tag{3.134}$$

where $\exp(i\boldsymbol{q} \cdot \boldsymbol{r})$ represents the wave of the LO phonons. As magnetophonon resonance occurs in the vicinity of $\boldsymbol{q} = 0$, M is described by an overlap integral between the two wave functions of Landau levels, $\langle 0, X, k_z | N, X', k_z' \rangle$. As overlap integrals between different harmonic oscillator wave functions of the Landau levels are zero for a common center coordinate ($X = X'$), the transition occurs only between states which have slightly different center coordinates ($X \neq X'$). Figure 3.39 (b) shows the magnetophonon resonance transition in high electric fields. For a low electric field, the X-dependence of the Landau levels is small, so that the transition between shifted X states takes place at fields only slightly shifted from the resonance condition. In a high electric field, however, the electron energy is significantly altered by E. The energy difference between the position X and X' is $eE\Delta X$ by a shift of the center coordinate $X - X' = \Delta X$. The magnetophonon resonance condition is thus expressed as

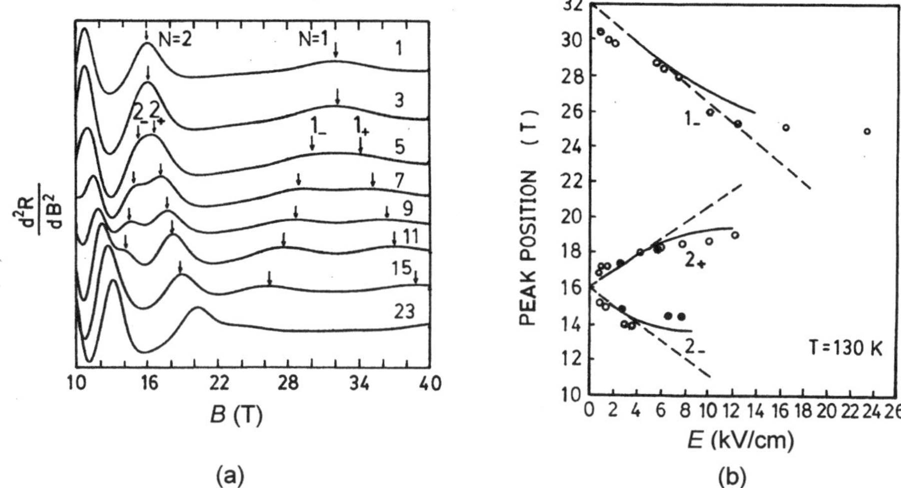

FIG. 3.40. (a) Theoretical curves of the oscillatory part of the magneto-conductivity in crossed magnetic and electric field for InP at different electric fields indicated on the right. (b) The experimentally observed magnetophonon resonance peak positions for $N = 1$ and $N = 2$ transitions [183]. Solid lines were theoretical curves numerically calculated from (3.137).

$$N\hbar\omega_c \pm eE\Delta X = \hbar\omega_o. \tag{3.135}$$

This implies that the magnetophonon resonance peaks appear on both sides of the magnetophonon resonance peak determined by (3.120). Hence apparently the resistivity minima are observed in the field positions represented by (3.120).

In InP, which has a larger effective mass and the larger LO phonon energy, magnetophonon resonance with the same harmonic number N occurs at higher fields than in GaAs. Yamada *et al.* measured the magnetophonon resonance in a short channel n^+nn^+ device of InP [183]. Fine magnetophonon resonance oscillation and the sign reversal of the peaks were observed. The results are in good agreement with the theoretical curve calculated following the formula obtained by Mori *et al.* [182],

$$\Delta\sigma_{xx} = \sum_{r=1}^{\infty} \frac{1}{r}\, exp(-2\pi r\gamma)\{\cos[2\pi r(\omega_o + \Delta\omega)/\omega_c] + \cos[2\pi r(\omega_o - \Delta\omega)/\omega_c]\} \tag{3.136}$$

$$\Delta\omega = \frac{3eEl}{2\hbar}[(\omega_o + \omega_c)/\omega_c]^{1/2}. \tag{3.137}$$

which is shown in Fig. 3.40 (a) for InP. The electric field dependence of the $N = 1$ and $N = 2$ peak positions are plotted in Fig. 3.40 (b). The experimental data are in good agreement with the theoretical curves calculated by using (3.137).

3.5.4 *Magnetophonon resonance in multi-valley semiconductors*

In multi-valley semiconductors such as n-type Si and Ge, the magnetophonon spectra are very complicated as the intervalley scattering take place as well as the intravalley scattering. These crystals are mono-polar, so that there are no polar optical phonons, but the magnetophonon resonance is still possible via the deformation potential coupling. In Si, for example, there are six valleys located in the $< 100 >$ and equivalent directions as shown in Fig. 2.8. In the case of Ge there are four valleys in the $< 111 >$ directions. In each valley, the intraband magnetophonon resonance occurs under the condition Eq. (3.120). In multi-valley semiconductors, there is also a possibility of intervalley transition. In the case of intervalley transition, depending on the magnetic field direction relative to the symmetry axis of each valley, different Landau level series are formed between the valley of the initial state and that of the final state. In such a situation, the magnetophonon resonance condition is written as

$$\left(N + \frac{1}{2}\right) \hbar\omega_{c1} - \left(M + \frac{1}{2}\right) \hbar\omega_{c2} = \hbar\omega_o, \tag{3.138}$$

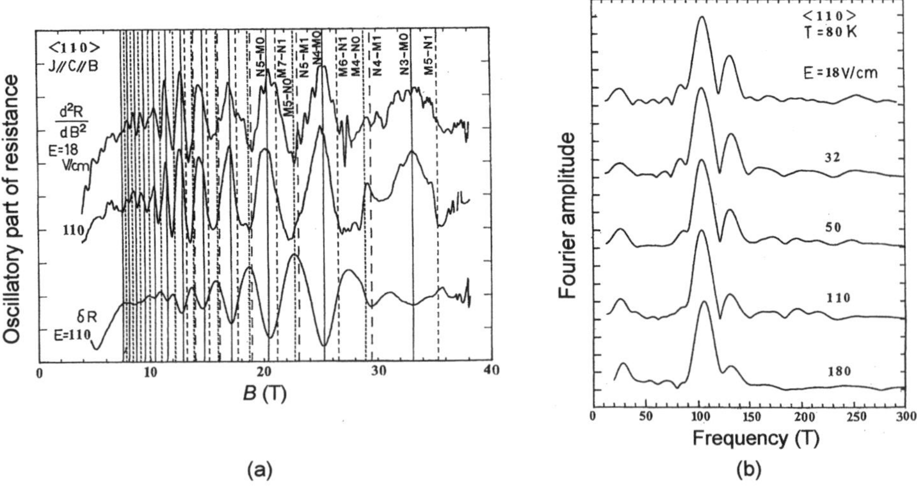

FIG. 3.41. Hot electron magnetophonon resonance spectra in n-type Si [187]. The magnetic field was applied parallel to the $< 110 >$ axis. $J \parallel B$. (a) Raw data of the oscillatory part of the magneto-resistance. The upper two curves represent the second derivative of the resistance with different bias voltages and the bottom curve represents the oscillatory part derived from the subtraction of the monotonic magneto-resistance. $T = 80$ K. The vertical lines indicating the resonance positions of the intervalley transition with different quantum numbers. (b) Fourier transformation of the d^2R/dB^2.

where N, M, $\hbar\omega_{c1}$, and $\hbar\omega_{c2}$ are the Landau quantum numbers and the Landau energy spacings in valleys 1 and 2, and $\hbar\omega_o$ is the involved LO phonon energy, respectively. Detailed studies of magnetophonon resonance have been made in n-type Ge [184] and n-type Si [185,186]. The magnetophonon resonance oscillation is very complicated because of the complexity in the combination of the sets of Landau levels and phonon branches. Figure 3.41 shows an example of the magnetophonon resonance spectra in n-Si observed in pulsed high magnetic fields up to 40 T [187]. The measurement was done in the longitudinal resistance and in the hot electron regime. The raw data of the oscillatory part is very complicated. Such complicated spectra involving oscillations with multiple frequencies can be analyzed by Fourier analysis with respect to $1/B$. The Fourier spectra for $B \parallel < 110 >$ axis are shown in Fig. 3.41 for different bias voltages. From the peaks in such Fourier spectra, the fundamental frequencies of different types of transitions can be obtained. The dominant series of the oscillation peaks are assigned as the intervalley transitions with the lighter electron mass by emitting a f110-Σ_3 phonons.

3.6 Angular dependent magneto-oscillation

As another type of quantum oscillation in high magnetic fields, the angular dependent magneto-oscillation (AMRO) is an important phenomenon which is useful for investigating the Fermi surface. The AMRO was first discovered by Kajita *et al.* and Kartsovnik *et al.* in an organic conductor θ-(BEDT-TTF)$_2$I$_3$ [188,189]. This material has a quasi-two-dimensional structure and its Fermi surface is represented as a warped cylinder. Kajita *et al.* found that when the magnetic field direction is changed from the symmetry axis by an angle θ, the conductivity changes as a function of θ. A remarkable feature of this phenomenon is that the oscillatory structure does not depend on the magnetic field. This is a completely different feature from the Shubnikov-de Haas effect. The oscillation was explained by Yamaji considering the warping of the Fermi surface which has a quasi-two-dimensional character [190, 191]. If the system is completely two-dimensional, the Fermi surface has a cylindrical shape, as shown in Fig. 3.42. When the magnetic field is applied with an angle θ from the cylinder axis, the extremal cross section of the cylinder perpendicular to the magnetic field has an area

$$S_k = \pi k_F^2 \frac{1}{\cos\theta}. \tag{3.139}$$

S_k does not depend on k_z. As shown in Section 2.3.2, the semiclassical quantization yields to the condition

$$S_k = (N+\gamma)\frac{2\pi}{l^2}, \tag{3.140}$$

where N is an integer and γ is a phase factor. As S_k depends on θ, the above quantization condition changes with θ. However, the change is just monotonic. If

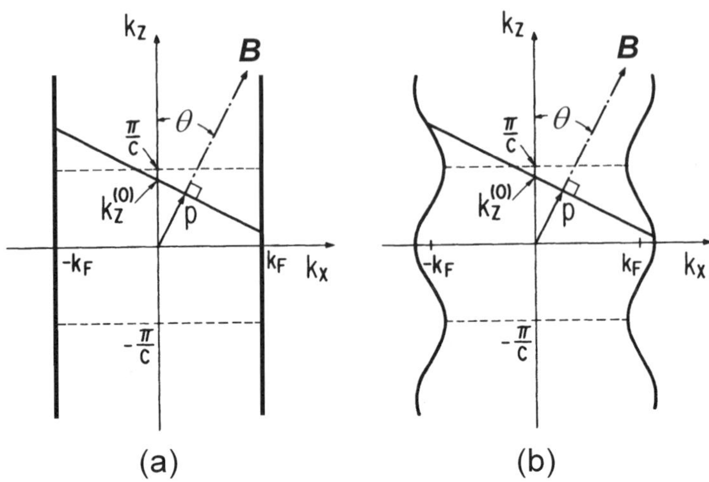

FIG. 3.42. (a) Cross-section of the Fermi surface of a two-dimensional system and the cross section perpendicular to the magnetic field applied with a tilted angle θ. (b) In the case of warped Fermi surface.

the system is quasi-two-dimensional with a small dispersion along the z-direction, the energy is represented as

$$\mathcal{E} = \frac{\hbar^2}{2m^*}\left(k_x^2 + k_y^2\right) - 2t\cos(ck_z), \tag{3.141}$$

then the extremal cross section is

$$S_k = \frac{1}{\cos\theta}\left[\pi k_F^2 + 4\pi m^* t \cos(ck_z^{(0)}) J_0(ck_F \tan\theta)\right], \tag{3.142}$$

where J_0 is the Bessel function of the zero-th order. Thus S_k has the $k_z^{(0)}$-dependence. In other words, the quantized energy depends on $k_z^{(0)}$. If the last term vanishes, however, with a condition,

$$J_0(ck_F \tan\theta) = 0, \tag{3.143}$$

S_k does not depend on $k_z^{(0)}$ any more and all the electrons on the Fermi surface obeys the same quantization condition. Under this condition, σ_{xx} or ρ_{xx} would show oscillation. As the Bessel function is approximated as

$$J_0(z) \sim (2/\pi z)^{1/2}\cos(N - \pi/4), \tag{3.144}$$

the condition (3.143) can be rewritten as

$$\tan\theta = \frac{\pi}{ck_F}N - \frac{\pi}{4ck_F}. \tag{3.145}$$

It was found that this condition almost coincides with the position where the conductivity maxima are observed [190].

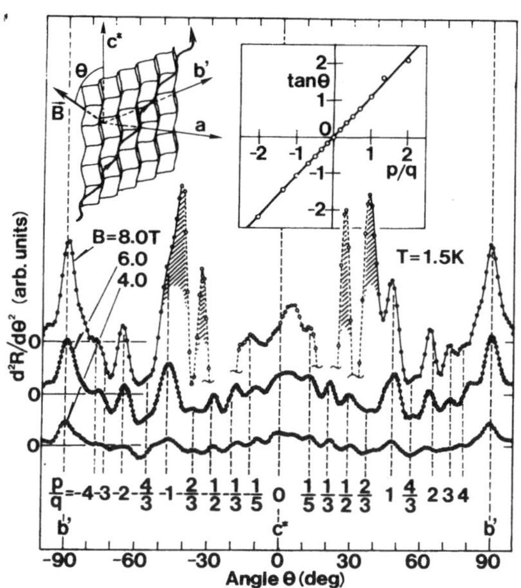

FIG. 3.43. Angular dependence of the second derivative of the magneto-resistance in $(TMTSF)_2ClO_4$ [193]. The let inset shows the Fermi surface. The peaks are denoted by fractions p/q.

A similar warped cylinder-shaped Fermi surface can be made in semiconductor superlattices. Yagi *et al.* observed the same kind of AMRO in doped AlGaAs/GaAs superlattices [192].

A different type of AMRO was observed in a quasi-one-dimensional organic conductor $(TMTSF)_2ClO_4$ [193]. Figure 3.43 shows experimental traces of AMRO in $(TMTSF)_2ClO_4$. The quasi-one-dimensional band of this crystal is represented as

$$\mathcal{E}(\boldsymbol{k}) = \frac{\hbar^2 k_x^2}{2m^*} - 2t_b \cos(bk_y) - 2t_c \cos(ck_z). \tag{3.146}$$

When the magnetic field is applied in the ab plane with an angle θ from the c-axis, the Hamiltonian is given by

$$\mathcal{H} = \frac{\hbar^2 k_x^2}{2m^*} - 2t_b \cos(G_b + bk_y)^2 t_c \cos(G_c x - ck_z), \tag{3.147}$$

where

$$G_b = \frac{beB}{\hbar}\cos\theta, \quad G_c = \frac{ceB}{\hbar}\sin\theta. \tag{3.148}$$

We can see that when a condition

$$\frac{G_b}{G_c} = \frac{q}{p} \quad (p, q = \text{integer}), \tag{3.149}$$

holds, the motion in the y, z-direction is periodic. Under such a condition, the magneto-conductivity should show maxima. This is exactly the phenomena observed in the experiment [193]. In organic conductors that have highly anisotropic Fermi surfaces, different types of AMRO have also been observed [194, 195].

4

CYCLOTRON RESONANCE AND FAR-INFRARED SPECTROSCOPY

4.1 Fundamentals of cyclotron resonance

Cyclotron resonance is a powerful tool to determine the effective mass of carriers, and to study energy band structure, electronic states, electron interactions with other elementary excitations, and electron-electron excitations. There are two ways to observe cyclotron resonance. One is by fixing the magnetic field constant and varying the wavelengh of the incident eletromagnetic radiation. The other method is to sweep the magnetic field at a constant wavelength. The latter is particularly suitable for pulsed high magnetic fields, since the magnetic field is automatically swept and hits the resonance when a monochromatic light is incident on a sample. High magnetic fields are advantageous in many ways for cyclotron resonance. Firstly, as the necessary condition of the cyclotron resonance $\hbar\omega_c > 1$ can be satisfied even in low mobility substances, the effective masses and the band structure can be accurately determined. Moreover, the cyclotron energy $\hbar\omega_c$ can exceed various excitation energies in semiconductors, such as LO phonons, plasmons, the energy band gap, or artificial quantum potentials, *etc.*, which allows us to study new interesting problems. Very high magnetic fields also cause the breakdown of the effective mass theory or new phenomena which arise from electron-electron interaction.

In high magnetic fields, the frequency for the resonance in most semiconductors is in the range 0.3–10 THz (teraherz), corresponding to the wavelength 30–1000 μm. Hence the far-infrared or teraherz spectroscopy technique is necessary. In this chapter, basic principles of the cyclotron resonance and different examples are presented, focusing on the effects of high magnetic fields.

4.1.1 *Observation of classical cyclotron resonance*

Cyclotron resonance in semiconductors was first observed by Dresselhaus, Kip, and Kittel in 1953 [196] and by Lax *et al.* in 1954 [197] for Ge and Si. When subjected to magnetic fields B, conduction electrons with effective mass m^* conduct cyclotron motion in the plane perpendicular to B with angular frequency

$$\omega_c = \frac{\mathrm{e}B}{m^*}. \tag{4.1}$$

If electromagnetic radiation is incident, the cyclotron motion resonantly absorbs the energy when the angular frequency of the radiation ω coincides with ω_c. This is the phenomenon called cyclotron resonance, which can be described well by a classical dynamics of electrons in static magnetic fields and electromagnetic radiation. The motion of electrons in a magnetic field $\boldsymbol{B}$ and an electric field $\boldsymbol{E}$ is represented by the following classical equation of motion,

$$m^* \frac{d\boldsymbol{v}}{dt} + \frac{m^* \boldsymbol{v}}{\tau} = -e\boldsymbol{E} - e(\boldsymbol{v} \times \boldsymbol{B}), \tag{4.2}$$

where the effect of the scattering of electrons is taken into account by the second term in the left side with relaxation time τ. Let us consider the case where a static magnetic field $\boldsymbol{B}$ is parallel to the z-direction, and the electromagnetic wave with an angular frequency ω propagates also along the z-direction. Then in most cases, the electric vector $\boldsymbol{E}$ is perpendicular to the z-axis. The carrier velocity v_x and v_y are obtained as follows,

$$v_x = \frac{e\tau}{m^*}\left[\frac{1 + i\omega\tau}{1 + (\omega_c^2 - \omega^2)\tau^2 + 2i\omega\tau}\right] E_x + \frac{\omega_c}{(\omega - i/\tau)^2 - \omega_c^2} E_y, \tag{4.3}$$

$$v_y = -\frac{\omega_c}{(\omega - i/\tau)^2 - \omega_c^2} E_x + \frac{e\tau}{m^*}\left[\frac{1 + i\omega\tau}{1 + (\omega_c^2 - \omega^2)\tau^2 + 2i\omega\tau}\right] E_y. \tag{4.4}$$

As the current is given by

$$\boldsymbol{J} = ne\boldsymbol{v}, \tag{4.5}$$

the conductivity tensor is obtained as

$$\boldsymbol{\sigma} = \frac{\sigma_0}{\tau}\begin{pmatrix} \dfrac{-i(\omega - i/\tau)}{(\omega - i/\tau)^2 - \omega_c^2} & \dfrac{\omega_c}{(\omega - i/\tau)^2 - \omega_c^2} & 0 \\ \dfrac{-\omega_c}{(\omega - i/\tau)^2 - \omega_c^2} & \dfrac{-i(\omega - i/\tau)}{(\omega - i/\tau)^2 - \omega_c^2} & 0 \\ 0 & 0 & \dfrac{-i}{\omega - i/\tau} \end{pmatrix}. \tag{4.6}$$

The diagonal conductivity is therefore

$$\sigma = \frac{j_x}{E_x} = \frac{Nev_x}{E_x} = \sigma_0\left[\frac{1 + i\omega\tau}{1 + (\omega_c^2 - \omega^2)\tau^2 + 2i\omega\tau}\right], \tag{4.7}$$

where σ_0 is the DC conductivity,

$$\sigma_0 = \frac{Ne^2\tau}{m^*}. \tag{4.8}$$

Equations (4.6) and (4.7) well known as the Drude model.

As the radiation proceeds in an absorbing medium in the z-direction, the intensity decreases with a dependence $\exp(-\alpha z)$. Here, α is called the absorption coefficient. When the radiation of intensity I_0, is incident to a semiconductor plate with a thickness d, the intensity I of the transmitted radiation through the plate is expressed using the above α in the following form,

$$I \approx I_0(1 - R)^2 \exp(-\alpha z), \tag{4.9}$$

where R is the reflectivity of the surface of the plate.

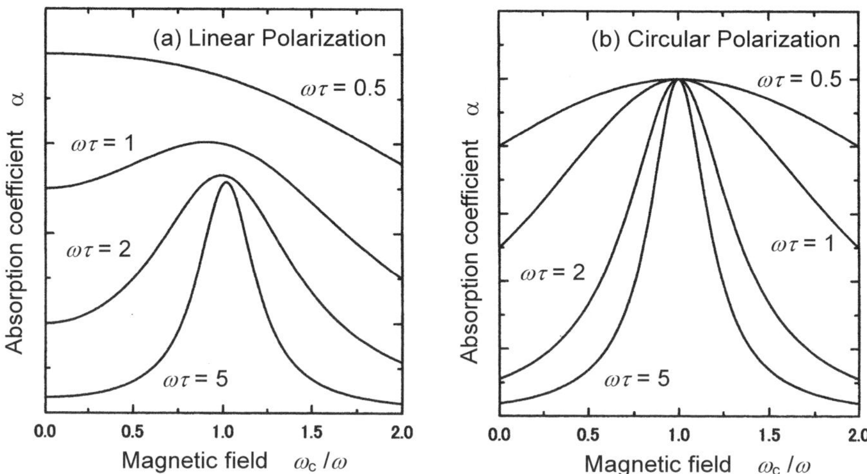

FIG. 4.1. Cyclotron resonance line-shape for at different values of $\omega\tau$. (a) Linear polarization. (b) Circular polarization.

The power absorption of the radiation per unit volume and per unit time is equal to the real part of $\boldsymbol{J} \cdot \boldsymbol{E}$. When the electric vector $\boldsymbol{E}$ is linearly polarized in the x-direction, it is proportional to

$$\Re(\sigma_{xx}) = \sigma_0 \frac{1 + (\omega\tau)^2 + (\omega_c\tau)^2}{[1 + (\omega^2 - \omega_c^2)\tau^2]^2 + 4(\omega\tau)^2}, \tag{4.10}$$

where $\Re$ denotes the real part. The absorption coefficient α is proportional to the power absorption, so that it is also proportional to $\Re(\sigma_{xx})$. It is depicted as a function of ω_c/ω in Fig. 4.1 (a) for the case of linear polarization. We can see that the absorption has a peak at $\omega_c/\omega = 1$, and the width of the peak is decreased as $\omega\tau$ is increased. The peak is visible only for $\omega\tau > 1$. This is the necessary condition for observing cyclotron resonance. The condition is equivalent to the requirement that the carriers should complete at least one cyclotron orbit between scattering.

The shape of the absorption curve in Fig. 4.1 (a) has a general background which decreases as the ω_c/ω increases. The background is more conspicuous for small $\omega\tau$. When we use the circularly polarized radiation, the peak is more clearly visible, because the fundamental propagation mode of the electromagnetic wave in magnetic fields is the circularly polarized one. The electric field of the circularly polarized radiation and the corresponding velocity of carriers are represented by

$$E_\pm = E_x \pm iE_y, \tag{4.11}$$

$$v_\pm = v_x \pm iv_y. \tag{4.12}$$

Here the + and − signs denote the right and left circular polarizations. The conductivity σ^+, σ^- for both circularly polarized radiation is obtained from an equation similar to (4.7), which is for the linear polarization,

$$\sigma_\pm = \sigma_{xx} \mp \sigma_{xy} = \sigma_0 \left[\frac{1}{1+(\omega \mp \omega_c)^2\tau^2} - i\,\frac{(\omega \mp \omega_c)\tau}{1+(\omega \mp \omega_c)^2\tau^2} \right]. \qquad (4.13)$$

The absorption of the radiation is proportional to the real part of the conductivity. For electrons, ω_c is positive (because we define the charge of electrons as $-e$), so that the resonance takes place for the right circularly polarized radiation for $\omega_c = \omega$ as the denominator takes the minimum. When we sweep the magnetic field fixing the photon energy $\hbar\omega$ of the radiation, the resonance occurs at a magnetic field B_r which satisfies,

$$\omega_c = \frac{\mathrm{e}B_r}{m^*} = \omega. \qquad (4.14)$$

From the magnetic field of the peak B_r, we can obtain the effective mass by

$$m^* = \frac{\mathrm{e}B_r}{\hbar\omega}. \qquad (4.15)$$

The curve has a Lorenzian shape as shown in Fig. 4.1 (b), and the half width of the half maximum ΔB is represented by

$$\frac{\Delta B}{B_r} = (\omega_c\tau)^{-1} = \frac{1}{\mu B}, \qquad (4.16)$$

Here $\mu = e\tau/m^*$ is the mobility of carriers. Using this relation, we can obtain information about the carrier scattering or mobility from ΔB. For holes, ω_c is negative (because e is negative for holes), and the resonance occurs for the left circular polarization. Therefore, from the sign of the polarization for which the resonance occurs, we can know whether the carriers are electrons or holes. Thus the right or left circular polarizations are often called "electron-active" or "hole-active" polarization, respectively. For both electrons and holes, the non-active polarization gives just a monotonic absorption spectrum. The origin of the background for the linear polarization as shown in Fig. 4.1 (a) is this non-active component, since the linear polarization is a sum of left and right circular polarizations. When $\omega_c\tau$ is not much larger than 1, the absorption is more clearly seen if we use a circular polarization. Therefore, the use of circular polarization is more advantageous than linear polarization. As the $\omega_c\tau$ becomes large, the absorption curve for the linear polarization approaches that for circular polarization because the contribution of the non-active component to the absorption becomes smaller.

4.1.2 *Quantum theory of cyclotron resonance*

Cyclotron resonance is regarded as a transition of electrons between Landau levels absorbing the energy of photons of the radiation. When the energy of one photon $\hbar\omega$ becomes equal to the spacing between Landau levels, resonant electron transition gives rise to the absorption of radiation. The transition probability of the resonance is calculated treating the electric field of the electromagnetic radiation

$$\boldsymbol{E}_R = \boldsymbol{E}_0 \exp[i(\omega t + \boldsymbol{q}\cdot\boldsymbol{r})], \tag{4.17}$$

as a perturbation. $\boldsymbol{E}_R$ is represented in terms of a vector potential

$$\boldsymbol{E}_R = -\frac{\partial \boldsymbol{A}_R}{\partial t}. \tag{4.18}$$

The magnetic field $\boldsymbol{B}$ is also represented by another vector potential $\boldsymbol{A}_B$,

$$\boldsymbol{B} = rot\boldsymbol{A}_B. \tag{4.19}$$

The Hamiltonian is given by

$$\begin{aligned}\mathcal{H} &= \frac{1}{2m}[\boldsymbol{p} + e(\boldsymbol{A}_B + \boldsymbol{A}_R)]^2 + U(\boldsymbol{r}) \\ &\sim \frac{1}{2m}(\boldsymbol{p} + e\boldsymbol{A}_B)^2 + U(\boldsymbol{r}) + \frac{e}{m}(\boldsymbol{p} + \boldsymbol{A}_B)\cdot\boldsymbol{A}_R.\end{aligned} \tag{4.20}$$

The first two terms of the right side compose the base Hamiltonian which forms the Landau levels. The third term can be treated as the perturbation which gives rise to the transition between the Landau levels. The absorption coefficient α is given by

$$\alpha = \frac{2\hbar\omega\mu_0}{ncE_0^2V}W, \tag{4.21}$$

where n is the refractive index, E_0 is the electric vector of the incident radiation. W is the transition probability between the n-th and the k-th Landau levels, which can be calculated by Fermi's golden rule,

$$W_{nk} = \frac{2\pi}{\hbar}\left|< k|\mathcal{H}'|n >\right|^2 \delta(\mathcal{E}_k - \mathcal{E}_n). \tag{4.22}$$

The matrix element is

$$\begin{aligned}< k|\mathcal{H}'|n > &= < k|\frac{e}{m}(\boldsymbol{p} + e\boldsymbol{A}_B)\cdot\boldsymbol{A}_R|n > \\ &= < k|\frac{e}{m^*}\boldsymbol{A}_R\cdot m\frac{d\boldsymbol{r}}{dt}|n > \\ &= \frac{i}{\hbar}\boldsymbol{A}_B(\mathcal{E}_k - \mathcal{E}_n) < k|\boldsymbol{r}|n > .\end{aligned} \tag{4.23}$$

According to the effective mass approximation, the wave functions of the n-th and k-th Landau levels are represented by

$$|n >= C_n u_{\nu 0} f_n, \quad |n >= C_n u_{\kappa 0} f_n, \tag{4.24}$$

where $u_{\nu 0}$ *etc.* are the Bloch function at $\boldsymbol{k} = 0$ and f_n *etc.* are the envelope functions. Thus the matrix element in (4.23) is

$$
\begin{aligned}
< k|\boldsymbol{r}|n > &= \int C_k^* u_{\kappa 0}^* f_k^* \boldsymbol{r} C_n u_{\nu 0} f_n \\
&\sim C_k^* C_n \left[\int_{cell} u_{\kappa 0}^* \boldsymbol{r} u_{\nu 0} d\boldsymbol{r} \int_{crystal} f_k^* f_n d\boldsymbol{r} \right. \\
&\left. + \int_{cell} u_{\kappa 0}^* u_{\nu 0} d\boldsymbol{r} \int_{crystal} f_k^* \boldsymbol{r} f_n d\boldsymbol{r} \right] . \qquad (4.25)
\end{aligned}
$$

To derive the second relation here, we used properties of the wave functions; namely, us are rapidly oscillating functions, whilst fs are slowly varying functions. The first term in the bracket is the interband term which is zero within the same band ($\kappa = \nu$). The second term is the intraband term which gives the contribution to cyclotron resonance. The selection rule is then obtained from the condition that $\int f_k^* \boldsymbol{r} f_n d\boldsymbol{r}$ is non-zero. When the Landau levels are represented by harmonic wave functions as in (2.60), the selection rule is obtained as

$$\Delta N = \pm 1 \qquad \text{(Landau level)}, \qquad (4.26)$$

$$\Delta s = 0 \qquad \text{(spin)}. \qquad (4.27)$$

From this selection rule, we can see that the carriers can undergo transition only to the adjacent Landau level. When there is a non-parabolicity, however, the Landau levels have some components of different harmonic oscillator functions, and there is some probability of transitions between remote levels, so that the above selection rule sometimes breaks down.

Quantum treatment of the Landau levels is important in the case where the spacing between the Landau levels is non-uniform due to the non-parabolicity or the degeneracy of the bands.

4.1.3 *Cyclotron resonance in anisotropic bands*

When the band extrema in the $\boldsymbol{k}$ space are located at points other than the Γ point, the effective mass is generally anisotropic even in cubic crystals. Good examples are Ge and Si, in which the conduction band minima are at the L-point and the Δ-point. In a band whose equi-energy surface is a rotational ellipsoid, the energy dispersion is represented by

$$\boldsymbol{E} = \frac{\hbar^2}{2m_t^*}(k_x^2 + k_y^2) + \frac{\hbar^2}{2m_l^*}k_z^2, \qquad (4.28)$$

where the longitudinal mass m_l^* is usually much larger than the transverse mass m_t^*. In such an anisotropic band, different effective mass is observed in the cyclotron resonance, depending on the direction of the magnetic field. Dresselhauss, Kip, and Kittel derived an expression of the apparent effective mass m_c^* observed in the cyclotron resonance for arbitrary directions of the magnetic field [196].

When magnetic field $\boldsymbol{B}$ is applied in the direction tilted from the z-direction by an angle θ in the $x-z$ plane,

$$\boldsymbol{B} = B(\sin\theta, 0, \cos\theta), \tag{4.29}$$

the effective mass m_c^* is represented by an equation

$$\left(\frac{1}{m_c^*}\right)^2 = \frac{\cos^2\theta}{m_t^{*2}} + \frac{\sin^2\theta}{m_l^* m_t^*}. \tag{4.30}$$

For $\boldsymbol{B}$ parallel to the z-direction, the effective mass is $m_c^* = m_t^*$ and for $\boldsymbol{B}$ parallel to the x- or y-direction, it is $m_c^* = \sqrt{m_t^* m_l^*}$.

In Ge, whose conduction band consists of four valleys, the principal axis of the ellipsoids (z-axis in Eq. (4.28)) is directed towards the $< 111 >$ direction, so that each curve should be calculated by different coordinate conversion to obtain the θ dependence. Four different curves are obtained in the general direction of $\boldsymbol{B}$. However, for the $\boldsymbol{B}$ direction in the (110) plane, two valleys are equivalent, so that three curves are obtained.

When the anisotropy is particularly large, the selection rule against the circular polarization is relaxed for magnetic fields applied in the direction off the axial symmetry axis, because then the cyclotron orbit has an ellipsoidal form rather than a perfect circle. In other words, the electron cyclotron resonance is also observed for "hole-active" polarization (left circular polarization) to some extent and vice versa. Let us assume that the magnetic field is applied in the x-direction for the conduction band with a dispersion represented by (4.28). The equation of motion in the $y-z$ plane is

$$m_t^* \frac{dv_y}{dt} + \frac{m_t^* v_y}{\tau} = -eE_y - ev_z B, \tag{4.31}$$

$$m_l^* \frac{dv_z}{dt} + \frac{m_l^* v_z}{\tau} = -eE_z - ev_y B. \tag{4.32}$$

$$\tag{4.33}$$

For circular polarization,

$$E_\pm = E_y \pm E_z, \tag{4.34}$$

$$v_\pm = v_y \pm v_z. \tag{4.35}$$

Therefore, for electrons, v_+ is obtained from the above equation of motion as

$$v_+ = \frac{1}{i(\omega - \tilde{\omega}_c) + 1/\tau} \left[\frac{\frac{1}{2}(i\omega + \frac{1}{\tau})(\frac{1}{m_t^*} + \frac{1}{m_l^*}) + \frac{i\tilde{\omega}_c}{\sqrt{m_t^* m_l^*}}}{i(\omega + \tilde{\omega}_c) + \frac{1}{\tau}} eE_+ \right] \tag{4.36}$$

$$+ \left[\frac{\frac{1}{2}(i\omega + \frac{1}{\tau})(\frac{1}{m_l^*} - \frac{1}{m_t^*})}{i(\omega + \tilde{\omega}_c) + \frac{1}{\tau}} eE_- \right], \tag{4.37}$$

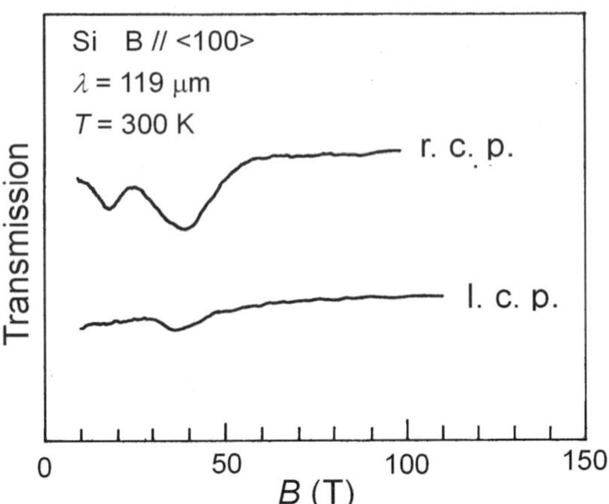

FIG. 4.2. Cyclotron resonance in n-Si for electron-active and hole-active circular polarizations. $B \parallel < 100 >$. The wavelength is 119 μm. $T = 300$ K.

where

$$\tilde{\omega}_c = \frac{eB}{\sqrt{m_t^* m_l^*}}. \tag{4.38}$$

The absorption is proportional to the real part of the conductivity $\sigma = Nev/E$. The ratio of the absorption for the left and right circular polarizations is

$$\frac{\Re\left(\sigma(E_+)\right)}{\Re\left(\sigma(E_-)\right)} \sim \left| \frac{\frac{1}{m_t^*} + \frac{1}{m_l^*}}{\frac{1}{m_l^*} - \frac{1}{m_t^*}} \right| = \frac{m_t^* + m_l^*}{m_l^* - m_t^*}. \tag{4.39}$$

An example is shown in Fig. 4.2 for n-type Si for right circular polarization (r.c.p.) and left circular polarization (l.c.p.) at a wavelength of 119 μm [198]. When we apply magnetic fields to the $< 100 >$ direction, two resonance peaks are observed corresponding to the two inequivalent valleys with effective cyclotron masses m_t^* and $\sqrt{m_l^* m_t^*}$, respectively, for the electron active mode (r.c.p.). The peak with m_t^* is not discernible for l.c.p. because the cyclotron orbit is circular, whilst the peak of $\sqrt{m_l^* m_t^*}$ is observed even for l.c.p., although the intensity is smaller than for r.c.p. This is because the cyclotron orbit is ellipsoidal.

In some crystals, effective masses are all different for three principal axes (no rotational symmetry), and expressed by

$$\mathcal{E}(\boldsymbol{k}) = \frac{\hbar^2 k_a^2}{2m_a^*} + \frac{\hbar^2 k_b^2}{2m_b^*} + \frac{\hbar^2 k_c^2}{2m_c^*}. \tag{4.40}$$

In such a case, the effective mass observed in cyclotron resonance is

$$\left(\frac{1}{m^*}\right)^2 = \frac{1}{m_a^* m_b^* m_c^*}(m_a^* \alpha^2 + m_b^* \beta^2 + m_c^* \gamma^2), \tag{4.41}$$

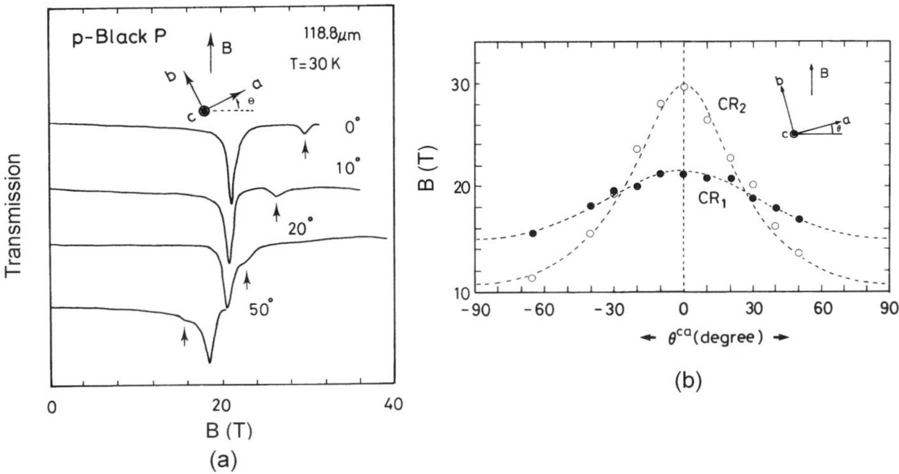

FIG. 4.3. Cyclotron resonance in *p*-type black phosphorus [200]. (a) Experimental traces at different angles when the ac-plane is rotated around the c-axis set perpendicular to B. Besides the main peak of hole cyclotron resonance, a small peak corresponding to the electron cyclotron resonance (indicated by arrows) is observed. (b) Angular dependence of the resonant field for the peak CR_1 (hole) and CR_2 (electron).

Table 4.1 Effective masses of electrons and holes in black phosphorus (in units of m.

	$\boldsymbol{B} \perp$ ab	$\boldsymbol{B} \perp$ bc	$\boldsymbol{B} \perp$ ac	Photon energy	Reference
	m^*_{ab}	m^*_{bc}	m^*_{ac}	(meV)	
holes	0.436	0.166	0.238	10.4	[200]
electrons	0.380	0.120	0.330		
holes	0.427	0.146	0.222	1.3–3.2	[199]
electrons	0.362	0.103	0.291		
	m^*_a	m^*_b	m^*_c		
holes	0.625	0.304	0.091	10.4	[200]
electrons	0.436	0.166	0.238		
holes	0.648	0.280	0.076	1.3–3.2	[199]
electrons	1.03	0.128	0.083		

where (α, β, γ) is a direction cosine of the magnetic field against the a, b, and c-axes. Black phosphorus is an example of such crystals. It is a rhombohedral layered structure stacked in the direction of the b-axis. Cyclotron resonance was measured by Narita *et al.* [199] in the low field range up to 7 T. Takeyama *et al.* extended measurements to higher fields up to 40 T, and determined the anisotropic effective masses [200]. Figure 4.3 shows the experimental traces of the cyclotron resonance in black phosphorus and the angular dependence of the effective mass [200]. The measurement was performed for p-type crystals grown under high pressure. As the samples also contain small n-type regions even in a p-type sample, two peaks were observed corresponding to holes (CR_1 peak) and electrons (CR_2 peak). Table 4.1 lists the effective masses of black phosphorus obtained from the cyclotron resonance experiments for all three directions.

4.1.4 *Cyclotron resonance in two-dimensional systems*

The extreme case of the anisotropic system is the two-dimensional (2D) electron systems in quantum wells or MOS-FET. The first cyclotron resonance experiments were reported by Allen, Tsui, and Dalton [201] and by Abstreiter *et al.* [202] for Si MOS-FET. The ground state of the two-dimensional electrons in Si MOS-FET fabricated on a [100] surface is quantized in such a way that the effective mass in the two-dimensional plane is the transverse mass m_t^*. Therefore, when we apply a magnetic field perpendicular to the plane, we can expect that the effective mass should not be very different from the value for the three-dimensional (3D) sample, $m_t^* = 0.1905m$. Actually, the observed mass for Si MOS-FET was $0.20m$ [201] and $0.21m$ [202] which are very close to the 3D values.

However, in the case of 2D systems, many different features are expected to occur in the cyclotron resonance. In fact, in more detailed studies, it has been revealed that the non-parabolicity, the polaron effect, the spin split cyclotron resonance, *etc.* all show properties different from the 3D case, Furthermore, the electron-electron interaction should be enhanced in the 2D systems at relatively low carrier densities since the interaction is more enhanced in 2D systems. Many experiments have been performed to explore the new effects in the cyclotron resonance of 2D carriers since the early days. Figure 4.4 shows the carrier density dependence of the cyclotron mass m_c^* for Si MOS-FET measured by Abstreiter *et al.* [203]. It was found that as the electron density n_s is decreased to below 1×10^{12} cm^{-3}, m_c^* starts increasing. With further decreasing n_s below 5×10^{11} cm^{-3}, m_c^* shows a drastic decrease. The first increase of the mass seems to arise from the electron-electron interaction. The large drop of the mass at very low carrier density is due to the localization of the carriers [203]. A similar increase of the effective mass with decreasing carrier concentration below 10^{12} cm^{-2} was also observed by Pan and Tsui [204].

Wilson, Allen, and Tsui investigated the cyclotron resonance at low electron densities while searching for the Wigner crystal [205]. In 2D systems, we can expect that the ground state of electrons in sufficiently high magnetic fields and low temperatures should be the Wigner crystal [119, 206]. Although the electron-electron interaction effect should not show up in the cyclotron resonance according to Kohn's theorem, as will be discussed in Section 4.4.1 [207], the rule is relaxed if the translational symmetry of the system is broken. Consid-

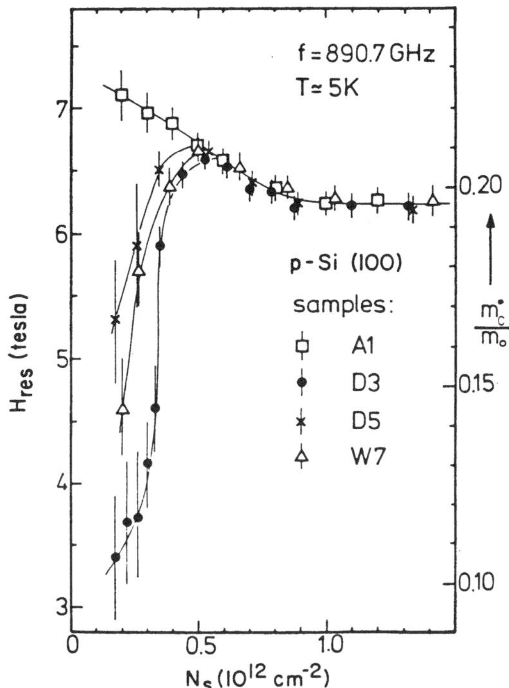

FIG. 4.4. Dependence of the cyclotron mass and the resonance magnetic field on the carrier density for different samples of Si-MOS-FET [203].

erable modification of the cyclotron resonance spectra will be observed if there are impurity potentials in a Wigner crystal [208]. Wilson *et al.* found that the cyclotron mass m_c^* and the line-width show significant changes near the filling factor ν crosses unity. The experiment in the low electron density, however, is difficult in the sense that the effect of the electron localization that becomes significant in the low densities obscures the effect of electron-electron interaction. Therefore, no conclusive result has been obtained regarding the Wigner crystallization. As regards the electron-electron interaction and its effect on cyclotron resonance, discussion will be made again in Section 4.4.

4.1.5 *Impurity cyclotron resonance and impurity transition*

In the quantum limit regime with high magnetic fields and low temperatures, we often observe subsidiary absorption peaks arising from the transitions between impurity states in the vicinity of cyclotron resonance absorption. As mentioned in Section 1.3.1, the donor states have hydrogen atom-like states such as $1s$, $2s$, $2p$, $3s$ $3p$, $3d$, $\cdots$. The absorption associated with the $1s \to 2p$ transition is fairly strong at low temperatures when electrons are captured in the donor states. The $2p$ state shows a Zeeman splitting to form $2p^-$, $2p^0$ and $2p^+$ states

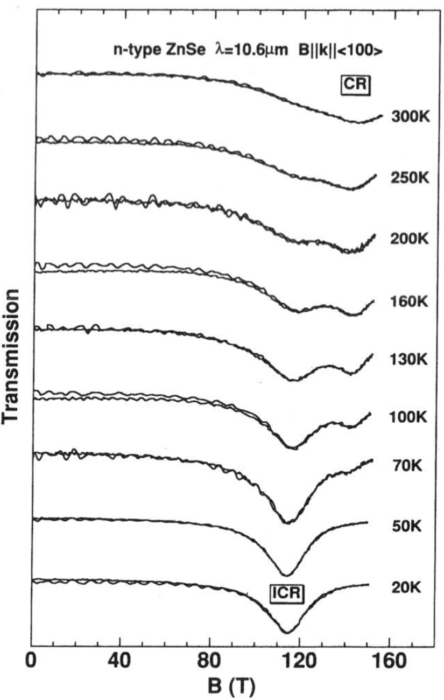

FIG. 4.5. Experimental traces of cyclotron resonance in n-type ZnSe at different temperatures [209].

in the low field range, and the transition occurs between the $1s$ state to $2p^-$ and $2p^+$ states. In the high field range, the $2p^+$ state tends to the (010) state which is associated with the $N = 1$ Landau level. Both the $1s$ state and the $2p^+$ state are almost parallel to the $N = 0$ and $N = 1$ Landau levels, respectively, but the binding energy of the $1s$ state is larger than the $2p^+$ state, because the wave function extension is smaller. Therefore the $1s \rightarrow 2p^+$ transition peak always appears in the low field side of the cyclotron resonance peak, and is often called the "impurity cyclotron resonance". As an example of such absorption peaks, we show in Fig. 4.5 cyclotron resonance spectra in n-type ZnSe [209]. Besides the cyclotron resonance peak, the impurity cyclotron resonance peak ($1s \rightarrow 2p^+$ or (000) $\rightarrow$ (010) transition) is seen in the low field side. Figure 4.6 (a) shows schematically the transitions of the cyclotron resonance and the impurity cyclotron resonance. Due to the very high field, the system is already in the high field range ($\gamma \sim 10$). At low temperatures, the impurity cyclotron resonance peak dominates the spectra because of the carrier freeze-out into the donor state, whilst the free carrier cyclotron resonance becomes prominent as the temperature is raised. From the temperature dependence of the intensity of the impurity cyclotron resonance peak as shown in Fig. 4.6 (b), we can determine the magnetic field dependence of the donor binding energy E_d [209].

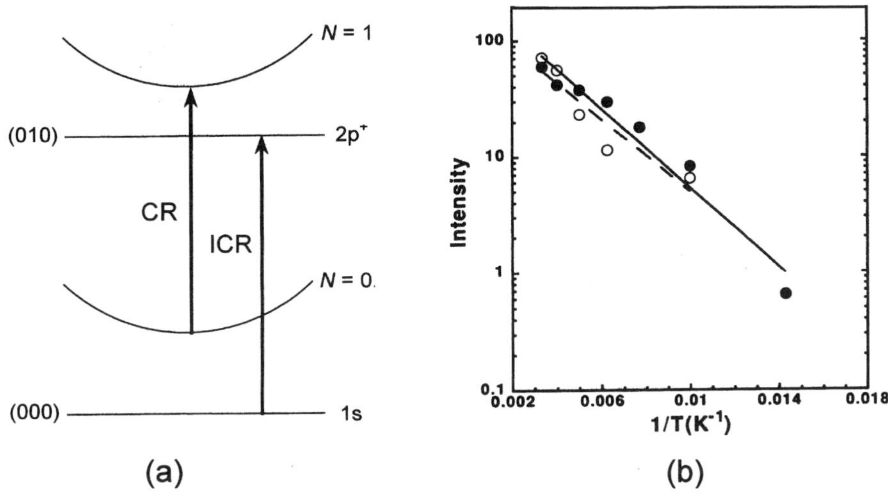

FIG. 4.6. (a) A schematic diagram of the transitions of cyclotron resonance (CR) and the impurity cyclotron resonance (ICR) [209]. (b) Temperature dependence of the integrated intensity of the cyclotron resonance absorption in ZnSe. The solid and the open circles indicate the data at wavelengths of 10.6 μm and 16.9 μm, respectively. The solid and broken lines stand for the best curve fitting with a function, $\exp(-E_d/2kT)$.

As already mentioned in Section 2.8.1, the $1s \rightarrow 2p$ transition is a useful means for studying the central cell correction or chemical shift among different spiecies of donor or acceptor impurities, and hence for characterizing high purity samples [59–61].

Another structure oberved in the far-infrared magneto-optical spectra in connection with impurity states is the absorption from the D^- state. The neutral shallow donor can accommodate a second electron to form a negatively charged ion, similar to the H^- state. This is called the D^- state [210]. The binding energy of the D^- state is very small. In the case of H^-, the binding energy is 5.5% of the Rydberg energy. Najda *et al.* observed the transition from the D^- state to a series of Landau levels in far-infrared magneto-photoconductivity for GaAs, InP, and InSb [211]. The binding energy of the D^- state was determined as a function of magnetic field.

4.2 Non-parabolicity and spin splitting

4.2.1 *Cyclotron resonance in non-parabolic bands*

As discussed in Section 2.4, the energy dispersion of electronic energy bands is parabolic only in the vicinity of their extrema, and as the energy becomes large, it deviates from the k^2 dependence. Generally, the effective mass increases with

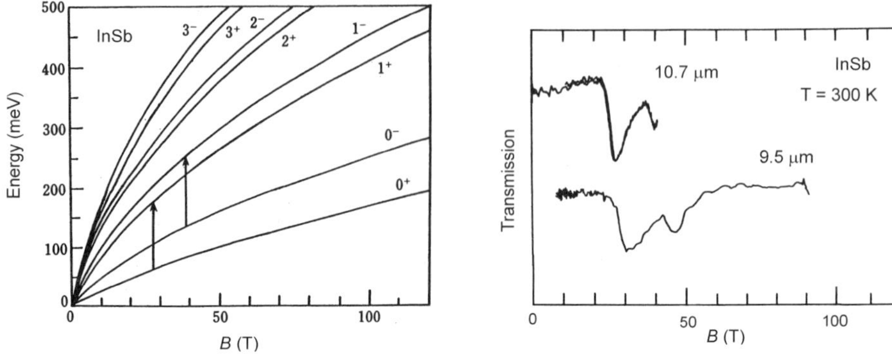

FIG. 4.7. (a) Landau levels and cyclotron resonance transitions in a non-parabolic band in InSb. (b) Spin-split cyclotron resonance in n-InSb at wavelengths of 10.7 μm and 9.5 μm [213]. $T = 300$ K.

increasing energy. The simplest case is the energy dispersion derived by the two-band model as discussed in Section 2.4.1. Based on the two-band model, the effective mass at m^* at an energy $\mathcal{E}$ is obtained from (2.123) as follows,

$$\frac{1}{m^*} = \frac{1}{m_0^*}\left(1 - \frac{2(N+1)\hbar\omega_c(0)}{\mathcal{E}_g}\right). \tag{4.42}$$

Equation (4.42) indicates that as the energy of the carriers increases and becomes non-negligible in comparison to the band gap, the increase of the effective mass is significant. The transitions between Landau levels ($0 \to 1$, $1 \to 2$, $\cdots$) are observed at different magnetic fields with a given $\hbar\omega$. The non-parabolicity appears also in the spin splitting. As is evident in Eqs. (2.125) or (2.127), the spin splitting is larger for the $N = 0$ Landau level than for the $N = 1$ level. Therefore, as shown in Fig. 4.7 (a), the transition of $(0+) \to (1+)$ occurs at lower magnetic field than that of $(0-) \to (1-)$ for a given $\hbar\omega$. The non-parabolicity in InSb and GaAs at very high magnetic fields was studied in cyclotron resonance by Herlach, Davis, and Schmidt [212] and Miura, Kido, and Chikazumi [213] in early days of the actual application of megagauss fields to semiconductor physics. Figure 4.7 (b) shows the experimental data for InSb obtained by Miura *et al.* The cyclotron resonance comprising two peaks due to transitions of different spins is clearly seen. Such a splitting between the transitions with different spin direction is sometimes called Δg splitting.

In the case of a two-dimensional electron system, the non-parabolicity is larger than in the three-dimensional case because the Landau levels are formed in the higher energy range due to the subband energy. In the actual band structures in Si or III-V, the valence bands are degenerate and the effect of other remote bands is significant. Therefore, the two-band model is not necessarily a good approximation to describe the conduction band. The three level model taking account of the degenerate valence bands or five level model taking account of the higher lying conduction bands in addition becomes necessary. Furthermore, the polaron effect also affects the non-parabolicity as will be discussed in the next section. To obtain a simple expression in spite of these effects, a non-dimensional parameter K is introduced to represent the non-parabolicity. Using the parameter K, the cyclotron resonance mass m_c^* corresponding to the transition from the $N = 0$ to the $N = 1$ Landau level is given by

$$\frac{1}{m_c^*} = \frac{1}{m_0^*}[1 + \frac{2K}{E_g}\{(N+1)\hbar\omega_c + < T >_z\}], \tag{4.43}$$

where $< T >_z$ is the kinetic energy in the z-direction (along the magnetic field) from the bottom of the energy band [214]. In three-dimensional systems, the density of states diverges at $k_z = 0$, so that $< T >_z = 0$. In two-dimensional electron systems confined in quantum wells, $< T >_z$ is regarded as the subband energy. In systems confined in a triangular potential as in Si MOS-FET, $< T >_z$ depends on the distance from the interface so that becomes a function of the electron density n_s.

The parameter K is different depending on substances. In the ideal two-band model, $K = -1$ irrespective of substances, as shown in (4.42). In bulk GaAs, $K = -0.83$ derived from the three-level model [215], and values $K = -1.1$ or -1.5 were obtained according to the five level model [216]. From the cyclotron resonance in GaAs heterostructures at 119 μm wavelength, a value of $K = -1.4 \pm 0.1$ was obtained [217]. In the case of two-dimensional systems, we have to keep in mind that m_c^* depends on $< T >_z$.

4.2.2 *Spin-split cyclotron resonance*

In the case of GaAs, the g factor is a small negative value. Therefore, the spin splitting (Δg splitting) is so small that it is not easily observed except by high resolution experiments or by experiments in very high magnetic fields. Sig *et al.* measured the cyclotron resonance of GaAs in a magnetic field up to 15 T and found that the spin splitting is larger than expected from the $\boldsymbol{k}\cdot\boldsymbol{p}$ theory based on the valence bands and the conduction band [218]. Normally, when we consider the band structure of zinc blende-crystals, we take account of three bands: the conduction band (Γ_6 band), the valence band (Γ_8 band), and the split-off valence band (Γ_7 band). However, when we measure cyclotron resonance in very high magnetic fields, the Landau levels are formed in the high energy range, so that the interaction with higher lying bands (Γ_{7c} and Γ_{8c}) has some significant effect.

Such an effect was actually observed in the cyclotron resonance in the megagauss range. Figure 4.8 (a) shows cyclotron resonance traces for bulk n-type GaAs at a CO_2 laser wavelength at different temperatures [219]. At low temperatures, four split peaks are observed. They correspond to the free carrier cyclotron resonance and the impurity cyclotron resonance as shown in Fig. 4.8 (b). In other words, the two peaks in the lower field side correspond to the transition

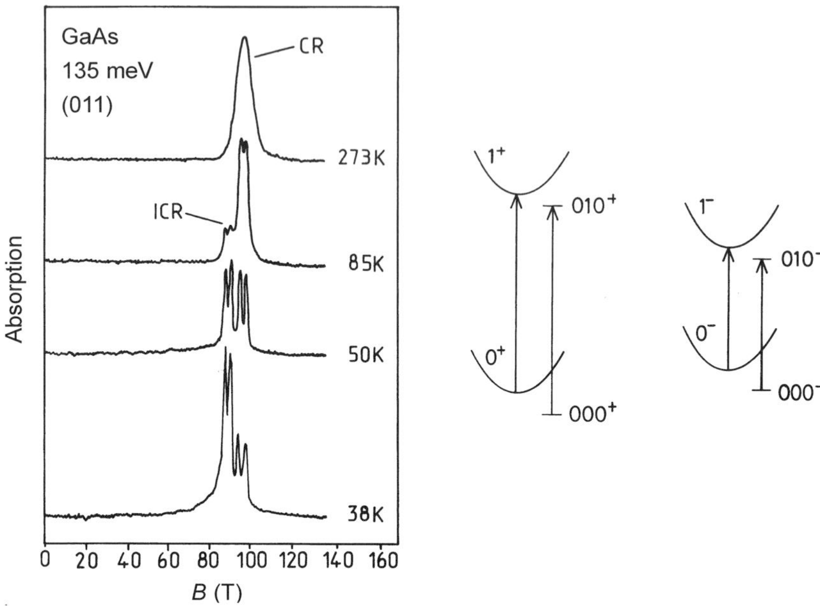

FIG. 4.8. (a) (Left) Cyclotron resonance traces in bulk n-type GaAs (epitaxial film) at different temperatures. (b) (Right) The Landau levels and the impurity states and the electron transitions responsible for the four observed absorption peaks.

between the donor states and the two peaks in the higher field side correspond to the real cyclotron resonance in the conduction band. Both transitions split due to the energy dependence of the g factor arising from the non-parabolicity. As the temperature is elevated, the impurity cyclotron resonance peaks diminish due to the decrease of the thermal population. Also, the peak of the spin up transition (lower field peak) is larger than that of the spin-down transition (higher field peak) due to the difference in the thermal population. From such measurements, we can obtain the dependence of the g-factor on energy and magnetic field. It was concluded that the $\boldsymbol{k}\cdot\boldsymbol{p}$ calculation with three bands (three-level model) is not enough. The five-level model including two higher lying conduction bands (Γ_{7c} and Γ_{8v} bands) as shown in Fig. 4.9 (a) is necessary to explain the magnitude of the Δg splitting.

Figure 4.9 (b) shows the magnetic field dependence of the g-factor for the $N = 0$ and $N = 1$ levels for two different directions of magnetic field in GaAs. As discussed in Section 2.4.2, the g-factor of the conduction band is a small value of -0.44 in GaAs, but it shows a change towards a positive value. It should tend to the free electron value $g = 2.0$ in the high field limit. The interesting point is that it crosses zero at some high magnetic field. In a high field where the sign of the g factor of the $N = 0$ level becomes positive, the spin splitting is larger in the $N = 0$ level than in the $N = 1$ level. Then the lower field peak among

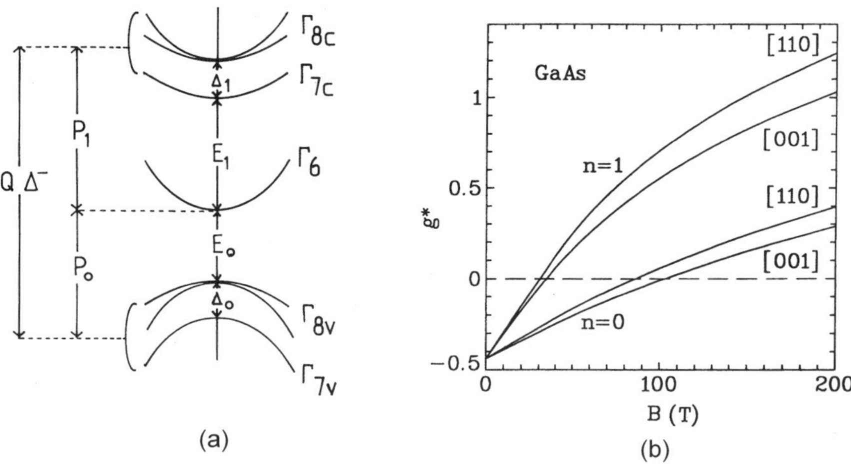

FIG. 4.9. (a) Five-level model. Not only the three levels (Γ_6, Γ_{8v}, and Γ_{7v} bands) but also the higher-lying conduction bands Γ_{7c}, Γ_{8c} are included. (b) Magnetic field dependence of the g-factor of the $N = 0$ and $N = 1$ levels in GaAs for $B \parallel< 100 >$ and $< 110 >$ direction.

the spin split peaks should become larger than the higher field peak. This was actually observed in the cyclotron resonance in very high magnetic field around $B = 240$ T using a CO laser and the electromagnetic flux compression as shown in Fig. 4.10 [220]. Only three peaks are observed as the peaks for the spin-down cyclotron resonance are merged with the spin-up impurity cyclotron resonance. However, we can see that the lower field peak is smaller than the higher field peak. From the analysis of such spin-split cyclotron resonance using the five-level model, not only the momentum matrix element between the conduction band $|S>$-like function and the valence bands $|X>$-like function, $P_0 \sim< S|p_x|X_v >$, but also the matrix elements associated with the higher-lying conduction bands, $P_1 \sim< S|p_x|X_c >$ and $Q \sim< X_v|p_y|Z_c >$ were obtained [220].

4.3 Resonance line-width

As discussed in Section 4.1.1, the classical treatment of the cyclotron resonance predicts that the half width ΔB of cyclotron resonance gives the carrier scattering time τ through the relation (4.16). Since early days, many studies have been made on the cyclotron resonance line-width to investigate the carrier scattering mechanism. The relaxation time τ is determined by various scattering mechanisms, such as acoustic phonon scattering, LO phonon scattering, ionized impurity scattering, neutral impurity scattering [221], carrier-carrier scattering [222], *etc.* In the low field range τ obtained from cyclotron resonance is usually in agreement with those obtained from the DC transport measurement, and thus

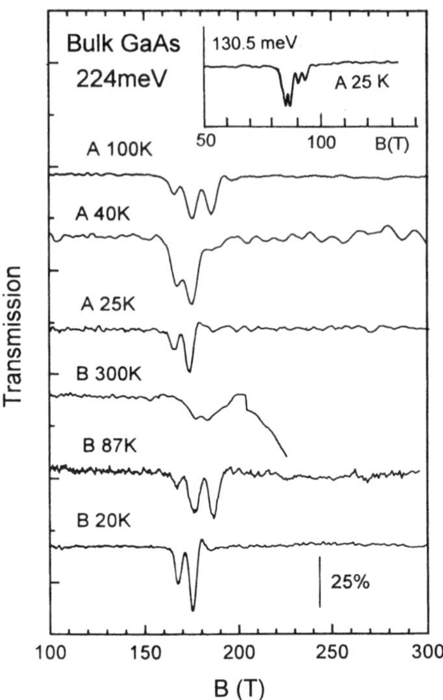

FIG. 4.10. Cyclotron resonance in the conduction band of GaAs at different temperatures in very high magnetic fields generated by electromagnetic flux compression [220]. Two different samples A and B were measured. The photon energy is 224 meV. The inset shows the data at a lower photon energy of 130.5 meV, where four peaks were resolved.

ΔB does not depend on B_r. As regards the acoustic phonon scattering, Bagguley, Stradling, and Whiting showed that in n-type Ge and Si, the line-width is proportional to $T^{-\frac{3}{2}}$ in the temperature range 30–100 K [223,224]. This temperature dependence is the same as the temperature dependence of the DC mobility when the acoustic phonon scattering is the dominant scattering mechanism.

In high magnetic fields when the quantum limit condition $\hbar\omega_c > kT$ holds, however, a quantum mechanical treatment of the line-width becomes necessary. Fink and Braunstein measured the cyclotron resonance in n-Ge up to 11 T where the condition $\hbar\omega_c \geq kT$ is satisfied, and found that there is scattering in addition to the acoustic phonon scattering in a classical form [225]. As regards the acoustic phonon scattering in the quantum limit, Meyer derived a quantum mechanical expression of the line-width of cyclotron resonance [226]. The relaxation time τ which determines the line-width is given from Fermi's golden rule by

$$\frac{1}{\tau} = \frac{2\pi}{\hbar} \int |< N = 1, k_z|\mathcal{H}'|N = 0, k_z - q >|^2 \, \delta(\mathcal{E}_1 \pm \hbar\omega_q - \mathcal{E}_0) d\boldsymbol{q}, \qquad (4.44)$$

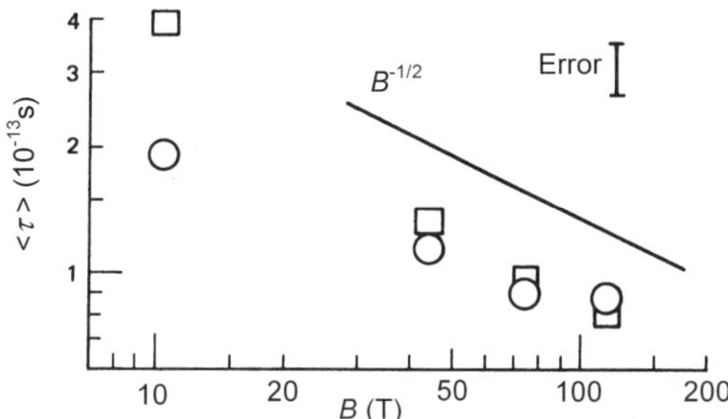

FIG. 4.11. Electron relaxation time $<\tau>$ deduced from cyclotron resonance line-width in n-Ge in very high magnetic fields [231]. The magnetic field was applied parallel to the $< 111 >$ axis. Data for two samples with different carrier concentrations are shown. Circle: 2.7×10^{16} cm^{-3}, Square: 2.2×10^{15} cm^{-3}. $T = 300$ K.

where k and q denote the wave vectors of electrons and phonons, and the matrix element is taken only between the $N = 0$ and $N = 1$ Landau levels. Taking the deformation potential for electron-phonon interaction as $\mathcal{H}'$, (4.44) is reduced to

$$\frac{1}{\tau} = \frac{E_1^2}{(2\pi)^2}\int \frac{q}{\rho v_s}\frac{r_c^2 q_\perp^2}{2}\exp\left(-\frac{r_c^2 q_\perp^2}{2}\right) \times \left(N_q + \frac{1}{2}\right)\delta\left(\hbar\omega_c + \frac{\hbar^2}{2m^*}[k_z^2 - (q_z + k_z)^2]\right) d\boldsymbol{q}, \qquad (4.45)$$

where v_s is the sound velocity, ρ is the density, r_c is the cyclotron radius and E_1 is the deformation potential constant. In high magnetic fields where the condition $\hbar\omega_c \gg kT \gg (\frac{1}{2}m^* v_s^2 \hbar\omega_c)^{\frac{1}{2}}$ is satisfied, we can obtain

$$\frac{1}{\tau} \propto T(\hbar\omega_c)^{\frac{1}{2}}. \qquad (4.46)$$

Ito, Fukai, and Imai investigated the line-width of the cyclotron resonance in n-Ge both theoretically and experimentally, and found that the line-width shows a $T^{-3/2}$ dependence in relatively high temperature and low frequency range, but it deviates from this dependence for $\hbar\omega_c > 1$ [227]. They showed that the experimentally observed line-width in the quantum limit is in good agreement with a theory based on the phonon induced transitions between Landau levels.

As regards the quantum calculation for the acoustic phonon scattering, Suzuki and Dunn [228, 229], and Kobori, Ohyama, and Otsuka [230] also obtained formula a little different from the results of Meyer.

Figure 4.11 shows the resonant photon energy dependence of the relaxation time in n-type Ge at room temperature in high magnetic fields [231]. It was found that $< \tau >$ is almost proportional to $\omega_c^{-\frac{1}{2}}$ in good agreement with the theory of Meyer.

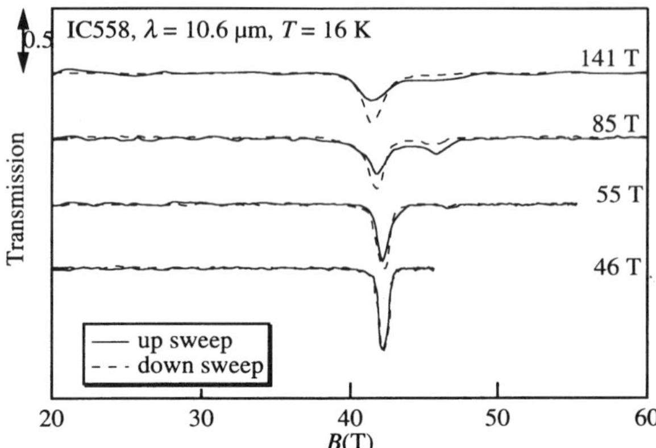

FIG. 4.12. Cyclotron resonance traces for InAs/AlSb quantum wells measured by using the single-turn coil technique [238]. The wavelength was 10.6 μm. T = 16 K. (a) As a function of time t. The magnetic field has nearly a sinusoidal form starting at t = 0 and has a maximum at about $t \sim 3$ μs. The sweep speed of the magnetic field is varied by changing the maximum magnetic field. (b) The same data replotted as a function of magnetic field.

In semiconductors containing appreciable concentrations of impurities, such as InSb, the ionized impurity scattering plays a predominant role in the cyclotron resonance width. The cyclotron resonance line-width arising from the ionized impurity scattering has been a controversial problem for many years, and there have been many reports [232–237].

4.3.1 *Spin relaxation*

As will be described in Chapter 7, the single turn coil technique provides a means to generate very high pulsed magnetic fields above 100 T. The rise time to the peak is 2–3 microseconds and the decay time is about 4 microseconds, and we can obtain signals twice on both slopes of the magnetic field (see Fig. 7.10 in Section 7.1.3). The short pulse duration is usually an obstacle for the experiments, as we need a fast detection technique. However, sometimes we can exploit the rapid variation of the magnetic field in a short time for studies of electron relaxation process. As an example of such a study, we show below the relaxation process of spin states in quantum wells.

Figure 4.12 shows cyclotron resonance traces for quantum wells of InAs/AlSb [238] measured by using the single-turn coil technique. Two spin-split-cyclotron resonance peaks are observed corresponding to the spin up transition ($|0\uparrow> \rightarrow |1\uparrow>$) and the spin down transition ($|0\downarrow> \rightarrow |1\downarrow>$). As the thermal population of electrons in the spin up state $|0\uparrow>$ (plus-spin state) is larger than in the spin down state $|0\downarrow>$ (minus-spin state), the peak corresponding to the spin up transition (the peak at the lower field side) is always larger than the other peak.

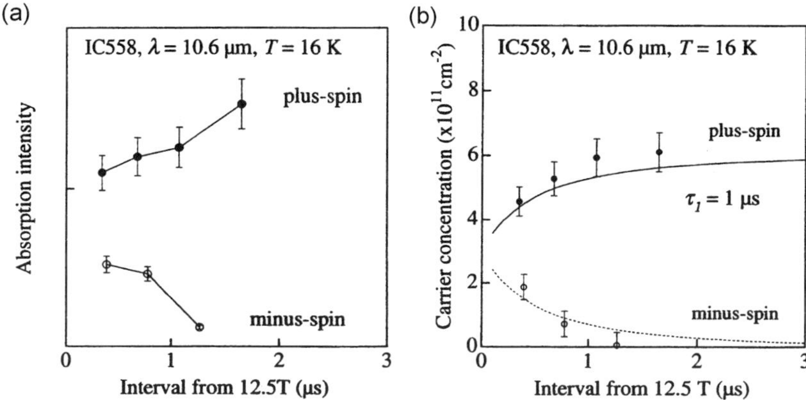

FIG. 4.13. (a) Integrated absorption intensity in the up-sweep as a function of the sweep rate (time from 12.5 T to the resonance field of each spin in the cyclotron resonance) in InAs/AlAs quantum well [238]. (b) Experimental and calculated carrier concentration of each spin in the up-sweep as a function of the sweep rate. The solid (dashed) line is the value of the carrier concentration of the plus (minus) spin calculated from the rate equation assuming $\tau_l = 1\mu$s.

However, we can see in Fig. 4.13 that there is a large hysteresis between the up sweep and the down sweep. There is a difference in the relative absorption intensity of the two peaks between the up and the down traces. The hysteresis is nothing to do with the artifact such as an insufficiently fast response time of the detection system, the heating of the carrier temperature, *etc.*, because in most other systems, traces in the up sweep and the down sweep almost perfectly coincide. The intensity in the down trace is considered to reflect the equilibrium thermal population of carriers in the initial states of the transition. However, the intensity of the up trace shows inequilibrium carrier population depending on the sweep rate. Figure 4.13 (a) shows the integrated intensity of the plus-spin peak and the minus-spin peak in the up-sweep as a function of the sweep rate. Here the sweep rate is varied by changing the maximum field. The horizontal axis represents the sweep speed in terms of the time from the 12.5 T to the resonant field of each peak. The hysteresis was ascribed to the slow relaxation time between the states with different spins. As the magnetic field is increased, the spin splitting is increased, and the electrons in the down spin state $|0\uparrow>$ should relax to the spin down state $|0\downarrow>$. The relaxation occurs via the electron-multiphonon interaction that takes a relaxation time of the order of 0.1–1 μs. Figure 4.13 (b) shows the carrier concentration of each spin in the up sweep. The population was obtained from the rate equation

$$\frac{df_-}{dt} = -\frac{f_-(1-f_+)}{\tau_l}, \tag{4.47}$$

where τ_l is the spin-flip relaxation time and f_+ (f_-) is the filling factor of the plus- (minus-) spin states. As shown in the figure, the theoretical curves assuming $\tau_l = 1\mu s$ are in good agreement with the experimental curve [238], This value is in good agreement with a theoretical value of the spin-flip relaxation time [239]. Such slow spin relaxation time was observed in the spin-flip Raman scattering [240] or photoluminescence experiments using circularly polarized radiation [241]. A similar slow relaxation time was also revealed in the cyclotron resonance in HgTe [242].

4.4 Electron-electron interaction

4.4.1 *Kohn's theorem*

As mentioned in Section 4.1.4, many studies have been made on the many body interaction among electrons in two-dimensional systems by cyclotron resonance. As regards the electron-electron interaction, however, there is a well-known Kohn's theorem that states that it should not appear in cyclotron resonance [207]. The Hamiltonian for the electron system under magnetic field is given by

$$\mathcal{H} = \frac{1}{2m}\sum_{i=1}^{N} P_i^2 + \sum_{i,j} u(\boldsymbol{r}_i - \boldsymbol{r}_j), \tag{4.48}$$

where

$$\boldsymbol{P}_i = \boldsymbol{p}_i + e\boldsymbol{A}_i = [p_{ix}, p_{iy} + eBx_i, p_{zi}], \tag{4.49}$$

and $u(\boldsymbol{r}_i - \boldsymbol{r}_j)$ is the Coulomb potential between electrons. If we take the momentum operator for all the electrons,

$$\boldsymbol{P} = \sum_{i=1}^{N} \boldsymbol{P}_i, \tag{4.50}$$

the equation of motion for $\boldsymbol{P}$ is represented by a commutation relation with $\mathcal{H}$, and we obtain

$$\frac{d\boldsymbol{P}}{dt} = \frac{i}{\hbar}[\mathcal{H}, \boldsymbol{P}] = -\frac{e}{m}\boldsymbol{P} \times \boldsymbol{B}. \tag{4.51}$$

It does not contain $u(\boldsymbol{r}_i - \boldsymbol{r}_j)$. It implies that the electron-electron interaction does not show up in cyclotron resonance. That is to say, the effective mass observed in cyclotron resonance does not depend on electron-electron interaction, as the cyclotron resonance sees just the center of mass motion, in which the internal force of electron-electron interaction does not have any effect. This is called Kohn's theorem. In fact, in organic conductors, it was found that the effective mass measured by cyclotron resonance is much less than that measured by the Shubnikov-de Haas effect [243], as the latter is affected by electron-electron interaction whereas the former is independent of it.

According to Kohn's theorem, it seems that the cyclotron resonance is useless for studying the effect of electron-electron interaction. However, Kohn's theorem holds for electrons moving in a potential with perfect translational symmetry. If there is an impurity potential, non-parabolicity, or electron-phonon interaction, or when there are two different kinds of carriers such as in the case of spin-split cyclotron resonance, the electron-electron interaction effect can be observed. Actually, many of electron-electron interaction in cyclotron resonance have been observed in different systems.

4.4.2 *Mode coupling in spin-split cyclotron resonance*

In spin-split cyclotron resonance, the effect of the coupling between the two cyclotron resonance modes of the spin up and down transitions is observed. Summers *et al.* [244], Nicholas *et al.* [245] and Engelhardt *et al.* [246] found that the intensity and peak positions of the spin-split cyclotron resonance of a GaAs/AlGaAs heterostructure at low temperatures show large dependences on temperature and the filling factor. The relative absorption intensity of the spin up and spin down transitions is not in agreement with that expected from the thermal population in the initial states, and the peak positions show a shift as a function of temperature and the filling factor. This phenomenon was studied theoretically by many authors, and it was revealed that the two cyclotron resonance modes with different spin directions are coupled together and form new modes. In other words, the cyclotron resonances of spin up electrons and spin down electrons interact with each other, and depending on temperature, magnetic field and relative carrier population, the position and the relative intensity change keeping the integrated intensity constant. Cooper and Chalker calculated the cyclotron resonance spectra in the case of the Wigner crystal, and obtained a good agreement with the experiments of Summers *et al.* for $\nu < 1/6$ [247]. They showed that the degree of the spin polarization is obtained from the center of gravity of the spectrum. However, the theory is valid only at $T = 0$ K and for $\nu \ll 1$.

Asano and Ando made a numerical calculation by exact diagonalization of the finite system [248,249]. The results are applicable for finite temperatures and arbitrary filling factor. According to their results, the absorption spectra (which are proportional to the conductivity) change as a function of the parameter $\mathcal{E}_c/\hbar\Delta$, where $\mathcal{E}_c$ is the magnitude of the interaction $e^2/\epsilon l$ (l is the magnetic length), and $\hbar\Delta$ is the energy difference between the up and down spin levels. There are three cases depending on the relative filling factors: (1) Positive mode repulsion; with increasing $\mathcal{E}_c/\hbar\Delta$, the up spin peak (higher energy peak) is pushed away to the higher energy side and the intensity is gradually concentrated into the down spin peak (lower energy peak), (2) Motional narrowing; with increasing $\mathcal{E}_c/\hbar\Delta$, the two peaks come close together, (3) Negative mode repulsion; with increasing $\mathcal{E}_c/\hbar\Delta$, the down spin peak is pushed away to the lower energy side and the intensity is gradually concentrated into the up spin peak. For example, in the case of $\nu < 1$, the spectra are that of positive mode repulsion, and in the case of $\nu > 1$, they are either that of motional narrowing or negative mode repulsion.

In early days, a slightly different type of mode coupling was reported in the cyclotron resonance in Si-MOS [250]. The conduction band minima in Si are located near the X points as shown in Fig. 2.8, and two inequivalent valleys should give two different cyclotron masses when we apply magnetic fields in the $< 100 >$ direction. As the A valleys have a larger effective mass in the $< 100 >$ direction (parallel to the confinement potential), the ground state of the A valleys is about 20 meV lower than that of the B valleys. Therefore, the effective mass observed in the cyclotron resonance is usually that of the A valleys. By increasing the electron density, the B valleys should be populated, so that the cyclotron resonance in the B valleys should be observable in heavily doped Si-MOS. However, the resonance corresponding to the B valleys has never been observed. This was attributed to the mode interaction between the valleys.

In very high magnetic fields in the megagauss range, a very prominent effect of the electron-electron interaction was observed in the cyclotron resonance of two-dimensional electrons in GaAs [251]. Figure 4.14 (a) shows experimental traces of cyclotron resonance in GaAs/AlGaAs multi-quantum wells with different electron concentrations n [251]. Clear cyclotron resonance peaks and the

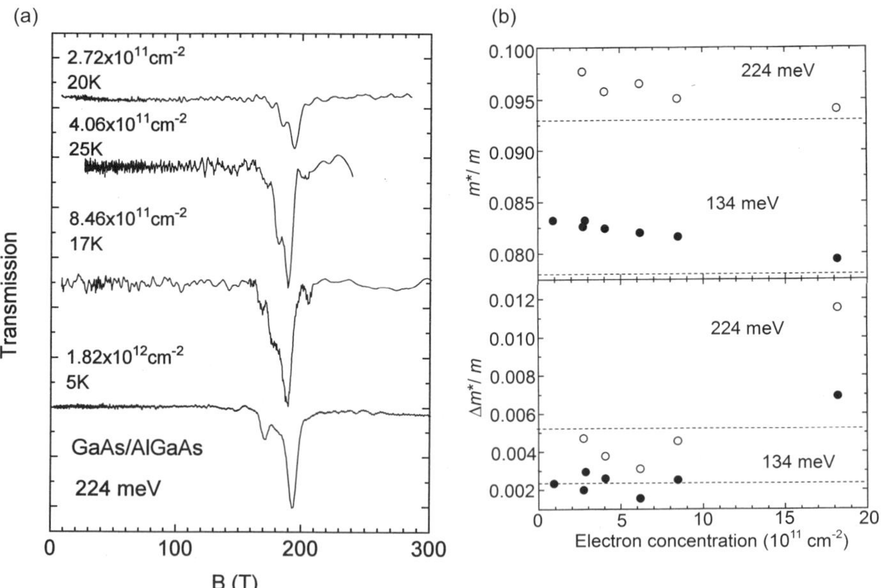

FIG. 4.14. (a) Cyclotron resonance traces for four different samples of GaAs/AlGaAs multi-quantum wells with different electron concentrations in very high magnetic fields up to 300 T [251]. The well width is 10 nm. The photon energy is 224 meV. $T =$ 5–25 K. (b) Top: Electron concentration dependence of the cyclotron effective mass of 2D electrons at photon energies of 224 meV (open circles) and 134 meV (closed circles). Bottom: Magnitude of spin splitting as a function of electron concentration. The broken lines denote the data for bulk GaAs samples.

spin splitting were observed. Here the higher field peak (– spin) is larger than the lower field peak (+ spin) because of the very large photon energy, similarly to the 3D case as shown in Fig. 4.10. A surprising result is that the spin splitting increases dramatically in a sample with a high electron concentration of 1.8×10^{12} cm^{-2}. The electron concentration dependence of the spin splitting $\Delta m^*/m$ is shown in Fig. 4.14 (b) together with the averaged effective mass. It was found that the averaged mass is a slowly decreasing function of n. We can see that the spin splitting has a prominent n-dependence. The $\Delta m^*/m - n$ curves take minima at around $n \sim 6\times6^{12}$ cm^{-2} for 224 meV at a slightly lower n for 134 meV. The electron concentration 6×6^{12} cm^{-2} corresponds to the filling factor $\nu \sim 0.13$. The pronounced electron concentration dependences of the spin splitting and the average mass are certainly a manifestation of some kind of electron-electron interaction. The temperature range of the experiment, 5–25 K, is much higher than the Wigner crystal regime, but there may by some kind of new phase. Michels *et al.* proposed a new liquid-like state for $\nu > 1/10$ in the high temperature range [252]. They assert that this phase is adjoining fractional quantum Hall liquid phase with a phase boundary near $\nu \sim 1/6$ which is close to the above value of $\nu \sim 0.13$. On the other hand, Engelhardt *et al.* suggested a transition from a bulk-like phase to a system with hybridized resonances caused by the effect of field-induced localization [246].

In InAs/AlAs quantum wells, prominent effects of the mode coupling were observed in the spin-split cyclotron resonance. Figure 4.15 shows the temperature dependence of the cyclotron resonance in InAs/AlAs quantum wells [238]. Both the plus-spin (up-spin) peak (lower field peak) and the minus-spin (down-spin) peak (higher field peak) show the temperature dependence. The plus-spin states are lower in energy than the minus-spin, so that the intensity of the plus-spin state should be always larger. As the temperature is increased, the electron population in the minus-spin states is increased, and thus the intensity of the minus-spin peak is increased. However, the experimentally observed increase of the minus-spin absorption intensity with temperature is much larger than expected from the level population, as shown in Fig. 4.15 (b). A remarkable result is that the minus-spin peak becomes even larger than the plus-spin peak at high temperatures. This anomalous temperature dependence is ascribed to the mode coupling. The difference from the case of GaAs system is that the effect is observed up to high temperatures. The major differences between InAs and GaAs are those in the spin splitting $\hbar\Delta$ and the coupling constant $\mathcal{E}_c/\hbar\Delta$. In InAs, $\hbar\Delta$ is very large, and therefore $\mathcal{E}_c/\hbar\Delta$ is small in comparison to GaAs. Therefore, the effect of the mode coupling can be observed up to room temperature.

Unusual temperature dependence of the cyclotron effective mass similar to the case of InAs/AlAs quantum well has also been observed in heavily doped InAs/GaSb quantum wells [253]. The InAs/GaSb system is known to be a type-II quantum wells, and electrons and holes usually coexist in the structure. In heavily n-doped samples ($> 10^{12}$ cm^{-2}), however, electrons are the dominant carriers, so that the electron CR is the only feature in magneto-optical spectra.

Figure 4.16 shows CR spectra for an InAs/GaSb quantum well, whose width is 15 nm. The total electron concentration is 3.56×10^{12} cm^{-2}. In this sample, not only the first subband but also the second subband is populated, so that four peaks are observed altogether. A surprising result is that the lowest field peak corresponding to the plus-spin transition of the first subband around 43 T is very weak. The tendency is more prominent as the doping is increased. Furthermore, the spin-splitting increases as the doping is increased. It should also be noticed that the peaks are shifted to the higher field side as the temperature is increased. This is in contrast to the ordinary cyclotron resonance in InAs where the effective mass should decrease as the temperature is increased due to the decrease of the band gap.

The effect of the mode coupling has also been observed in a p-type system as well. Cole *et al.* observed the effect in p-GaAs/AlGaAs heterojunctions as a significant decrease of the effective mass with increasing temperature [254].

4.4.3 *Filling factor dependence of effective mass and line-width*

Besides the remarkable effect in the spin-split cyclotron resonance, the electron-electron interaction effects are generally observed as the variation of the effective mass and the line-width as a function of the filling factor in 2D electron systems.

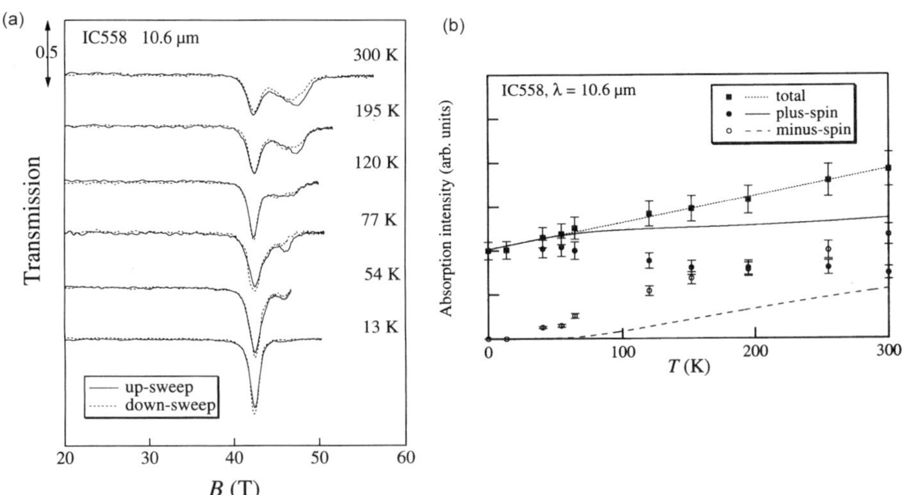

FIG. 4.15. Cyclotron resonance traces for InAs/AlSb quantum wells [238]. The well width is 15 nm, and the electron concentration is 10^{11} cm^{-2}. (a) Experimental traces of the transmission at different temperatures. (b) Temperature dependence of the absorption intensity of the two peaks. The total intensity shows a slight increase with temperature due to the slight increase of the electron concentration. The solid and broken lines are theoretical lines for the plus-spin and the minus-spin resonances calculated from the electron population of the states.

As described in Section 4.1.4, the effective mass should vary as a result of the formation of the Wigner crystal in the limit of $\nu \gg 1$.

For $\nu \geq 1$, the observed effective mass varies as a function of the position of the Fermi level relative to the Landau levels. Abstreiter *et al.* found that the cyclotron resonance peak in a Si-MOS-FET contains a small oscillating structure and the peak position of the small oscillation corresponds to the field where the Fermi level coincides with the Landau levels [203]. The phenomenon is similar to the Shubnikov-de Haas oscillation. An interesting point is that the peaks appear as maxima in the low field side of the main cyclotron resonance peak and as minima as in the high field side. This phenomenon was explained by Ando in terms of the oscillatory change of the Landau level width which depends on the position of the Fermi level [255]. In two-dimensional systems, the impurity scattering is screened by the presence of the carriers, the screening is varied depending of the position of the Fermi level. The oscillatory structure within the cyclotron resonance peak [203] is for the case when the resonance line-width is fairly large and contains several transitions from different Landau levels.

When the line-width is narrower, the cyclotron resonance peak contains just one initial state N just below the Fermi level and one final state $N+1$. In the GaAs system, such a condition is fulfilled in high magnetic fields. Englert *et al.* found that the line-width ΔB for the cyclotron resonance in GaAs heterostruc-

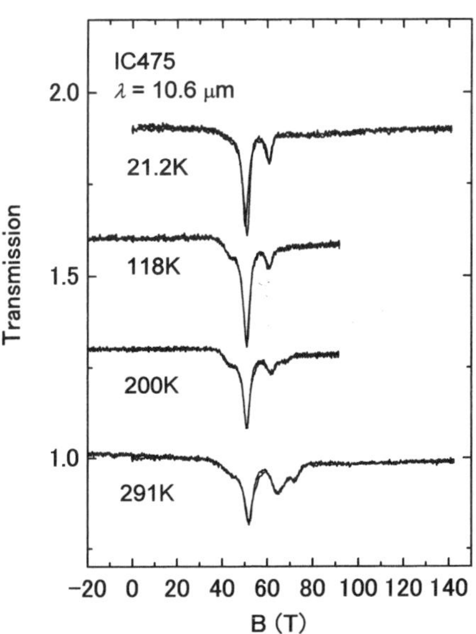

FIG. 4.16. Cyclotron resonance in heavily doped InAs/GaSb quantum well at different temperatures. The well width is 15 nm and the carrier concentration is 3.56 $\times 10^{12}$ cm^{-3}. The photon energy is 117 meV [253].

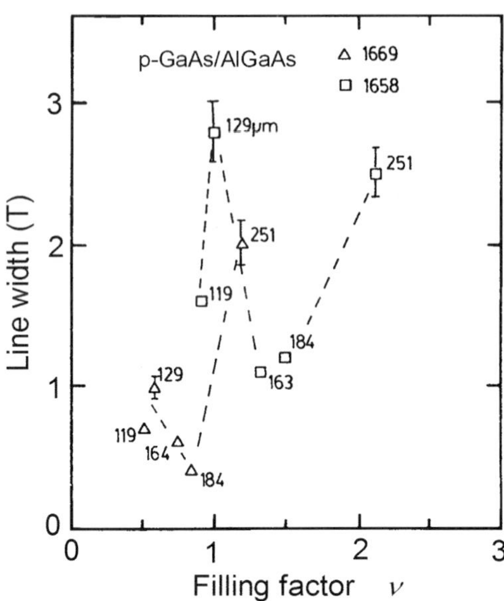

FIG. 4.17. Dependence of the cyclotron resonance line-width on the filling factor in p-GaAs/AlGaAs teterostructure for 2 different samples 1669 and 1658 [257]. The wavelength of the measurement is shown for each point.

tures shows an oscillation as a function of the electron density n_s [256]. The width takes peaks at filling factors $\nu = 2$ and 4. The origin of the oscillation is the screening of the impurity potential by free carriers which becomes maximum at the maximum density of state of the Landau levels. The factor 2 is due to the two spin states, which are almost degenerate. A similar oscillation of the line-width was also observed in a p-channel of the GaAs system by Staguhn, Takeyama, and Miura, as shown in Fig. 4.17 [257].

Kaesen *et al.* observed the oscillation of the effective mass extracted from the resonance frequency as a function of the filling factor in Si-MOS [258]. Figure 4.18 shows the experimental results obtained by Kaesen *et al.* The effective mass takes prominent maxima filling factors $\nu = 4n$, as the field and the gate voltage varied. It implies that the maxima occur when the Fermi level lies between distinct Landau levels in Si-MOS. The effect was ascribed to the impurity mediated electron-electron interaction. Similar oscillations of the cyclotron resonance effective mass have been observed in GaAs/AlGaAs [259–262], InAs/AlGaSb systems [263], CdTe/CdMgTe [264], and Si/SiGe [265]. In most cases, the oscillation was also observed for the line-width.

Figure 4.19 shows the effective mass deduced from the cyclotron resonance in a high mobility Si/SiGe quantum well [265]. The effective mass showed maxima of the effective mass appeared at filling factors $\nu \sim 2$ (not shown in the figure)

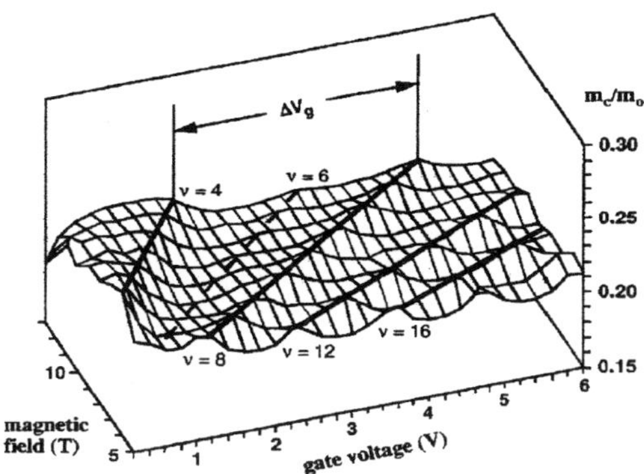

FIG. 4.18. Change of the effective mass as a function of magnetic field and the gate voltage observed by Kaesen *et al.* [258]. The effective mass takes maxima at filling factors $\nu = 4n$.

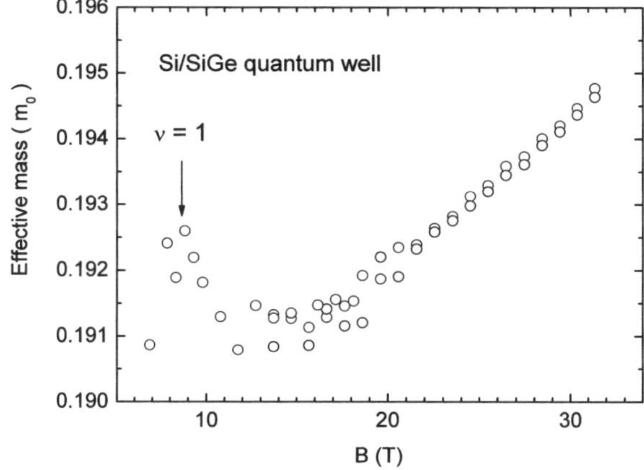

FIG. 4.19. Magnetic field dependence of the effective mass deduced from cyclotron resonance in Si/SiGe quantum well [265]. The well width = 10 nm. The carrier density and DC mobility are 2.0×10^{11} cm^{-2}, and 5.0×10^4 cm^2/V·s at 20K.

and 1, and a remarkable increase at higher fields. The effective mass at 30 T is 0.1942 m which is about 2.2% larger than at the low field limit 0.1900 m. The mass increase was found to persist in higher fields almost linearly up to 60 T. The band gap at the Δ point is 4 eV, so that the mass increase at 30 T by non-parabolicity is only 0.9% at most. Therefore, the observed mass enhancement at $\nu < 1$ is considered to be due to the electron-electron interaction [265].

4.5 Polaron effects

The non-parabolicity of energy bands is brought about not only by the $\boldsymbol{k}\cdot\boldsymbol{p}$ interaction between the bands but also by the electron-phonon interaction. As mentioned in Section 3.5, electrons interact strongly with LO phonons in polar crystals such as II-VI or III-V semiconductors, and move as polarons which have heavier mass than bare electrons. In relatively low magnetic fields, the effective mass which is measured by cyclotron resonance is this polaron mass, which is larger than the bare mass m^* by a factor of about $1+\alpha/6$. When $\hbar\omega_c$ approaches $\hbar\omega_o$ in high magnetic fields, the interaction between electrons and LO phonons resonantly increases and a large break takes place in the magnetic field dependence of the energy of the Landau levels. The effect of the large electron-LO phonon interaction near the resonance condition is called the resonant polaron effect, and was first observed by Johnson and Larsen in n-type InSb in which such a resonance can be readily achieved due to the low effective mass. The energy of the Landau levels can be calculated as the second order perturbation energy for the Fröhlich Hamiltonian (3.118). Larsen derived the energy of the $N = 0$ and $N = 1$ Landau levels as follows for the phonon emitting process

$$\mathcal{E}(0) = \frac{1}{2}\hbar\omega_c - \frac{\alpha(\hbar\omega_o)^2}{2\pi^2}\mathcal{P}\int d\boldsymbol{k}\sum_N \frac{|\mathcal{H}'_{N0}(\boldsymbol{k})|^2}{(N+\frac{1}{2})\hbar\omega_c + \hbar\omega_o + \frac{\hbar^2 k_z^2}{2m^*} - \mathcal{E}(0)}, \tag{4.52}$$

$$\mathcal{E}(1) = \frac{3}{2}\hbar\omega_c - \frac{\alpha(\hbar\omega_o)^2}{2\pi^2}\mathcal{P}\int d\boldsymbol{k}\sum_N \frac{|\mathcal{H}'_{N1}(\boldsymbol{k})|^2}{\mathcal{E}(0) + N\hbar\omega_c + \hbar\omega_o + \frac{\hbar^2 k_z^2}{2m^*} - \mathcal{E}(1)}, \tag{4.53}$$

where $\mathcal{H}'_{N0}$ and $\mathcal{H}'_{N0}$ denote the matrix element for the interaction of the $N = 0$ and $N = 1$ Landau levels with the N-th Landau levels, and are given by

$$|\mathcal{H}'_{N0}|^2 = \frac{r_o}{N!}(ak_\perp)^{2N}\frac{e^{-a^2k_\perp^2}}{k^2}, \tag{4.54}$$

$$|\mathcal{H}'_{N1}|^2 = \frac{r_o}{N!}\frac{(N - a^2k_\perp^2)^2}{(ak_\perp)^{2-2N}}\frac{e^{-a^2k_\perp^2}}{k^2}. \tag{4.55}$$

Here the parameters a and r_0 are defined to represent the cyclotron orbit radius and polaron radius, respectively, with the following relation,

$$a = \left(\frac{\hbar^2/2m^*}{\hbar\omega_c}\right)^{1/2} = \frac{l}{\sqrt{2}}, \tag{4.56}$$

$$r_o = \left(\frac{\hbar^2/2m^*}{\hbar\omega_o}\right)^{1/2},$$
$$\left(\frac{r_o}{a}\right)^2 = \frac{\hbar\omega_c}{\hbar\omega_o}. \tag{4.57}$$

The energies of the $N = 0$ and $N = 1$ Landau levels are shown in Fig. 4.20. We can see that under the resonance condition $\hbar\omega_c = \hbar eB/m^* = \hbar\omega_o$, there is a

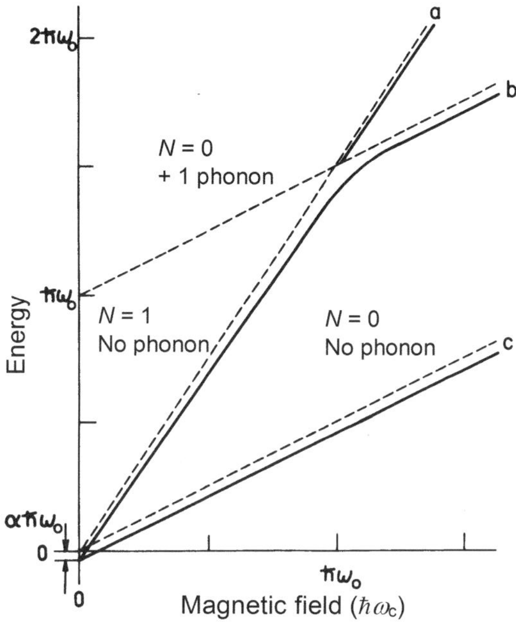

FIG. 4.20. Landau levels of $N = 0$ and $N = 1$ for a polar crystal. Broken lines show the unperturbed Landau levels with or without addition of LO phonon energy.

large splitting in the $N = 1$ Landau level. This resonant effect can be regarded, in other words, as the anti-cross-over effect between the $N = 0$ Landau levels with one LO phonon and the $N = 1$ Landau level with no LO phonon.

The resonant polaron effect was first observed by Dickey, Johnson, and Larsen [266] for n-InSb and analyzed in detail by Larsen [267]. In n-InSb, the effect is small because of the relatively small coupling constant α. The effect of the resonance is also observed in many other transitions [268] and the broadening of the line-width of the resonance peak near $\hbar\omega_c \approx \hbar\omega_o$. In III-V crystals with larger effective mass and larger phonon frequency, higher magnetic fields are necessary. The resonant polaron effect was investigated in GaAs [219] and InP [269] using a very high magnetic fields generated by the single turn coil technique.

In II-VI compounds, α is significantly larger than in III-V compounds, and the resonant polaron effect can be seen more conspicuously, although higher magnetic fields are needed to achieve the resonance condition due to the larger effective mass and the larger LO phonon energy in general. High magnetic fields are also useful for measuring the resonance in II-VI crystals with low mobility. Miura, Kido, and Chikazumi first observed a large resonant polaron effect in n-type CdS and CdSe [270, 271]. Figure 4.21 demonstrates the cyclotron resonance traces observed in II-VI compounds at very high magnetic fields [272]. In spite of relatively low mobilities in II-VI compounds, we can observe well-defined

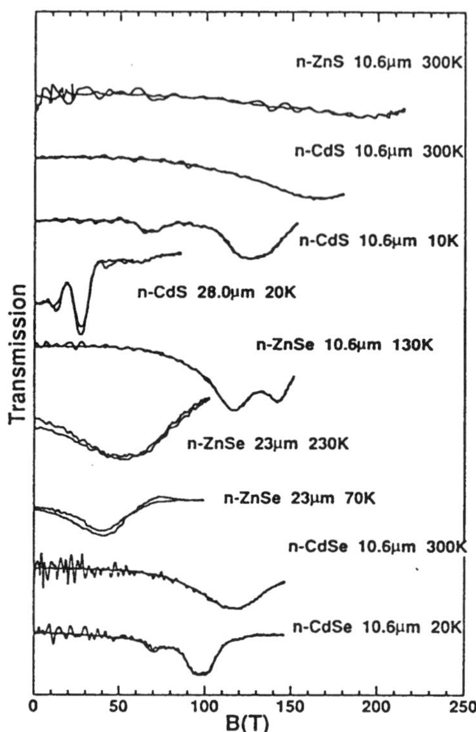

FIG. 4.21. Cyclotron resonance traces in II-VI compounds at high magnetic fields [272].

resonance peaks in very high magnetic fields. Figure 4.22 shows resonant photon energy as a function of B in n-ZnSe as compared with the theoretical curve calculated from (4.52) and (4.53). In Fig. 4.22, we see various characteristic features of the polaron cyclotron resonance: firstly, the effective mass observe in cyclotron resonance (cyclotron mass) m_c^* is larger than for the bare mass m^* in the low field range ($\hbar\omega_c \ll \hbar\omega_o$), as is evident from the smaller gradient of the $\hbar\omega$–B curve. This mass corresponds to the polaron mass $m_p^* = (1 + \alpha/6)m^*$. As $\hbar\omega_c$ approaches $\hbar\omega_o$, the effective mass becomes larger and larger, and finally the $\hbar\omega$–B curve is flattened near $\hbar\omega_c = \hbar\omega_o$. In the higher energy range, there appears another branch, where the m_c^* is smaller than m_p^*. In the vicinity just above $\hbar\omega_o$, the effective mass becomes very small due to the resonance effect. As the energy $\hbar\omega_c$ is further increased, m_c^* tends to approach the bare electron mass. The mass around this range is larger than the band edge mass due to the band non-parabolicity. The decrease of m_c^* for $\hbar\omega_c > \hbar\omega_o$ can be explained by considering the role of the phonon cloud in the polaron effect. Electrons are moving in crystals accompanying the phonon cloud. That is the reason why the

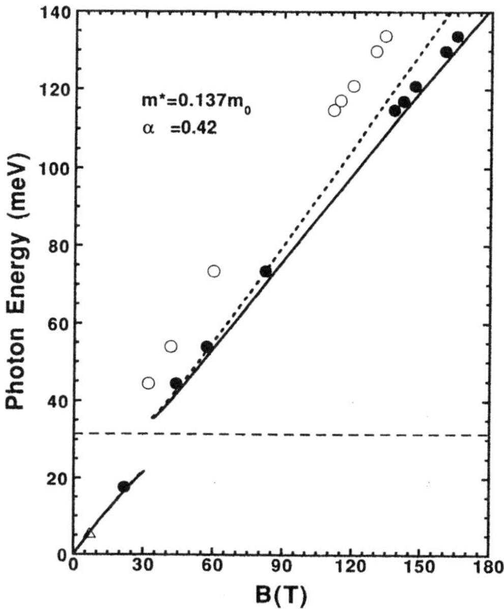

FIG. 4.22. Resonant photon energy as a function of B for n-type ZnSe [209]. Closed points and open points indicate the cyclotron resonance and impurity cyclotron resonance, respectively. The solid line and the broken line show calculated lines using the second order perturbation theory with and without non-parabolicity.

polaron mass is larger than the bare mass. In the high field region where $\hbar\omega_c > \hbar\omega_o$ holds, the cyclotron motion of electrons is so fast that the phonon cloud cannot follow the motion of electrons, and the observed mass becomes the bare mass.

By choosing proper values of m^* and α, we can obtain a good agreement between the experimental results and the calculation. In the calculation, we have to take into account the band non-parabolicity as it becomes significant in the high energy region. From such a comparison between experiments and calculation, we can obtain m^* and α for the crystals. Table 4.2 lists the values of m^* and α of typical II-VI compounds in such a manner [272, 273].

In the calculation of the polaron Landau levels, the second order perturbation theory provides only a rather crude approximation, when α is not small. However, it is difficult to obtain more accurate results by any other theoretical methods. The above results obtained by the second order perturbation method are considered to be reasonable. Larsen considered the problem by using a variational method [267].

In alkali-halide crystals, α is even larger, usually more than 1, so that large polaron effects should be observed. Experiments of the cyclotron resonance in

Table 4.2 The bare mass m^*, polaron mass m_p^* and the coupling constant α in typical semiconductors

Material	m^*/m	m_p^*/m	m_{CR}^*/m (117meV)	α	$\hbar\omega_o$ (meV)	B_r (T)	Reference
ZnO	0.25		0.286			280	[274]
ZnS (ZB)	0.200	0.221	0.188	0.630	31.4	54.3	[275]
ZnSe	0.137	0.1451	0.141	0.420	43.4	51.4	[209]
CdS ($\boldsymbol{B}\parallel$c)	0.168	0.188	0.162	0.514	37.8	54.9	[270, 271]
CdS ($\boldsymbol{B}\perp$c)	0.159	0.180	0.157	0.555	37.8	51.9	[270, 271]
CdSe($\boldsymbol{B}\parallel$c)	0.118	0.130	0.118	0.419	26.2	26.7	[270, 271]
CdTe	0.092	0.098	0.104	0.4	21.3	18.0	[276, 277]

KCl, KBr, KI, RbhCl, AgCl, AgBr and TlCl have been performed by Hodby *et al.* [154,278–280]. As the carrier concentration in alkali-halides is very small, they excited carriers by repeated pulses of a flash lamp and the resulting transient currents were measured.

In polar crystals, phonon assisted cyclotron resonance is sometimes observed. The resonance occurs at

$$\hbar\omega_c = \hbar\omega + N\hbar\omega_o \tag{4.58}$$

in addition to the main peak at the usual cyclotron resonance. Such a series of resonance was observed in InSb [281, 10].

Figure 4.23 shows a cyclotron resonance in ZnS [275]. A phonon assisted cyclotron resonance peak for $N = 1$ is observed as a shoulder in the low field side of the main cyclotron resonance peak.

In two-dimensional electron systems, the polaron effect is different from the three-dimensional case because the electron-phonon interaction takes place in a space with different dimensions. According to Das Sarma, the polaron mass of two-dimensional electrons is [282],

$$m_p^* = m_b^* \left(1 + \frac{\pi\alpha}{8} + \frac{9\pi\alpha}{64}\frac{\omega_c}{\omega_o}\right), \tag{4.59}$$

For low magnetic field range where $\hbar\omega_c \ll \hbar\omega_o$, (4.59) tends to

$$m_p^* = m_b^* \left(1 + \frac{\pi\alpha}{8}\right). \tag{4.60}$$

Comparing this expression with (3.119), we can see that the polaron effect is about 2.3 times larger than in the three-dimensional case. In two-dimensional electrons, however, the screening effect by electrons is larger, so that the polaron effect becomes smaller than expected from (4.60).

In 2D GaAs systems, however, the carrier screening effect of the electron-phonon interaction is important, and therefore the polaron effect is obscured [178,

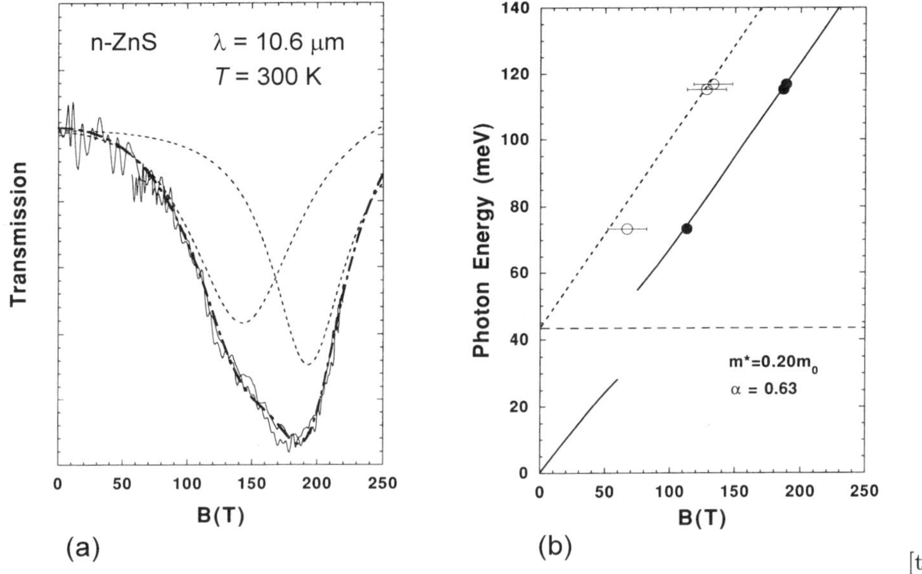

[t]

FIG. 4.23. (a) Cyclotron resonance in n-type ZnS at a wavelength of 10.6 μm at room temperature [275]. The absorption spectrum consists of two components as shown by dashed lines (Lorentzian curve). (b) Photon energy of the resonance peaks as a function of magnetic field. The solid line is calculated by the Brillouin-Wigner perturbation theory. The open points denote the phonon-assisted cyclotron resonance. The dashed line is a guide for the eyes representing $\hbar\omega = \hbar\omega_c + \hbar\omega_o$.

283]. In GaAs with high electron concentration, m_c^* is increased with increasing electron density n_s. This is because $< T >_z$ becomes larger as n_s increases. At high temperatures above 100 K, m_c^* decreases with increasing temperature due to the temperature dependence of the effective mass. It should be noted that m_c^* decreases with decreasing temperature below 100 K. This is because the screening of the polaron effect by electrons becomes more efficient as the temperature is lowered and the Lanudau level width is decreased.

4.6 Temperature dependence of effective mass

As already mentioned in Section 3.5.1, the carrier effective mass m^* has generally a significant temperature dependence as the band gap varies with temperature. In most semiconductors, as the temperature is increased, the band gap decreases so that the effective mass decreases. For example, the bare mass at the band edge of InSb and GaAs decreases from $0.0133m$ and $0.0665m$ at low temperatures of 40–60 K to $0.0121m$ and $0.0636m$ at nearly room temperature [159]. Such a temperature dependence has been observed in many semiconductors.

In Pb chalcogenides, PbTe, PbSe, and PbS, the band gap shows an anomalously large increase with increasing temperature contrary to normal semicon-

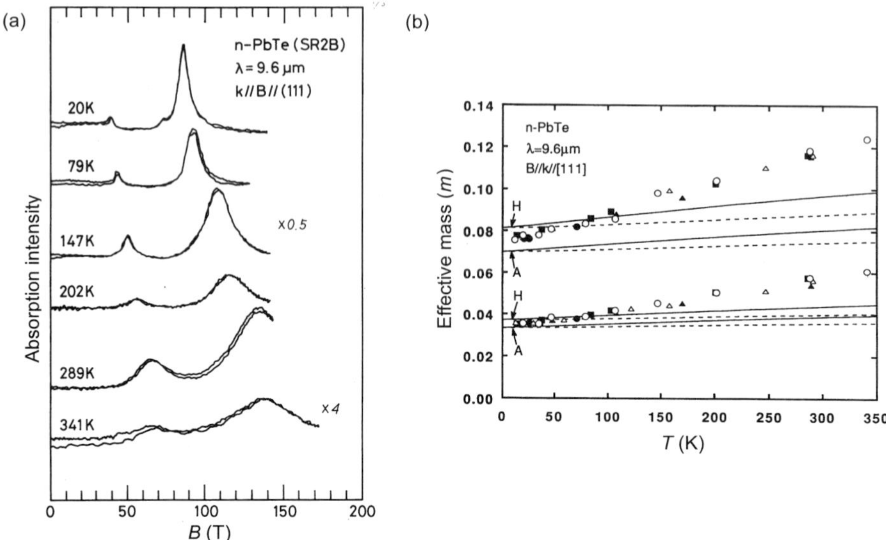

FIG. 4.24. (a) Cyclotron resonance absorption spectra in n-type PbTe [285]. Magnetic field was applied parallel to the $< 111 >$ axis. (b) Temperature dependences of the light and heavy effective masses of in PbTe at a wavelength of 9.6 μm ($\hbar\omega = 129$ meV). $B \parallel < 111 >$. Different data marks correspond to different samples. Solid and broken lines represent theoretical calculation by the six-band model taking account of total change of the band gap and only dilational part of the band gap change, respectively. P and H denote calculations using the parameters of Pascher *et al.* [286] and Hewes *et al.* [287]

ductors [284]. The temperature dependence of the carrier effective masses is actually very large in these crystals. It is an intriguing problem how this anomalous temperature dependence of the band gap affects that of the effective mass. Figure 4.24 (a) shows the cyclotron resonance spectra for an energy of 129 meV at different temperatures [285] when a magnetic field is applied parallel to the $< 111 >$ direction. Two sharp resonance peaks are observed corresponding to a light electron valley whose principal symmetry axis is directed towards the magnetic field and three degenerate heavy electron valleys which are oriented in a oblique direction relative to the magnetic field by about 70°. Both peaks shift to higher fields with increasing temperature. The effective mass increase between $T = 15$ K and 300 K amounts to nearly 50%. As the band gap of PbTe shows a dramatic increase with increasing temperature, the large change of the effective mass is partly explained as due to the increase of the band gap. In Fig. 4.24 (b), the experimental points are compared with the theoretical lines calculated by the $\boldsymbol{k}\cdot\boldsymbol{p}$ theory and the band parameters determined by Pascher *et al.* [286] and Hewes, Adler, and Senturia [287]. The experimental data show a much larger temperature dependence than the theoretical lines. A similar anomalously large temperature dependence of the effective mass was also observed in the valence band of PbTe and in both the conduction and the valence bands on PbSe [288]. The large temperature dependence of the effective mass in Pb chalcogenides is not fully understood, but the possibility of the contribution of the temperature dependences of the momentum matrix elements and the spin orbit interaction was suggested [285].

4.7 Spin resonance and combined resonance

The selection rule of cyclotron resonance is such as shown in Eq. (4.27) for the electric dipole transition. So the spin resonance or combined resonance as shown in Fig. 4.25 which requires $\Delta s = \pm 1$ are forbidden for the electric dipole transition. However, the wave function of a Landau level of the conduction band with a certain spin contains some component of the opposite spin through the mixing with the valence band due to the spin-orbit interaction. Therefore, a small transition probability exists for the electric field-excited spin resonance and combined resonance. McCombe *et al.* were the first to observe the combined resonance in InSb [289, 290]. The resonance was observed by the electric field polarization parallel to the magnetic field ($\boldsymbol{E} \parallel \boldsymbol{B}$). They found that the absorption intensity is proportional to the square of the field, and concluded that the non-parabolicity is the dominant mechanism of the resonant transition. The electric dipole-excited spin resonance was also observed by McCombe and Wagner [291]. The resonance was observed by the cyclotron resonance-inactive circular polarization. These transitions are ordinarily forbidden transitions, so that the intensity is very weak.

In Pb chalcogenides, a different type of combined resonance was observed prominently in the cyclotron resonance spectra [292]. At a wavelength of 16.9 μm (73.4 meV), two resonance peaks were observed corresponding to the light hole valley (L) and the heavy hole valley (H) for a magnetic field parallel to $< 111 >$ direction. At higher photon energies, however, a new peak was observed. Figure 4.26 shows cyclotron resonance spectra for p-type PbTe in very high

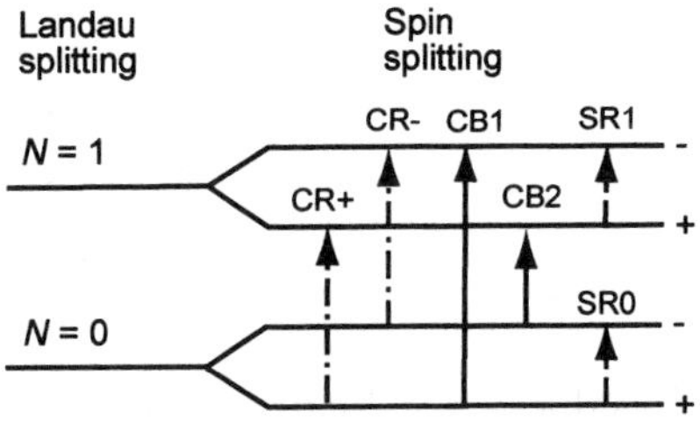

FIG. 4.25. Cyclotron resonance (CR), spin resonance (SR), and combined resonance (CB) between the spin split Landau levels.

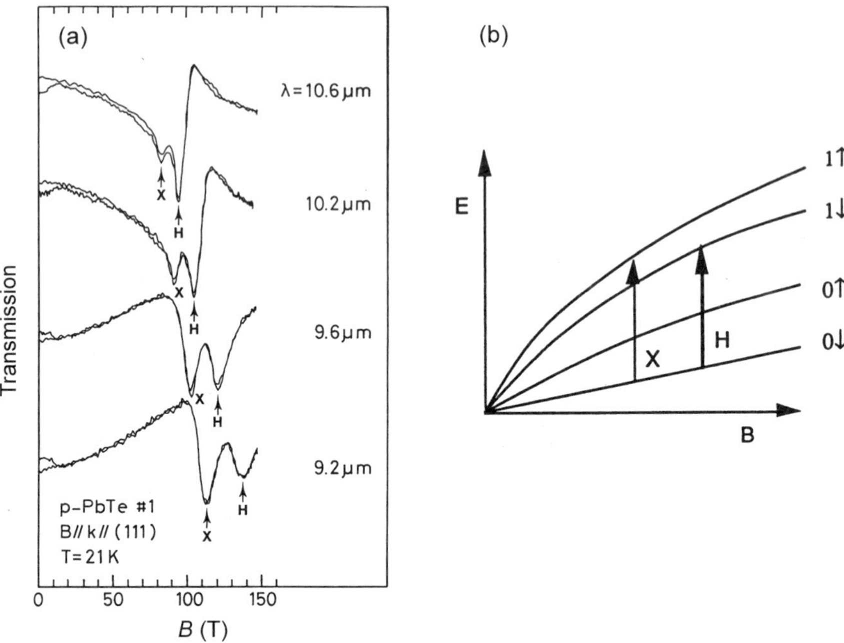

FIG. 4.26. New type of combined resonance observed in p-type PbTe [292]. (a) Cyclotron resonance spectra for $B \parallel < 111 >$ at different wavelengths. $T = 21$ K. (b) Model of the transitions causing the resonance peak.

magnetic fields at various wavelengths near 10 μm. In addition to the two peaks L and H, we can see a new peak "X"in between. In the figure, the peak L is seen at 36 T, but it is very small because of the small population in the valley responsible to the peak. On the other hand, the peak X is a strong feature and grows with increasing photon energy (shorter wavelengths). It exceeds even the main cyclotron resonance line H for $\lambda \leq 9.6$ μm. The strikingly intense absorption at the combined resonance condition is ascribed to the mixing of the wave functions of the $|1 \downarrow>$ and the $|1 \uparrow>$ levels due to the large spin-orbit interaction. According to the six band model calculation using the existing band parameters, the two levels approach very close to each other at about 100 T, and repel in higher fields. As a result, a strong mixture and the interchange of the character of the energy levels take place near the crossing point. Therefore, the peak "X"becomes the dominant cyclotron resonance line in magnetic fields higher than the crossing point.

4.8 Structural phase transitions

Lead chalcogenides SnTe, GeTe, SnSe, and GeSe are known to be unstable against the structural phase transition. At high temperatures above the critical field T_c, they have a NaCl type (cubic) crystal structure. Below T_c, they undergo a phase transition to an As type (rhombohedral) structure. In alloys doped with Sn or Ge to PbTe, $Pb_{1-x}(Sn,Ge)_xTe$, the structural phase transition is also expected to occur if x is sufficiently high ($x > 0.5$ at. % in the case of Ge and $x > 35$ at. % in the case of Sn [293]). As the structural phase transition is caused by the zero center TO phonon softening [294, 295] which is governed by the interband interaction, we can expect a large effect of the magnetic field to the structural phase transition. The renormalized TO phonon frequency $\tilde{\omega}_{TO}$ is given by

$$\tilde{\omega}_{TO}^2 = \omega_o^2 + \omega_A^2(T) - \frac{2}{MNa^2} \sum_k |\Xi_{cv}(k,k)|^2 \frac{f_v(k) - f_c(k)}{\mathcal{E}_{ck} - \mathcal{E}_{vk}}, \tag{4.61}$$

where ω_o is the uncoupled TO phonon energy, $\omega_A(T)$ is the anharmonicity term representing the phonon-phonon interaction. The third term represents the electron-phonon interaction, where M, N, a and k are the reduced mass of the ion pair, the number of atoms in the unit cell, the lattice constant and the wave-number. $\Xi_{cv}(k,k)$, $f_{v,c}(k)$ and $\mathcal{E}_{ck}$ are the interband optical deformation potential, the Fermi distribution function, and the band energy, respectively.

PbTe is a crystal for which high quality crystals have been grown and the band structures are well known. We can also obtain good alloy crystals, (PbGe)Te and (PbSn)Te, which show structural phase transitions. It is an interesting question how the phase transition should be modified by high magnetic fields. Figure 4.27 (a) show the experimental traces of cyclotron resonance for p-type $Pb_{1-x}Ge_xTe$ ($x = 0.0096$) in a magnetic field parallel to the $< 111 >$ direction [296]. The term $\omega_A(T)^2$ contributes positively to $\tilde{\omega}_{TO}^2$, but the third term is negative and the structural phase transition takes place when $\omega_A(T)^2$ becomes zero. In PbTe, the band gap is small and the energy denominator in the third term is small. Moreover, as the energy spectrum is modified drastically due to the small effective mass, we can expect that the transition temperature should be changed drastically by the external magnetic field. Moreover, the magnetic field-induced structural phase transition can be expected even in crystals with small x which do not show the phase transition in zero field. Actually, in PbGeTe and PbSnTe which exhibit the structural phase transitions at finite temperatures, it was found that the transition temperature is varied by magnetic fields [294, 295].

Figure 4.27 shows the cyclotron resonance spectra in p-type $Pb_{1-x}Ge_xTe$ (x=0.0096) at different temperatures [296]. As mentioned in the previous subsections, two cyclotron resonance peaks are observed for $\boldsymbol{B} \parallel < 111 >$. The peaks 1 (L) and 2 (H) in Fig. 4.27 (a) correspond to these two peaks. The peak X corresponds to the combined resonance as discussed above. As can be seen in Fig. 4.27 (b), the effective masses increase significantly as the temperature increases. It was found that the temperature gradient of the effective masses changes abruptly at $T = 42$ K and 48 K for the light mass peak (Peak 1) and the heavy mass peak (Peak 2), respectively. These temperatures are regarded as the transition temperature T_c where the structural phase transition occurs. In this sample, we

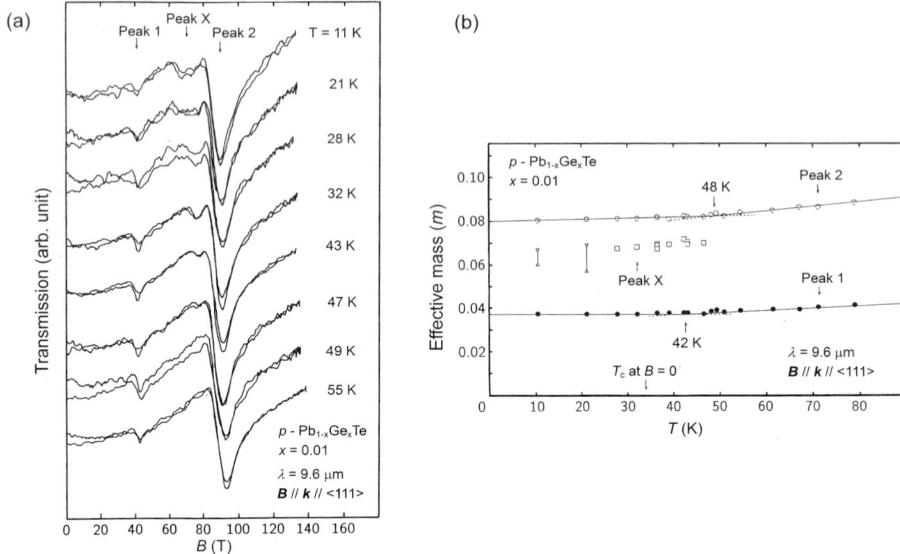

FIG. 4.27. (a) Experimental traces of cyclotron resonance of p-type $Pb_{1-x}Ge_xTe$ ($x = 0.0096$) [296]. The photon energy is 134 meV. (b) Temperature dependence of the effective mass measured by the cyclotron resonance.

found that $T_c = 34$ K at zero magnetic field, but it was increased as the magnetic field was increased. The transition is the second order and the change of displacement of the atomic arrangement is not so large that the phase transition shows up not as a sudden change of the effective masses but as the change of the temperature coefficient. By extrapolating such magnetic field dependence of T_c as a function of x, we can predict that even in pure PbTe which does not show any structural phase transition at zero field, the transition should occur at about 400 T. In the rhombohedral phase below T_c, the crystals show ferroelectricity. Thus such a structural phase transition caused by magnetic field can be called the magnetic field-induced ferroelectric transition, which is one of the unique effects of high magnetic fields.

Such a phenomenon has been explored by means of cyclotron resonance in n-type PbTe [251]. Figure 4.28 shows the experimental traces of cyclotron resonance in n-type PbTe at a photon energy of 224 meV ($\lambda = 0.553$ μm). At low temperatures below 40 K, many prominent absorption peaks were observed in the low magnetic field range. These peaks are due to the interband magneto-absorption. As the photon energy exceeds the band gap at low magnetic fields, the interband transitions become possible. It is interesting that both interband transition and cyclotron resonance (intraband transition) are observed in one spectrum. The absorption jumps up at some field as the band gap opens up in the high field. Then, with further increasing field, the traces at low temperatures

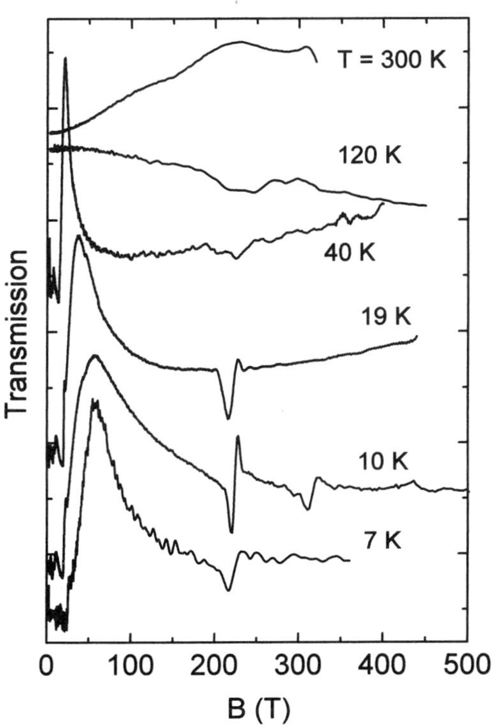

FIG. 4.28. Cyclotron resonance in n-type PbTe in very high magnetic fields at different temperatures [251]. The photon energy is 224 meV. $B \parallel < 111 >$.

show a gradual decrease. The decrease tends to saturate around nearly 100 T. The cyclotron resonance peak is observed after this saturation of the absorption at around 217 T, and the position does not change much between 7 K and 19 K. The effective mass is estimated as $0.112m$ from the resonance field. This mass is corresponding to the heavy mass for $\boldsymbol{B} \parallel < 111 >$. It is much larger than the value $0.0723m$ at 117 meV. The difference is due to the non-parabolicity. Unfortunately, it was not easy to obtain any evidence of the structural phase transition in PbTe at low temperatures from this experiment. However, a number of intriguing new phenomena were observed in this experiment using very high magnetic fields; the decrease of the transmission below 100 T, the splitting of the cyclotron resonance absorption peak at 40–120K, *etc.* The origin of these phenomena are not totally known yet.

4.9 Cyclotron resonance in valence bands

4.9.1 *Hole cyclotron resonance in bulk crystals*

As discussed in Section 2.4.3, the Landau levels in the valence band in crystals with diamond or zinc blende structure, such as Si or GaAs, have complicated dispersions because of the degeneracy of the bands at the Γ point. For small Landau level quantum numbers N, the spacing of the adjacent Landau levels is not uniform and each Landau level is not expressed by a single harmonic oscillator function but contains multiple functions. Therefore, in the cyclotron resonance for p-type samples of these crystals, many different transitions are allowed between different levels with different intervals. Thus cyclotron resonance often shows complicated spectra comprising many peaks under nearly a quantum limit condition $kT < \hbar\omega_c$ at low temperatures and high magnetic fields. At high temperatures or in low magnetic fields, on the other hand, holes are populated up to large N levels. Then most of the electron transition for cyclotron resonance occurs between the uniformly spaced Landau levels with large N. Two absorption peaks are observed corresponding to the heavy holes and light holes [196]. Cyclotron resonance under the latter condition is called "classical cyclotron resonance"being distinguished from the former case called "quantum cyclotron resonance".

The quantum effect in hole cyclotron resonance was first observed for Ge by Fletcher, Yaeger, and Merritt [297]. Subsequent experiments by Hensel and Suzuki for p-Ge revealed many features of quantum cyclotron resonance [298]. Many absorption peaks are observed and spectra show a large dependence on the factor $\eta = \hbar\omega/kT$ ($\hbar\omega$ is the photon energy for the measurement) representing the thermal population of carriers in the Landau levels. Further complexities occur in quantum cyclotron resonance due to the "k_H effect"(k_H is the k-vector along the field direction). In the case of usual cyclotron resonance, the transition occurs mostly near $k_H = 0$ as the density of states is the largest there. However, the low lying Landau levels have a complicated k_H dependence so that the joint density of states sometimes takes an extremum at a finite k_H. Therefore, in order to obtain the absorption peak position theoretically, we have to calculate the transition probability over the entire range of k_H.

Figure 4.29 shows the cyclotron resonance absorption spectra observed for p-type Ge in pulsed high magnetic fields [299]. The values of $\eta = \hbar\omega/kT$ for the three traces in each figure are about the same. Although the wavelength of the radiation has a large variation, the spectra show nearly the same shape for $\eta = \hbar\omega/kT$. There is a tendency that for greater resonance energy $\hbar\omega$, the mass m^* at the peaks is larger. This dependence is due to the interaction of the valence band with the conduction band and the split-off band.

Figure 4.30 shows the spectra of the quantum cyclotron resonance of p-type GaSb at different temperatures [299]. In each field direction, three absorption peaks were resolved (A, B and C in the figure). Comparing Fig. 4.29 and Fig. 4.30, we can see that there is a similarity in the spectra between Ge and GaSb. For example, for $\boldsymbol{B} \parallel < 111 >$, the peaks B and C for GaSb seem to correspond to

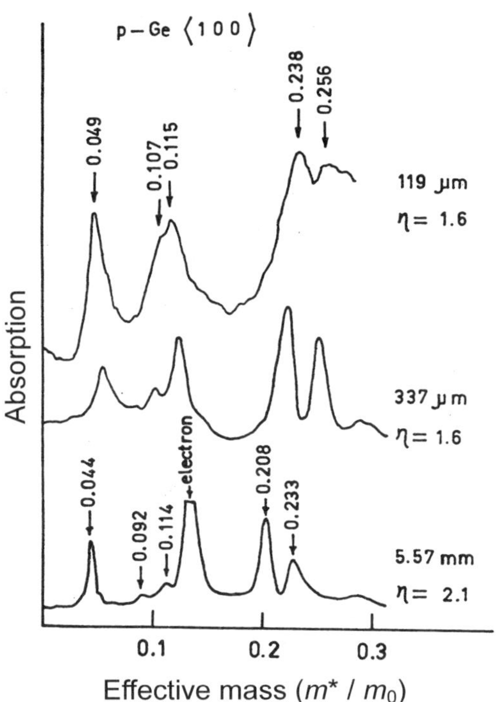

FIG. 4.29. Cyclotron resonance in the valence band of Ge at different photon energies [299]. $B \parallel < 100 >$. The abscissa is scaled by the effective mass $m^*/m = eB/\omega m$ to stand for the magnetic field in order to compare the peaks for different ω. The data for 5.57 mm, 337 μm and 119 μm are by Hensel [300], by Bradley *et al.* [301] and by Suzuki and Miura [299], respectively.

the peaks in Ge with effective mass $0.125m$ and $0.250m$, respectively. Moreover, the temperature dependence of the absorption intensity of each corresponding peak also shows a similarity. On referring to a detailed analysis of Ge [300], the peaks B and C for GaSb can be assigned as the allowed transitions $1_0 \rightarrow 2_1$ and $3_2 \rightarrow 4_3$ at $k_H = 0$, respectively. Once the assignment is done we can obtain the Luttinger parameters and the classical masses of heavy and light holes. The values are listed in Table 4.3. The classical masses and the Luttinger parameters deduced from the quantum cyclotron resonance are in reasonably good agreement with those obtained from the classical cyclotron resonance [302] and the theory of Lawaetz [12].

4.9.2 *Hole cyclotron resonance in two-dimensional systems*

In two-dimensional systems the Landau level structure is even more complicated than in bulk systems because of the presence of quantum potentials. Cyclotron resonance spectra should show features different from the three-dimensional case.

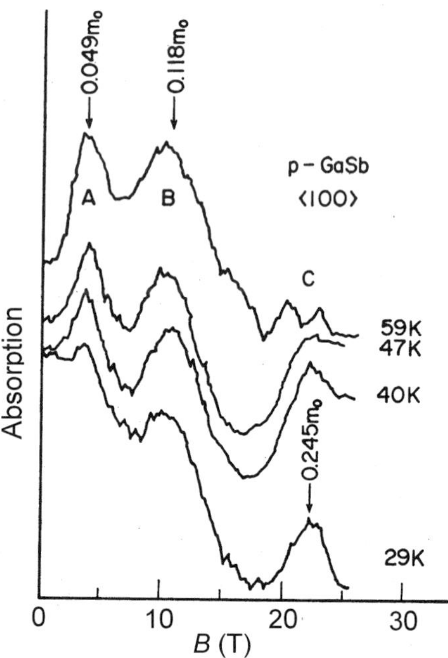

FIG. 4.30. Cyclotron resonance in p-type GaSb [299]. $B \parallel < 100 >$.

Table 4.3 Comparison of the values of the valence band parameters in GaSb [299]

Band parameter	Quantum cyclotron resonance [299]	Classical Cyclotron resonance [302]	Theory [12]
m_{lh}^*/m	0.046± 0.002	0.052 ± 0.004	0.046
$m_{hh}^*/m < 111 >$	0.43± 0.12	0.36 ± 0.03	0.529
$m_{hh}^*/m < 100 >$	0.29 ± 0.05	0.26 ± 0.04	0.321
γ_1	12.4 ± 0.4		11.80
γ_3	5.4 ± 0.2		5.26
$\gamma_3 - \gamma_2$	1.5 ± 0.1		1.23
κ	3.2 ± 0.1		3.18

Cyclotron resonance of holes in two-dimensional systems was first measured by Kotthaus and Ranvaud for p-type Si MOS-FET [303]. They observed cyclotron resonance on the three high symmetry surfaces ($< 100 >$, $< 110 >$, and $< 111 >$) of Si at a wavelength of 333 μm. The effective masses increased as the carrier concentration was increased in magnetic fields up to 6 T. They found that the effective masses for all the directions obtained by the cyclotron resonance were in good agreement with the data of the Shubnikov-de Haas effect [304] and the theory [305, 306].

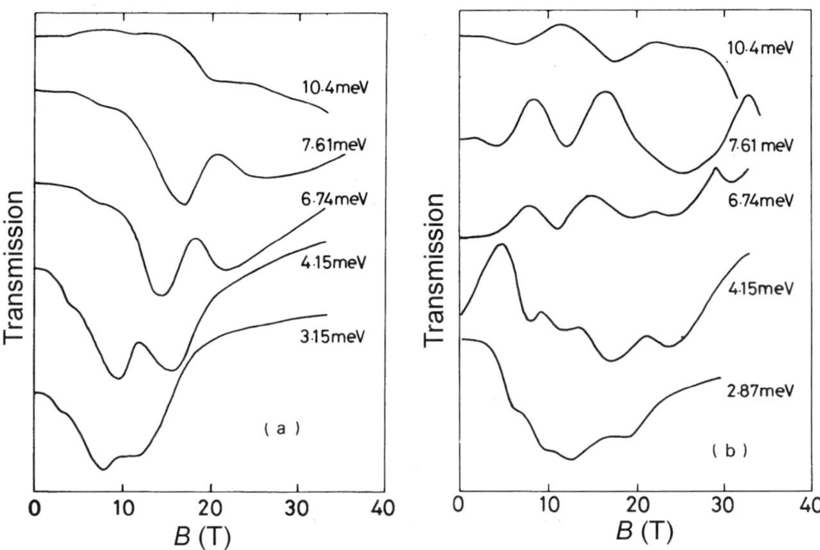

FIG. 4.31. Cyclotron resonance in the valence band of two different samples of GaAs/AlGaAs quantum well [307]. (a) L_z = 5.0 nm, $p = 9.8\times10^{11}\text{cm}^{-2}$. (b) L_z = 12.3 nm, $p = 1.1\times10^{12}$ cm^{-2}.

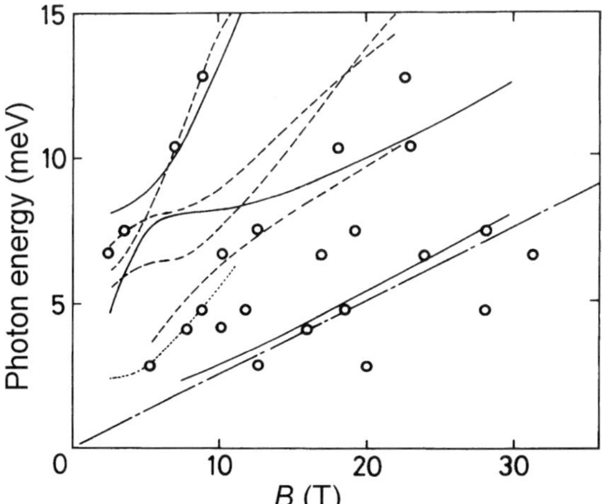

FIG. 4.32. Cyclotron resonance energy as a function of magnetic field for the valence band of a GaAs/AlGaAs quantum well [307]. L_z = 10.85 nm. $p = 4.5\times10^{11}$ cm^{-2}. The solid, broken, and dotted lines are the calculated transition energy from different initial states. The dash-dotted line denotes the transition for $m^* = 0.45\ m$.

Iwasa *et al.* measured the cyclotron resonance in p-type modulation doped GaAs/AlGaAs quantum wells [307, 308]. Figure 4.31 shows the cyclotron resonance traces in p-type quantum wells with a well width of $L_z = 12.3$ nm at different wavelengths [307]. The cyclotron resonance spectra consist of many absorption peaks. The photon energy of the resonance peaks is plotted as a function of magnetic field in Fig. 4.32 for a GaAs/AlGaAs quantum well with a well width of $L_z = 10.8$ nm. The data show a very complicated field dependence of the resonance energy. However, they are in good agreement with a theoretical curves for various transitions described in Section 2.5.2.

4.10 Unusual band structures and phenomena

In this section, we describe some examples of cyclotron resonance in unusual band structures.

4.10.1 *Camel's back structure*

When two bands cross each other at a symmetry point in the Brillouin zone with k-linear terms, the two bands may show an anti-crossing effect if there is an interaction between them in crystals that do not have an inversion symmetry. The camel's back structure in the band extrema is caused by such a mechanism. The first example of the camel's back structure observed in cyclotron resonance was in Te. Te crystals are always grown as p-type in the extrinsic range at low temperatures. The Landau level structure in the valence band of Te is very complicated due to the camel's back structure, as we have already seen in Fig. 3.34. Couder *et al.* [169] and Yoshizaki *et al.* [170] found that the cyclotron resonance lines as functions of magnetic field are very non-linear and show splitting at some field, corresponding to such a complicated structure. Very similar phenomena are also observed in the conduction band at the X point in GaP.

Figure 4.33 shows the structure of the band minimum with and without the inversion symmetry when there is a k-linear term at a symmetry point X. Examples of the former and latter cases are Si and GaP, respectively. In Si, there is an inversion symmetry at the X point so that degeneracy is preserved at the X point. The conduction band minima are located at a point on the Δ axis. In III-V compounds, the degeneracy is lifted because of the lack of inversion symmetry. In the latter case, the $\boldsymbol{k}\cdot\boldsymbol{p}$ Hamiltonian for the two bands is written as

$$\mathcal{H} = \begin{pmatrix} Ak_\perp^2 + Bk_\parallel^2 - \frac{\Delta}{2} & Pk_\parallel \\ Pk_\parallel & Ak_\perp^2 + Bk_\parallel^2 + \frac{\Delta}{2} \end{pmatrix}, \tag{4.62}$$

where the first and the second lines designate the X_1 and X_3 bands, respectively, and $Pk_\parallel$ is the matrix element connecting the two bands. (In Si this k-linear term is zero). Solving the secular equation of (4.62), we obtain the energy dispersion,

$$\mathcal{E}(k) = Ak_\perp^2 + Bk_\parallel^2 - \left[\left(\frac{\Delta}{2}\right)^2 + P^2k_\parallel^2\right]^{1/2}. \tag{4.63}$$

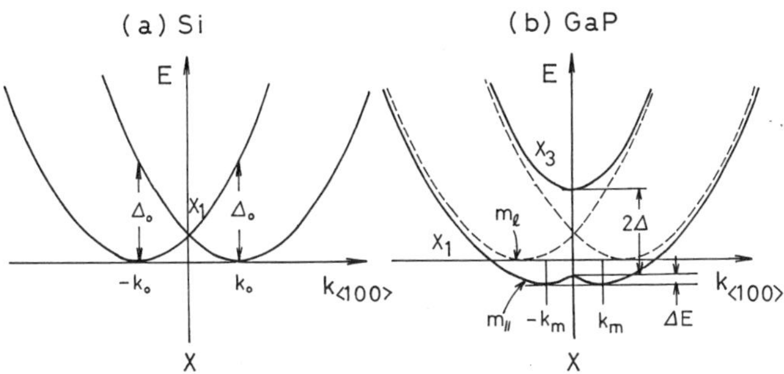

FIG. 4.33. Comparison of the conduction band structure near X point between (a) Si (with the inversion symmetry) and (b) GaP (without the inversion symmetry).

The dispersion is parabolic in the direction of $\boldsymbol{k}_{\perp}$ but has double minima in the direction of $\boldsymbol{k}_{\parallel}$. The minima occur at

$$k_m = \frac{P}{2B}\left[1 - \left(\frac{\Delta}{\Delta_0}\right)^2\right]^{\frac{1}{2}} \quad (\Delta \leq \Delta_0),$$
$$= 0 \quad (\Delta \geq \Delta_0). \tag{4.64}$$

$$\Delta_0 = \frac{P^2}{B}. \tag{4.65}$$

In the case $\Delta \leq \Delta_0$, the $\boldsymbol{k} = 0$ point is a saddle point. Such a band dispersion is called the camel's back structure or dumbbell structure. III-V compounds that have conduction band minima have such a structure in principle. However, if $\Delta \geq \Delta_0$ holds, the two minima overlap at $k = 0$ (X point), and the double minima disappear. Also, if the energy undulation $\Delta\mathcal{E} = \Delta_0(1 - \Delta/\Delta_0)^2/4$ is very large, the effect is not visible. It is an intriguing situation when $\Delta\mathcal{E}$ is comparable with the cyclotron frequency.

The cyclotron resonance in the conduction of GaP was studied by Leotin *et al.* [309] and Miura *et al.* [310]. Figure 4.34 shows the cyclotron resonance spectra in n-type GaP for $\boldsymbol{B} \parallel < 100 >$ [310]. When we apply magnetic fields along the $< 100 >$ direction, two resonance peaks corresponding to the transverse mass m_t^* and the geometrical average of the transverse and longitudinal mass $\sqrt{m_t^* m_l^*}$ should be observed, if we stand on an ellipsoidal band model. As the mobility is not so high in GaP and the latter mass is very large, we need a high magnetic field for observing both peaks. Reflecting a large anisotropy, the latter peak is observed with the hole-inactive circular polarization of the radiation as well (*cf.* Section 4.1.3). Table 4.4 lists the effective masses of the conduction electrons in GaP obtained by cyclotron resonance. We can see that the m_t^*/m does not depend so much on the photon energy, but there is a large

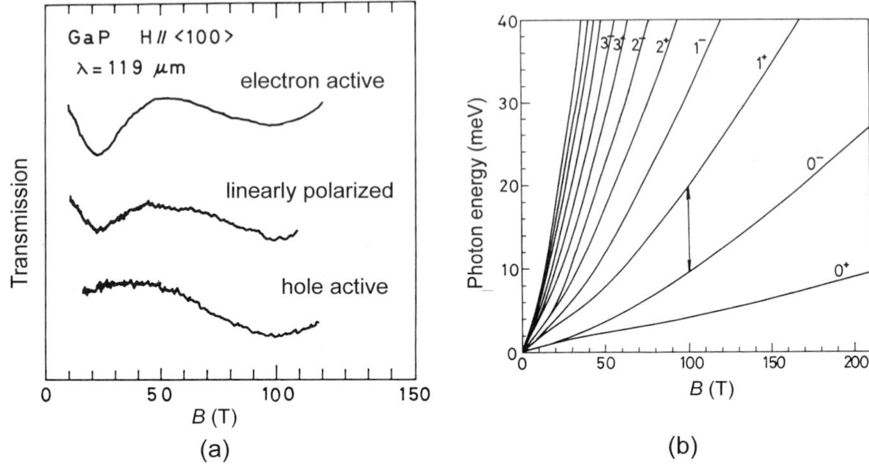

FIG. 4.34. (a) Cyclotron resonance in the conduction band of GaP [310]. Spectra for two different circularly polarized radiation (electron-active and hole-active) and linearly polarized radiation are shown. $\lambda = 119\ \mu$m. $T = 152$ K (Top), $T = 152$ K (Middle), $T = 152$ K (Bottom). (b) Calculated energies of the Landau levels. The arrow indicates the main transition which gives the dominant contribution to the absorption peak for the heavier mass.

Table 4.4 Cyclotron effective mass of the conduction band of GaP

Wavelength	m_t^*/m	$K = m_l^*/m_t^*$	m_l^*/m
337 μm	0.252±0.003	28±7	6.9±1.7
119 μm	0.254±0.004	19±2	4.8±0.5

energy dependence in the anisotropy factor K or m_l^*/m. This is considered to be the manifestation of the camel's back structure. The Landau level energies and the transition probabilities were calculated by expanding the wave functions into harmonic oscillator functions based on the Hamiltonian (4.62) [310]. Figure 4.34 (b) shows calculated results of the Landau level energies using suitable parameters. As the spacing between the levels is non-uniform and non-linear, and also many transitions become allowed due to the mixing of the wave function, the absorption peak with the heavier mass consists of several transitions. Adjusting the parameters and comparing the experimental data with the calculation, it was found that the transition $(0^- \rightarrow 1^+)$ gives the dominant contribution to the absorption peak for the heavier mass.

Other candidates which might have the camel's back structure are AlAs and SiC which have the conduction band minima close to the X point.

AlAs is an important material for the barrier layers in GaAs/AlAs quantum

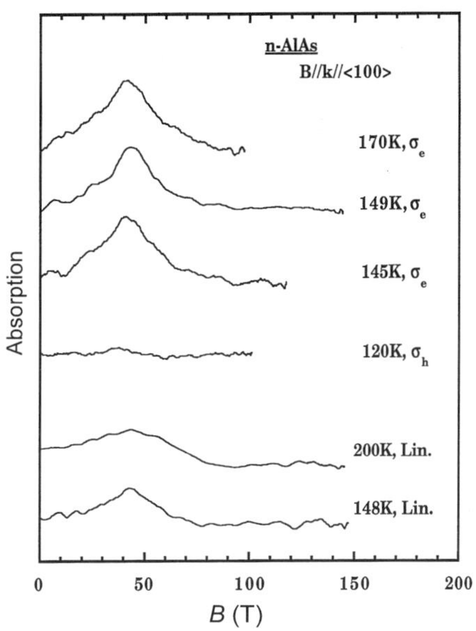

FIG. 4.35. Cyclotron resonance spectra in n-type AlAs at different temperatures [312]. $\lambda = 119\ \mu$m.

wells or heterostructures. However, in comparison to the counterpart GaAs, not much is known about AlAs, because of the difficulty in preparing bulk AlAs with high quality. Recently, very high mobility was realized in the quantum well of AlAs sandwiched in between AlGaAs layers [311], and cyclotron resonance and quantum Hall effect was observed. However, for measuring cyclotron resonance in bulk AlAs, we need a high magnetic field as the mobility is quite low. Figure 4.35 shows high field cyclotron resonance spectra in n-type AlAs grown by MBE on a GaAs substrate [312]. A prominent peak at 42 T and a shoulder-like structure at 22 T were observed at a wavelength of 119 μm. From the resonance positions, the effective masses $\sqrt{m_t^* m_l^*}$ and m_t^* at the X point were obtained. Table 4.5 lists the effective mass of AlAs obtained by various experiments. As the anisotropy factor K is considerably smaller than in GaP, it is unlikely that the camel's back structure has a prominent effect.

Similar measurements were made on n-type SiC which has also the conduction band minima near the X point [316]. Figure 4.36 shows cyclotron resonance spectra in n-type SiC. Two peaks corresponding to the effective masses m_t^* and $\sqrt{m_t^* m_l^*}$ are clearly seen at $\lambda = 119\ \mu$m and 36 μm, and a peak corresponding to m_t^* at shorter wavelengths, $\lambda = 28\ \mu$m and 23 μm. Table 4.6 lists the effective masses obtained by cyclotron resonance experiments at different photon energies. The effective masses are almost independent of photon energy although there is a slight non-parabolicity. As there is no apparent non-linearity in m_l^*/m, we can

Table 4.5 Cyclotron effective mass of the conduction band of AlAs at X point

Experiment	m_t^*/m	m_l^*/m	$\sqrt{m_t^* m_l^*}/m$	$K = m_l^*/m_t^*$
Faraday rotation [313]	0.19	1.1	0.46	5.8
Shubnikov-de Haas [314]			0.55±0.05	
Theory [315]	0.24	1.48	0.60	6.2
Cyclotron resonance [312]	0.25±0.01	0.88±0.07	0.47±0.01	3.5±0.4
Cyclotron resonance (2D) [311]			0.46	

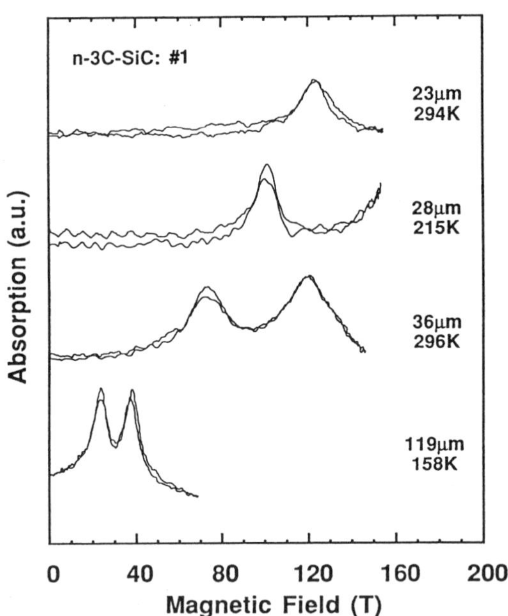

FIG. 4.36. Cyclotron resonance spectra in n-type SiC at different wavelengths [316].

conclude that there is no signature of the camel's back structure and the X point is probably located at the X point [316].

4.10.2 *Semiconducting diamond*

Diamond has recently attracted much attention as a promising material for semiconductor devices that can operate at high temperatures due to its large band gap. SiC or BN are interesting materials for the same reason. However, the mo-

Table 4.6 Cyclotron effective mass of the conduction band of SiC at X point

Photon energy (meV)	m_t^*/m	m_l^*/m	$\sqrt{m_t^* m_l^*}/m$	$K = m_l^*/m_t^*$	Reference
3.15	0.247	0.667	0.406	2.70	[317]
4.23	0.247				[317]
10.4	0.25	0.67	0.41	2.68	[316]
34.4	0.25	0.67	0.41	2.68	[316]
44.2	0.26	0.68	0.42	2.62	[316]
53.8	0.27				[316]

bility of carriers in diamond is rather low, and the condition $\omega_c\tau > 1$ is hardly fulfilled, especially at high temperatures of interest. Recently, doped-diamond single crystals have been fabricated by artificial crystal growth. Although high quality single crystals have been synthesized in case of p-type, it is still very difficult to grow n-type crystals.

Diamond has a band structure similar to those for typical semiconductors Ge or Si. However, the spin-orbit interaction is much smaller than in Ge and Si, and the splitting of the split-off band from the degenerate heavy hole and the light hole bands is less than 13 meV [318–320]. It is interesting how this situation shows up in the effective masses of the three bands. First attempts for the cyclotron resonance measurement in p-type diamond have been made at weak magnetic field and low temperature range. Rauch made the measurement in the microwave wavelength range at low temperatures exciting carriers by light illumination [321]. Complicated cyclotron resonance peak structure was observed. There has been controversy concerning the origin of the observed resonance absorptions because the experiment was made using a sensitive cavity with light illumination, where absorptions may be brought about by impurities or defects. It would be very desirable for experiments to be performed on thermally excited carriers at high temperatures. Then, because of the low mobility at high temperatures, the cyclotron resonance experiment is not an easy task.

High magnetic fields in the megagauss range are very useful for measuring cyclotron resonance in such samples. The condition $\omega_c\tau > 1$ can be rewritten as $\mu(\mathrm{cm}^2/\mathrm{V{\cdot}s})B(\mathrm{T})\times 10^{-2} > 1$ using the practically used units. High magnetic fields allow us to satisfy this condition in low mobility crystals. For example, for carriers with a mobility of 100 $\mathrm{cm}^2/\mathrm{V{\cdot}s}$, a field of 100 T is sufficient for the condition.

The first experiment of cyclotron resonance in thermally excited carriers in p-type diamond was done by Kono *et al.* [322]. Figure 4.37 (a) is an example of the experimental traces of the cyclotron resonance in p-type diamond. The experiment was done at T = 90–100 °C. A broad peak was observed in different directions of the magnetic field [322]. At a temperature of T = 90 °C, the mobility estimated from the line-width is $\mu \sim 700$ $\mathrm{cm}^2/\mathrm{V{\cdot}s}$. It was found that the peak can be decomposed to three peaks. They showed an angular dependence as shown in

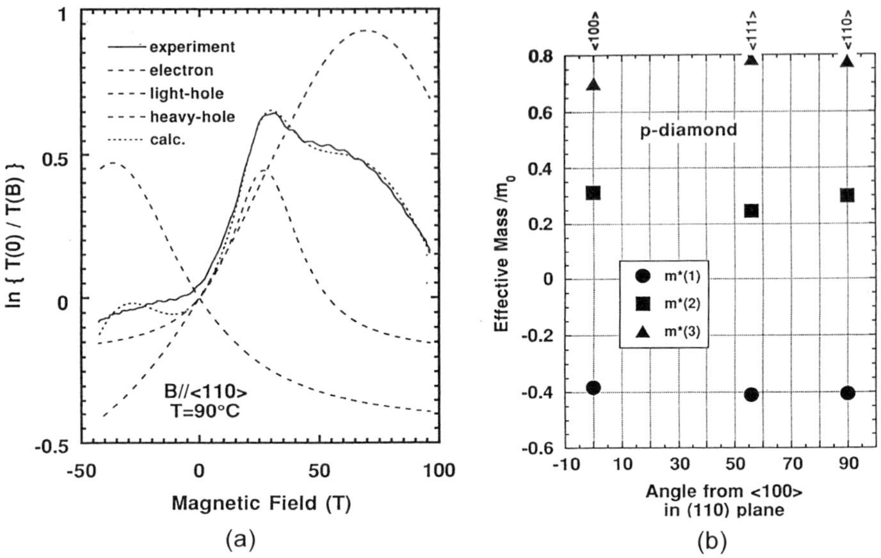

FIG. 4.37. (a) Cyclotron resonance spectra in p-type diamond [322]. $B \parallel< 110 >$. $T = 90$ °C. Measuring wavelength was 119 μm. The solid line shows the experimental trace, and the broken lines show the calculated resonance lines decomposed for the heavy hole, the light hole, and the split-off band holes. The positive magnetic field stands for the hole-active circular polarization of the radiation and the negative field stands for the electron-active one. (b) Angular dependence of the effective masses of three different holes responsible for the cyclotron resonance.

Fig. 4.37 (b). The three peaks probably arise from the three hole bands. In Ge or Si, the cyclotron resonance for the split off band has never been observed, as it is well separated from the top of the valence bands. In diamond which has a very small spin orbit splitting, however, it should be quite possible to be observed. However, a remarkable point is that one of the resonance peaks was observed for the electron-active polarizations. The origin of the electron-active resonance peak is not fully understood yet, but it is probably due to the heavy hole band as the cyclotron motion of the heavy holes may become peculiar due to the complicated interaction between the nearly degenerate three bands. In fact, the Landau level structure should be quite complicated in such a situation. From the experimental data as shown in Fig.4.37 (b), the valence band parameters were determined [322].

4.10.3 *Graphite*

Graphite is a semimetal which exhibits many interesting properties in high magnetic fields. A drastic jump of the magneto-resistance was discovered when a high magnetic field of 20–50 T was applied parallel to the c-axis (perpendicular to the stacking layers) at low temperatures (below 10 K) [323]. Since the transition field depends on temperature, it was assigned by Yoshioka and Fukuyama [324] as the magnetic field-induced electronic phase transition forming a charge density wave or spin density wave. According to the Yoshioka-Fukuyama theory, the transition is brought about by a nesting of Fermi-points on the Landau levels in the quantum limit. As graphite has a layered crystal structure, the energy

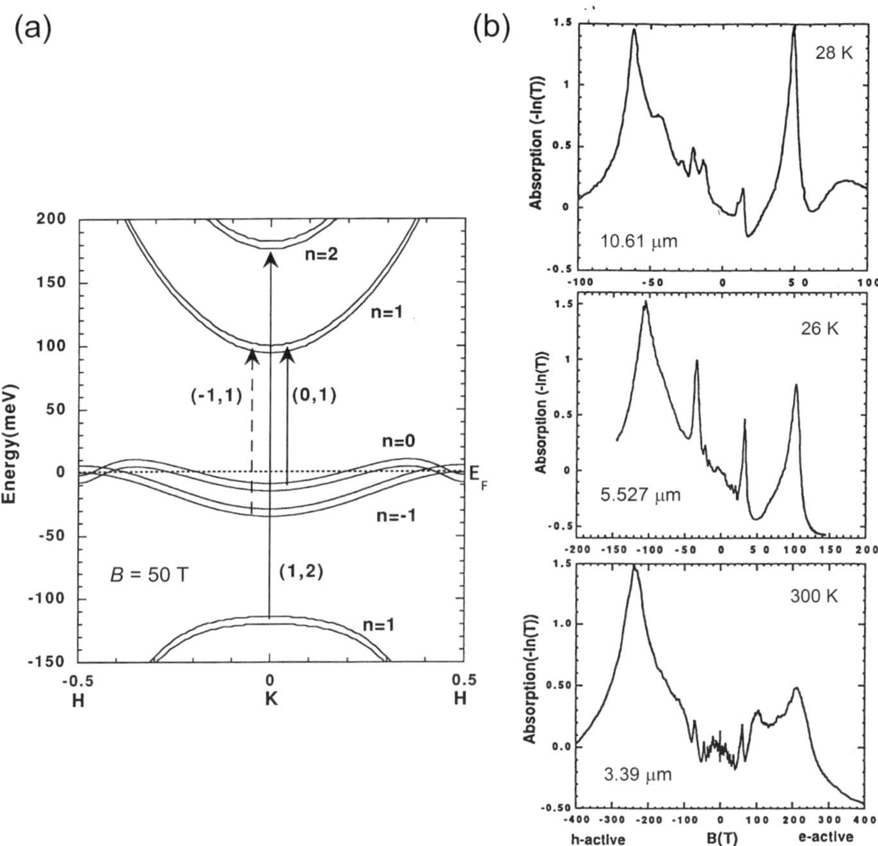

FIG. 4.38. (a) Landau level energy dispersion when a magnetic field of 50 T is applied perpendicular to c-axis [330]. Four Landau levels remain below the Fermi level at this field. The solid and broken lines with arrows indicate the allowed transitions for electron-active and hole-active circular polarizations, respectively. (b) Cyclotron resonance spectra in graphite for different wavelengths [330]. The positive and negative values of the magnetic fields on the abscissa indicate the electron-active and hole-active circular polarizations of the radiation, respectively. In order that the peaks from the same origin can be seen nearly at the same position, the scales of the horizontal axis are adjusted.

dispersion is very anisotropic; it is very conducting with a small effective mass in the layers but the effective mass in the perpendicular direction to the layers is fairly large. Therefore, the density of states at the Fermi-points when the field is applied perpendicular to the layers is very large. This is a favorable point for forming a charge density wave or spin density wave. The phenomena were investigated in detail in high magnetic fields [325, 326]. The theory predicts that the phase transition should be re-entrant with respect to temperature. The re-entrant feature of the phase transition was actually observed in a high field of around 50 T [326, 327].

In highly quantized electron states, more phase transitions of different types are expected to occur in higher magnetic fields. A number of cyclotron resonance studies in graphite have been made in lower fields [328,329], but the investigation in very high magnetic fields is of interest, since new features may show up in relation to the electronic phase transition. Figure 4.38 shows the Landau level energy dispersion of graphite and the data of cyclotron resonance when very high magnetic fields up to 400 T were applied.

Many absorption peaks are observed corresponding to different transitions between Landau levels with non-uniform spacing. Interesting features are the peaks associated with the transitions involving 0± levels. The (1+)→(0+) transition is observed at 62 T for the hole-active polarization with 10.61 μm. This fact implies that the (0+) state is already depopulated below 62 T. On the other hand, the observation of the peak at 208 T with 3.39 μm for the electron active polarization (corresponding to (0–)→(1–) transition) indicates that the (0–) level is below the Fermi level and should cross the Fermi level at a higher field. This is a striking result, if we consider that the spin splitting should be small as expected from a small value of g-factor which should be nearly 2. In other words, if the (0+) state crosses the Fermi level below 62 T, (0–) level should cross it at a field not so much higher than this field. This contradiction is reconciled by a theory of Takada and Goto who predicted the enhancement of the spin-splitting due to the exchange interaction [331].

4.10.4 *Bi_2Se_3 family*

Layered semiconductors Bi_2Te_3, Sb_2Te_3, Bi_2Se_3, and their solid solutions are known to have extraordinarily large thermoelectric coefficients. They are interesting materials also from the viewpoint that the lattice constant in the c-axis is very large (2.6–2.8 nm). Many investigations have been made on p-type Bi_2Te_3 and n-type Bi_2Se_3. However, not much is known concerning the energy band structures in these materials.

It is known that these materials are narrow gap semiconductors. According to tunneling measurements, the band gap of Bi_2Te_3 is $E_g \sim 0.20$ eV at room temperature and increases up to 0.25 eV when the temperature is decreased down to 4.2 K [332]. The valence band in Bi_2Te_3 consists of an upper valence band (UVB) and a lower valence band (LVB) [333, 334] as shown in Fig. 4.39. Both bands have many-valley structure consisting of six ellipsoids, centered on a mirror plane of the Brillouin zone. The conduction band also consists of two bands UCB and LCB. In contrast to Bi_2Te_3 in which the conduction band has two extrema with six-ellipsoid valleys, the conduction band of Bi_2Se_3 has its principal minimum located at the center of the Brillouin zone (Γ point), and therefore, it can be represented by one ellipsoid [335, 336].

Cyclotron resonance was observed in single crystals of n-type $Bi_{2-x}Sb_xSe_3$, p-type Bi_2Te_3, $In_xBi_{2-x}Te_3$ and p-type $Sb_2Te_{1.2}Se_{1.8}$ under high pulsed magnetic fields up to 150 T generated by the single turn coil technique, and the effective masses were determined [337]. Figure 4.39 shows the magnetic field dependence of infrared transmission in an n-type $Bi_{1.48}Sb_{0.52}Se_3$ sample at laser wavelengths of $\lambda = 10.6$ μm and 9.25 μm in different temperatures [337]. At low temperatures, the traces shown in Fig. 4.39 exhibit two peaks (marked by vertical arrows). The low field peak (peak 1) has a larger intensity than the high field peak (peak 2). As the temperature increases, the peak 2 diminishes rapidly and becomes almost indiscernible at 180 K. These two peaks were observed in all samples with different Sb content.

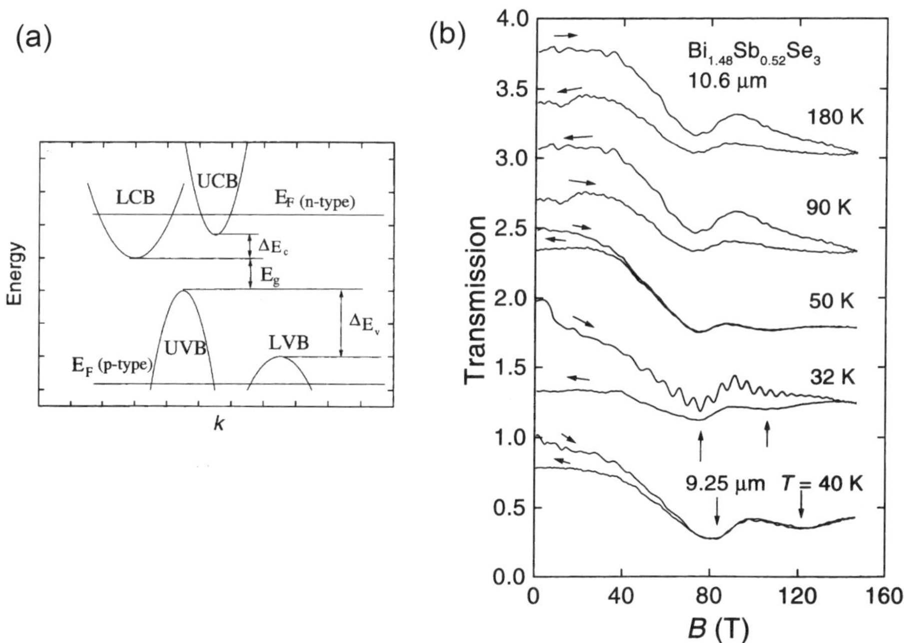

FIG. 4.39. (a) Schematic band structure of Bi_2Te_3 (p-type) and Bi_2Se_3 (n-type). (b) Transmission traces for n-type $Bi_{1.48}Sb_{0.52}Se_3$ sample versus magnetic field at different temperatures. Wavelength of radiation was 10.6 μm except for the lowest trace where it was 9.25 μm. Arrows show magnetic field sweeps in the up and down directions. Two minima are shown by vertical arrows.

At $T = 32$ K, the effective mass is estimated as $m^* = 0.075m$ for peak 1 and $m^* = 0.105m$ for peak 2. It was found that these effective masses have almost no dependences on temperature and electron concentration. The position of the peaks is almost independent of x, and of electron concentration. The larger mass $m^* = 0.105m$ for peak 2 is slightly smaller than but close to the effective mass obtained by the Shubnikov-de Haas effect $m^*_{\rm SdH} = 0.12m$ [335,338], so that it was assigned as the electron mass in the lower conduction band (LCB). The smaller effective mass $m^* = 0.075m$ for peak 1 is then considered to be attributed to the upper conduction band (UCB) [337]. The effective masses in other samples of the crystal family were also obtained [337].

4.11 Interplay between magnetic and quantum potentials

4.11.1 *Short period superlattices with parallel magnetic fields*

As regards two-dimensional systems, we have considered so far, cyclotron resonance when the magnetic field is applied in the direction perpendicular to the two-dimensional plane. The cyclotron motion itself is not affected by the quantum potential in this case. What will happen then if the field is applied parallel to the plane? In quantum wells, the magnetic field effect becomes conspicuous when the magnetic field is high enough to reduce the cyclotron radius to a size smaller than the quantum well width. However, at a lower field, cyclotron resonance is not observed in the quantum well since the cyclotron orbit is not closed within the quantum well. In short-period superlattices with thin barrier layers, however, the electron wave functions penetrating into the barrier layers from adjacent wells are connected with each other, such that a narrow band is formed in the direction perpendicular to the layers (k_x direction). In other words, the energy band in the three-dimensional Brioullin zone is folded back due to the periodic superlattice structure to form a mini-band. Thus the energy dispersion is generated in the k_x direction, and the perpendicular transport becomes possible. Then the cyclotron motion across the barrier layer causes the cyclotron resonance.

In the presence of a potential $U(x)$ in the plane of the cyclotron motion, the Schrödinger equation is given by

$$\left[-\frac{\hbar^2}{2m^*}\frac{\partial^2}{\partial x^2} + \frac{e^2B^2}{2m^*}(x-X)^2 + U(x)\right]\Phi(x) = \mathcal{E}\Phi(x), \qquad (4.66)$$

similarly to (3.113), as we have seen in Section 3.4.3, This is a one-dimensional equation with respect to x, adding a quantum well potential to a parabolic potential due to the magnetic field. The solution of this equation derives eigenenergies which depend on the center coordinate X. This indicates that the degeneracy of the Landau level energies with respect to X is lifted by the quantum potential $U(x)$. Equation (4.66) can be solved numerically by a computer calculation. Figure 4.40 (a) shows the results of the energy of the Landau levels as a function of X for a $(\mathrm{GaAs})_7(\mathrm{AlGaAs})_5$ superlattice (superlattice with seven GaAs

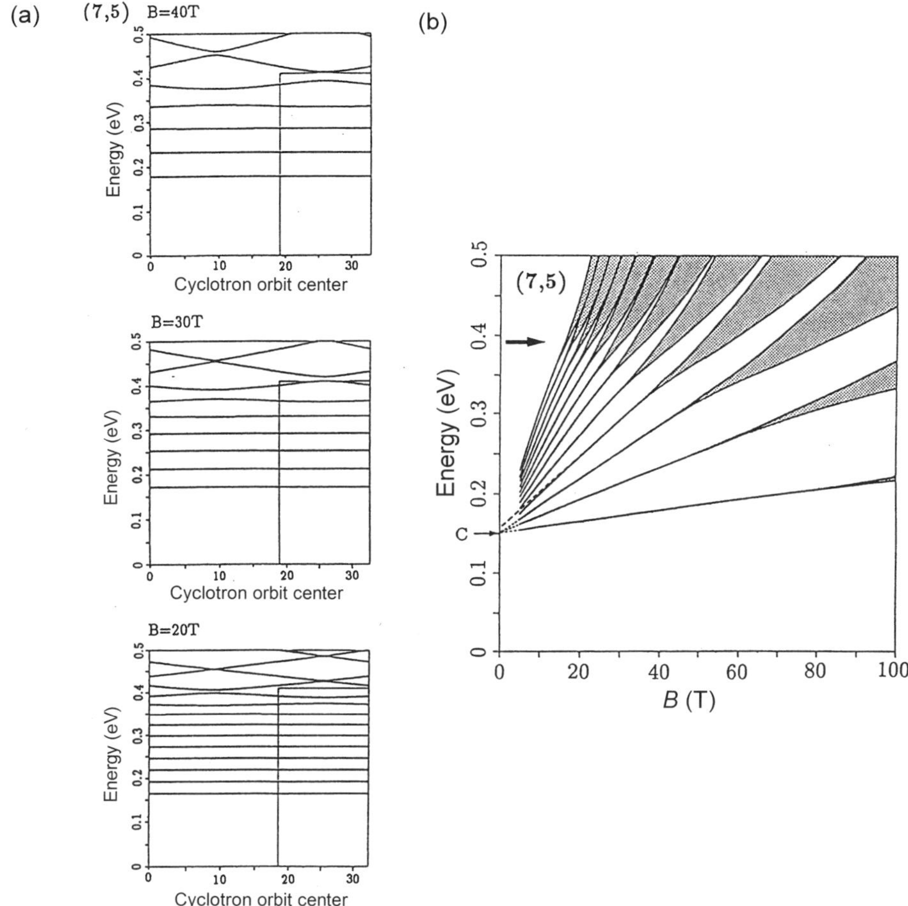

FIG. 4.40. (a) Landau level energies as a function of the center coordinate of the cyclotron motion for a GaAs/AlGaAs superlattice (7, 5) when the magnetic field is applied parallel to the superlattice layers [339]. The potential is shown in the figure. (b) Landau level energies as a function of magnetic field. Above the potential height shown by the arrow, the Landau levels are broadened.

monolayers and five AlGaAs monolayers, hereafter denoted as (7,5)) when the magnetic field is applied parallel to the superlattice layers [339]. The dispersionless Landau levels are formed in the low energy range when their energy is within the subband. However, in the region exceeding the mini-band width, the Landau levels have a dispersion whose energy is dependent on X. In other words, the Landau levels have a finite energy width in this region. This is a fundamental problem of solid state physics concerning the energy spectrum when the Landau level energy exceeds the band width.

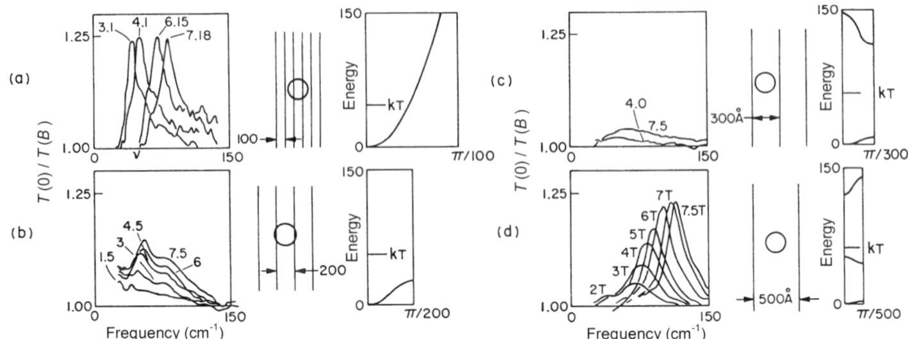

FIG. 4.41. Cyclotron resonance spectra in GaAs/AlGaAs short-period superlattices with different well width L_w when the magnetic field is applied parallel to the layers [340]. The field strength is shown for each graph. (a) L_w = 10 nm, (b) L_w = 20 nm, (c) L_w = 30 nm, (d) L_w = 50 nm. The energy dispersion is shown on the right-hand side of each figure.

If we measure cyclotron resonance in such a system, the broadening of the Landau level is actually observed. In low magnetic fields and low energy range, the width of the Landau level is small, so that sharp cyclotron resonance is observed. This corresponds to the situation in which the cyclotron motion is extended to many periods of the superlattice and the energy does not depend so much on X. In higher magnetic fields and higher energy range, the Landau levels are broadened and cyclotron resonance line-width is increased. This corresponds to a situation in which the Landau level energy strongly depends on X. When the magnetic field is further increased, the cyclotron resonance line-width becomes narrow again, because a cyclotron orbit is completed within the well layers. This corresponds to a situation in which the energy change with X occurs only near the interfaces and the density of states is concentrated mostly in the interior of the well layer.

This behavior of the cyclotron resonance line-width was actually observed by Allen *et al.* [340] for short-period superlattices of GaAs/AlAs. Instead of varying magnetic field in a wide range, they measured the cyclotron resonance for superlattices with different periods. As shown in Fig. 4.41, they found that the line-width is broadened when the cyclotron radius (or magnetic length) l is close to the well width L_w [340]. At $B \sim 8$ T, the cyclotron radius is ~9.2 nm. A sharp cyclotron resonance peak is observed when the well width is much smaller or larger than 9.2 nm but it is broadened for $L_w \sim 10$ nm.

4.11.2 *Angular dependence of the effective mass in quantum wells*

The effective magnetic field on the motion of the in-plane motion of two-dimensional electrons is the field perpendicular to the 2D plane. Therefore, when we tilt the magnetic field $\boldsymbol{B}$ from the normal to the 2D plane by an angle θ, the effective field is $B\cos\theta$, and the cyclotron frequency is given by

$$\omega_{c\perp} = \frac{eB}{m^*}\frac{1}{\cos\theta} \tag{4.67}$$

The cyclotron resonance peak should shift to higher magnetic field with increasing θ. This is actually observed in the normal case. However, at very high magnetic fields, if the Landau levels associated with the cyclotron resonance transition cross the subbands, the situation is very different since the Landau level-subband coupling (LSC) occurs in tilted magnetic fields. In other words, when the $\hbar\omega_{c\perp}$ becomes equal to the interval between the lowest two subbands $\mathcal{E}_1 - \mathcal{E}_0$, the $N = 1$ Landau level of the $\mathcal{E}_0$ ground subband crosses the $N = 0$ Landau level of the $\mathcal{E}_1$ subband, the anti-crossing behavior is observed because of the LSC. Such an anti-crossing was first observed by Beinvogl and Koch on the accumulation layer on Si [341]. Subsequently, many experiments have been done on GaAs/AlGaAs heterostructure [342–344], coupled quantum wells of GaAs/AlGaAs [345], InAs/AlSb quantum well [346] and GaAs/AlGaAs superlattice [347].

When the magnetic field is applied at an angle θ from the normal direction to the layers (z-direction) in the $x - z$ plane, the magnetic field and the vector potential can be expressed as $\boldsymbol{B} = (B_\parallel, 0, B_\perp)$ and $\boldsymbol{A} = (0, xB_\perp - zB_\parallel, 0)$. The Hamiltonian for electrons is then

$$\mathcal{H} = \mathcal{H}_\perp + \mathcal{H}_\parallel + \mathcal{H}', \tag{4.68}$$

$$\mathcal{H}_\perp = -\frac{\hbar^2}{2m^*}\frac{\partial^2}{\partial z^2} + U(z) + \frac{m^*}{2}\omega_\parallel^2 z^2, \tag{4.69}$$

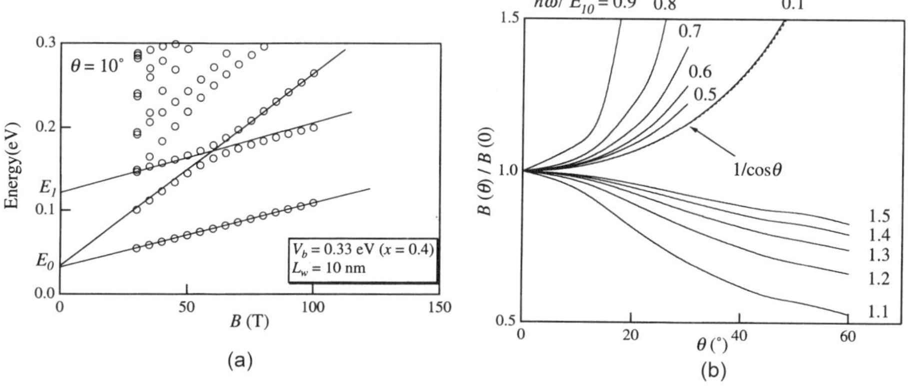

FIG. 4.42. (a) Calculated results of the Landau level energy for $\theta = 10°$ [347]. Energies without the LSC are shown by solid lines. (b) Cyclotron resonance field as a function of the tilt angle θ for different values of $\hbar\omega/\mathcal{E}_{10}$. The broken line shows the curve of the $1/\cos\theta$ dependence.

$$\mathcal{H}_{\parallel} = -\frac{\hbar^2}{2m^*}\frac{\partial^2}{\partial x'^2} + \frac{m^*}{2}\omega_\perp^2 x'^2, \tag{4.70}$$

$$\mathcal{H}' = -\frac{\omega_\parallel \omega_\perp}{m^*} x' z. \tag{4.71}$$

Here, $x' = x + \hbar k_y / eB_\perp$, $\omega_\parallel = eB_\parallel / m^*$, and $\omega_\perp = eB_\perp / m^*$. $\mathcal{H}_\perp$ is a term to represent the motion of electrons perpendicular to the two-dimensional plane (z-direction) in which the effect of the parallel field is added to the term giving the subband energies. $\mathcal{H}_\parallel$ is a term representing the motion of electrons within the two-dimensional plane, giving the Landau quantization by the perpendicular field $B_\perp$. $\mathcal{H}'$ is the term that represents the coupling between the Landau level and the subbands. As the cyclotron energy $\hbar\omega$ approaches the energy difference between the two subbands $\mathcal{E}_{01} = \mathcal{E}_1 - \mathcal{E}_0$, the coupling is resonantly enhanced and a large cross-over effect occurs. Figure 4.42 (a) shows the Landau level energies for $\theta = 10°$ [347]. Figure 4.42 (b) shows the cyclotron resonance field $B(\theta)/B(0)$ $\hbar\omega$ as a function of the angle θ at a photon energy. When $\hbar\omega_c/\mathcal{E}_{01}$ is small, the resonance field $B(\theta)$ increases with increasing θ with nearly the $1/\cos\theta$

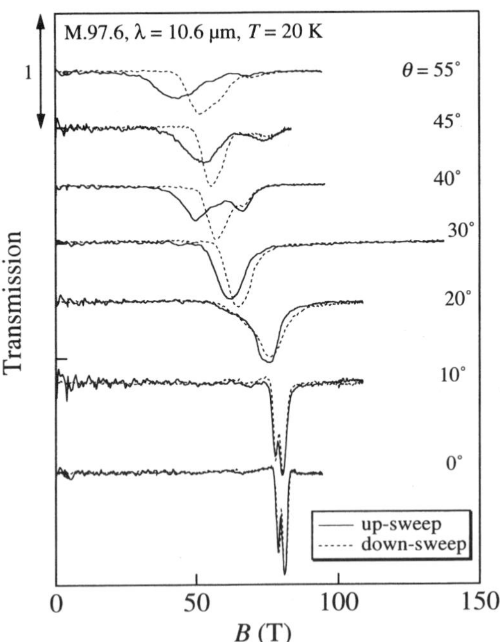

FIG. 4.43. Cyclotron resonance traces for a GaAs/AlGaAs quantum well at tilted magnetic fields [347]. The widths of the well and the barrier layers are 10 nm and 60 nm, respectively. The wavelength is 10.5 μm. The sample satisfies the relation $\hbar\omega/\mathcal{E}_{01} = 1.60$.

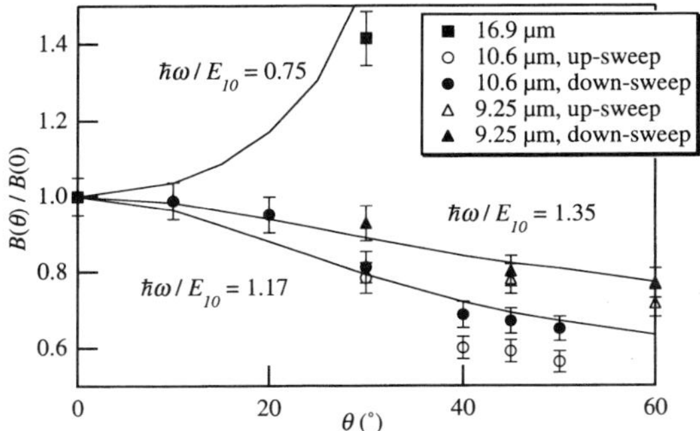

FIG. 4.44. Cyclotron resonance spectra in GaAs/AlGaAs multiple quantum wells at different angles θ of magnetic field relative to the growth direction [347].

dependence. As $\hbar\omega_c/\mathcal{E}_{01}$ becomes large, approaching 1, the angle dependence increases and as soon as it exceeds 1, the angular dependence changes its sign. In other words, the resonance field decreases as the tilt angle increases.

As an example of such a dependence, Fig. 4.43 shows the cyclotron resonance traces for GaAs/AlGaAs multiple quantum wells [347]. Spin splitting is observed at $\theta = 0°$. We can see clearly that the resonance field decreases with increasing angle. A large hysteresis is observed between the up-sweep and down-sweep of the magnetic field. The reason for the hysteresis is not known at this moment, but it suggests that in the up-sweep there must be some kind of relaxation phenomenon in the electronic state with a response time of about 1 μs. Figure 4.44 shows the observed angular dependence of the resonance field $B(\theta)$ for different $\hbar\omega/\mathcal{E}_{01}$ as compared with theoretical calculation. It is clearly seen that the $B(\theta)$ is an increasing function of θ for $\hbar\omega/\mathcal{E}_{01} < 1$, but it decreases with θ for $\hbar\omega/\mathcal{E}_{01} > 1$.

4.11.3 *Cyclotron resonance in quantum dots*

As discussed in Section 2.5.4, recent advances in microfabrication techniques have enabled us to make lateral potentials on a two-dimensional plane to confine electrons in a narrow region. Quantum wires and quantum dots are fabricated by different techniques. Quantum dots are the system in which the electron wave functions are confined in all three directions. The energy levels of the quantum dots in a magnetic field can easily be obtained as in (2.169) if we approximate the potential in the lateral direction by a parabolic potential. The uniformly spaced energy levels at $B = 0$ are split by magnetic field due to the angular momentum represented M. The transition energy between these energy levels is represented as

$$\hbar\omega_{\pm} = \sqrt{\left(\hbar\frac{\omega_c}{2}\right)^2 + \omega_0^2} \pm \frac{\hbar\omega_c}{2}, \tag{4.72}$$

for allowed transitions, $\Delta M = \pm 1$. This gives an energy ω_0 at $B = 0$ splitting to two branches in magnetic fields. Here $\hbar\omega_+$ and $\hbar\omega_-$ correspond to the transition

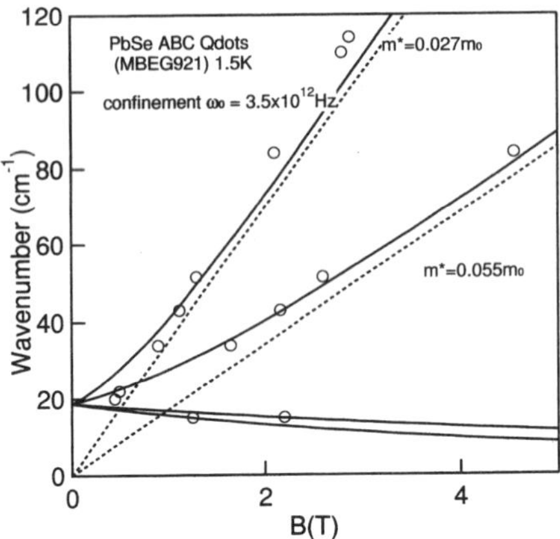

FIG. 4.45. Cyclotron resonance in a PbSe/PbEuTe quantum dot sample in a low field range [352].

energies for right circular and left circular polarizations, respectively. In the limit of large B where a relation $\omega_c \gg \omega_0$ holds, $\hbar\omega_+$ tends to the cyclotron resonance energy $\hbar\omega_c$.

In the experiments of cyclotron resonance in quantum dots, such a dispersion was actually observed in GaAs/AlGaAs dots [348–350], in InSb dots created by a gate voltage [351], and self-organized dots of PbSe/PbEuTe [352]. As an example of cyclotron resonance in quantum dots, we show the data for PbSe/PbEuTe quantum dots in Fig. 4.45 [352]. Springholz *et al.* succeeded in fabricating high quality self-assembled quantum dots of PbSe embedded in $Pb_xEu_{1-x}Te$ matrix [353, 354]. By stacking many layers of PbSe and PbEuTe on a $< 111 >$ surface of BaF_2 alternately by MBE, self-assembled quantum dots of PbSe are formed arranged densely in each PbSe layer due to the difference in the lattice constant between the two layers. The dots have a trigonal pyramid shape and regularly arranged with a trigonal symmetry on each plane and stacked with an ABCABC$\cdots$ order to form an fcc-like dot crystal. By choosing a proper parameter, an AAA$\cdots$ like order is also possible. It is of interest to investigate the regular arrangement of the quantum dots. Owing to the large dot density, cyclotron resonance can easily be observed. When we apply magnetic field $\boldsymbol{B}$ perpendicular to the layers, In Fig. 4.45, it can be seen that there are two sets of lines corresponding to the effective masses of $0.025m$ and $0.055m$. They correspond to the cyclotron resonance in two inequivalent valleys of the conduction band for magnetic field $\boldsymbol{B} \parallel < 111 >$. Each set of lines splits into two lines, one goes up and the other down with magnetic field. This is a typical dispersion of the cyclotron resonance mode in quantum dots.

Cyclotron resonance traces in higher magnetic fields are shown in Fig. 4.46. The measurements were performed using a CO_2 laser varying the wavelength

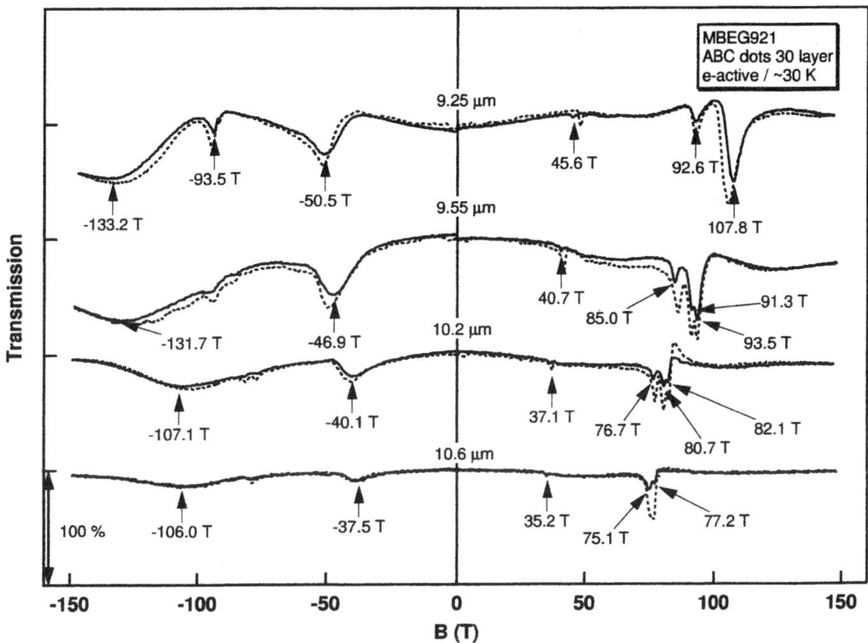

FIG. 4.46. Cyclotron resonance traces for PbSe/PbEuTe quantum dots [352]. The right and left panels show the data for electron-active and electron-inactive circular polarizations, respectively, at different wavelengths. The solid and broken lines represent the up-sweep and down-sweep of the magnetic field.

between 9.25 μm and 10.6 μm. Radiation was circularly polarized and electron-active (right panel) and hole-active (left panel) modes can be distinguished. In the hole-active mode, peaks arising from the PbEuTe matrix are observed. The absorption peaks in the electron-active mode are the cyclotron resonance peaks of the electrons in the PbSe quantum dots. The resonance peaks of the light electron valley (35.2–45.6 T) and the heavy electron valley (75.1–107.8 T) are observed. The heavy electron peaks show a splitting. This can be explained as the lift of the degeneracy of three equivalent valley for $\boldsymbol{B} \parallel < 111 >$ by a strain introduced around the dots. The interesting point is that there is a remarkable wavelength dependence of the absorption intensity. This was explained as due to the interference effect of the incident radiation in the sample [352].

4.12 Magnetic field-induced band cross-over

4.12.1 Γ-*L cross-over in GaSb*

When a magnetic field is applied, the band edge shifts to a higher energy by $\hbar\omega_c/2$. In very high fields, the shift becomes so large that the band edge may

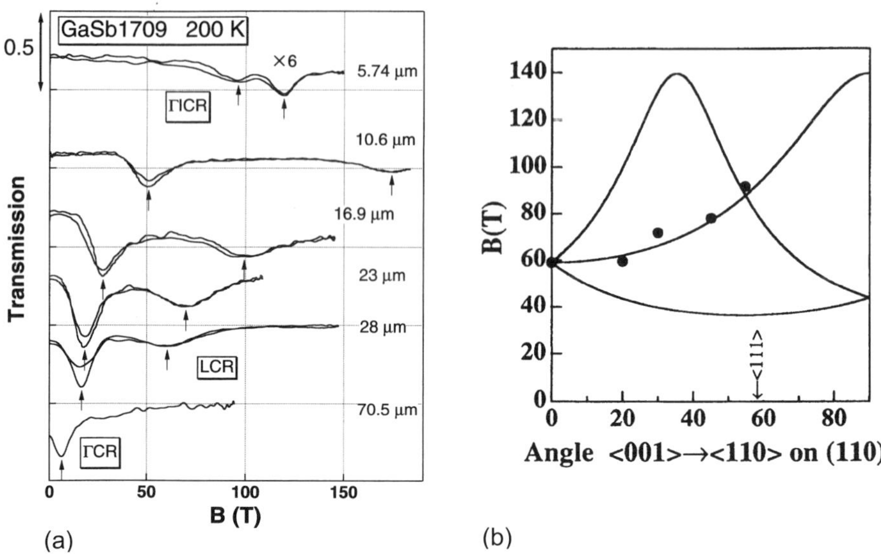

FIG. 4.47. (a) Cyclotron resonance traces for GaSb at different wavelengths [359]. $B \parallel < 100 >$. ΓCR, ΓICR, and LCR denote the cyclotron resonance at Γ point, impurity cyclotron resonance at Γ point and cyclotron resonance at L point, respectively. (b) Angular dependence of the resonance field. The direction of magnetic field is varied from $< 100 >$ to $< 110 >$ axis on the (110) plane.

take over a higher lying band edge which has a larger effective mass at some critical fields. This actually takes place in III-V compounds whose conduction band minimum is located at Γ point and there are higher lying minima at X points or L points. As an example of such field-induced band cross-over, we present below the Γ−L cross-over in GaSb. GaSb has a conduction band minimum at the Γ point, but the minima at the L point are very closely located just above the Γ minimum. The energy difference is $\Delta\mathcal{E}$(L-Γ) $\sim$ 86 meV at 200 K. The effective mass at the L point is expected to be larger than that at the Γ point. In sufficiently high magnetic fields, therefore, the $N = 0$ Landau level at the L point would become the lowest level above the Γ-L cross-over and the effective mass parameters at the L point would be revealed. Many experiments have been carried out to determine the longitudinal and transverse masses m_l and m_t at the L point such as the Hall effect [355], piezo-resistance [356], Faraday rotation [357], and the Shubnikov-de Haas effect [358]. However, no conclusive results have been obtained, as signals are a mixture of two series arising from the Γ and the L points.

A cyclotron resonance experiment was performed by Arimoto *et al.* [359]. Figure 4.47 (a) shows typical cyclotron resonance traces for $\boldsymbol{B} \parallel < 100 >$ at different wavelengths. At 28 μm and $T = 204$ K, two peaks are observed corresponding

Table 4.7 Effective masses of the conduction band of GaSb at the Γ and the L point

Minimum	direction	λ (μm)	$\hbar\omega$ (meV)	m^*/m
Γ		Band edge	0	0.039 ± 0.001
Γ		28.0	44.3	0.044 ± 0.001
Γ		10.6	117	0.050 ± 0.001
Γ		5.53	22.4	0.067 ± 0.001
Γ		88	14.1	0.0412
L	$m^* < 100 >$	Band edge	0	0.14 ± 0.01
L	$m^* < 100 >$	28.0	44.3	0.150 ± 0.002
L	$m^* < 100 >$	10.6	117	0.170 ± 0.002
L	$m^* < 111 >$	Band edge	0	0.21 ± 0.01
L	$m^* < 111 >$	16.9	73.4	0.243 ± 0.002
L	m_t	Band edge	0	0.085 ± 0.006
L	m_l	Band edge	0	1.4 ± 0.2

to the cyclotron resonance at the Γ point (ΓCR) and at the L point (LCR), from which we can determine the effective masses for the both conduction bands. At 9.24 μm and $T = 25$ K, the impurity cyclotron resonance peaks at the Γ point (ΓICR) and at the L point (LICR) are observed as the electrons freeze out in the donor states due to the low temperature. At 5.53 μm and $T = 150$ K, small peaks of ΓICR and ΓCR are observed. Larger peaks corresponding to LICR and LCR should be observed at higher fields. The dependencies of the cyclotron resonance on the field angle and the photon energy were investigated in detail as shown in Fig. 4.47 (b). From this analysis, the effective masses of the both conduction bands were determined as listed in Table 4.7.

A similar phenomenon, the Γ−X cross-over in GaAs was investigated by cyclotron resonance using explosive-driven flux compression technique [360,361].

4.12.2 Γ−*X cross-over in GaAs/AlAs superlattices*

In the case of GaAs/AlAs superlattices, the conduction band minimum in the GaAs layer is located at the Γ point while the band minimum in the AlAs layer is at the X point. At the X point, the conduction band minimum is lower in the AlAs layer than in the GaAs layer. Therefore, the quantum well is located in the AlAs layer and the GaAs layer plays a role as a barrier layer, contrary to the case of the Γ point. The situation is schematically illustrated in Fig. 4.48. GaAs/AlAs short-period superlattices are often denoted as $(\mathrm{GaAs})_m/(\mathrm{AlAs})_n$, with integer parameters m and n designating the monolayer numbers of each layer. In $(\mathrm{GaAs})_m/(\mathrm{AlAs})_n$, mini-bands are formed and the wave functions are extended in the both layers through tunneling. The lowest mini-band mainly consists of the GaAs Γ band (Γ-like band), and the mini-band consisting mainly of the AlAs X band is lying higher when the GaAs monolayer number m is larger than some value. As m is decreased, the energy of the Γ-like band is increased and the X-like band becomes the lowest mini-band. Thus short-period superlattices with small m should have a type II band line up where the conduction band minima are located at the X point in the AlAs layer while the valence band top is located at the Γ point in the GaAs layer. Many investigations have been made on the type I to type II transition in short-period superlattices of $(GaAs)_m/(AlAs)_n$ when m and n were varied [362–367]. There is a critical value of the monolayer number m_{I-II} at around 12–14 that divides the range of type I and type II superlattices.

Cyclotron resonance in short-period superlattices with m above and below m_{I-II} should reveal the effective mass of the Γ-like band and the X-like band, respectively. This was actually observed in the cyclotron resonance in a very high magnetic field. Figure 4.49 shows the cyclotron resonance spectra in n-type $(GaAs)_n/(AlAs)_n$ short-period superlattices for different n [368, 369]. For samples with $n = 16$, a relatively sharp resonance peak arising from the Γ point is observed with an effective mass of $0.053m$ as peak A. This effective mass is close to the value of the band edge mass of bulk GaAs $0.067m$. In a higher field a small peak B was also observed corresponding to the resonance in the X-band

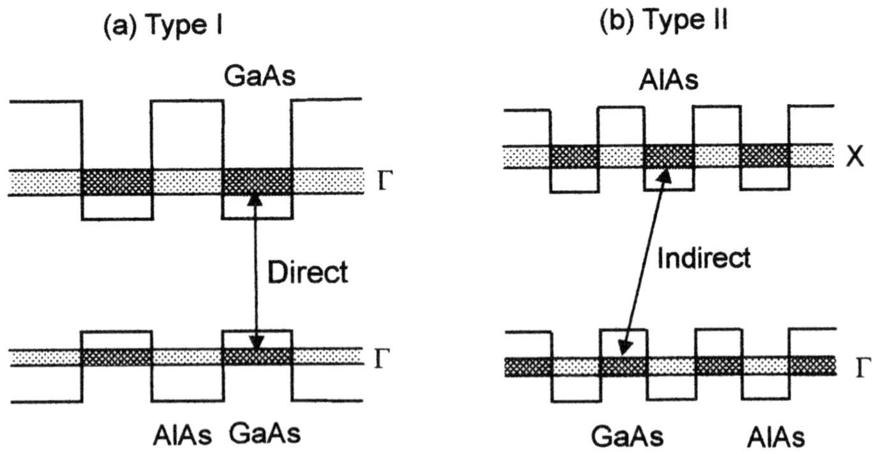

FIG. 4.48. Schematic energy diagram of the (a) type I superlattice and (b) type II superlattice of the GaAs/AlAs system. The density of the shadow region indicates the extent of the position probability density of electons and holes in the mini-bands. Regarding the Γ point, the well layer for the conduction band is the GaAs layer, while at the X point the well layer is the AlAs layer. For small width of the GaAs layer, the mini-band energy of the conduction band at the Γ point is increased in the GaAs layer and the conduction band minimum is located in the AlAs layers; thus the system has indirect nature.

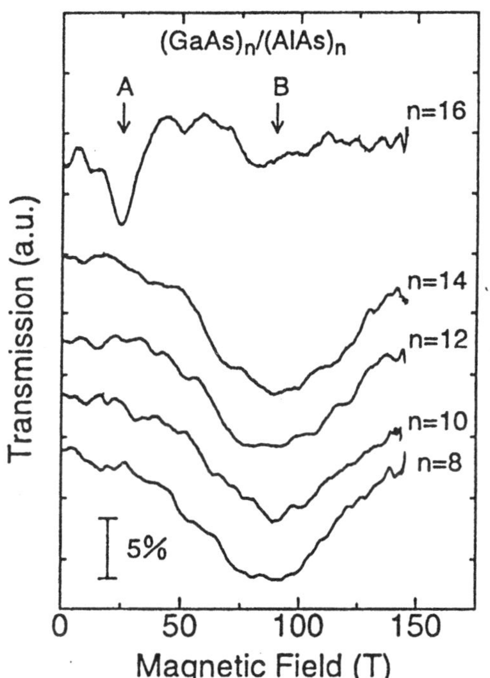

FIG. 4.49. Cyclotron resonance traces for $(\mathrm{GaAs})_n/(\mathrm{AlAs})_n$ short-period superlattices with $n = 8$–16 [368]. $\lambda = 23$ μm. $B \parallel < 100 >$.

at around 90 T. For $n \leq 14$, a broad peak near 90 T is the main feature and the peak A is not observed. This broad peak corresponds to the transverse mass m_t at the X-like band. From the resonance field, the effective mass is obtained as $0.193 \pm 0.006m$. This was the first observation of the effective mass of the short-period superlattice. It should be noted that the above value of m_t is close to that obtained from magneto-photoluminescence, $0.16m$ [370], and that obtained from cyclotron resonance in bulk AlAs, $0.25m$ [312]. The experimental data show that the effective mass change from that in type I to that in type II abruptly at $n = 14$. This is in contrast to the smooth change of the effective mass in GaAs/AlGaAs quantum wells [371].

Figure 4.50 shows cyclotron resonance traces for $(\mathrm{GaAs})_n/(\mathrm{AlAs})_n$ short-period superlattices at $\hbar\omega = 129$ meV ($\lambda = 9.61$ μm) in higher fields generated by electromagnetic flux compression [372, 373]. In a sample of $n = 16$ (Type I sample), only a peak from the Γ point was observed, giving the effective mass of $0.0809m$ at 76 K, and $0.0756m$ at 34 K. In a sample of $n = 12$, in addition to a small peak corresponding to the Γ point, a peak arising from the X point was observed at a higher field. The effective mass was obtained as $0.244m$ at 104 K,

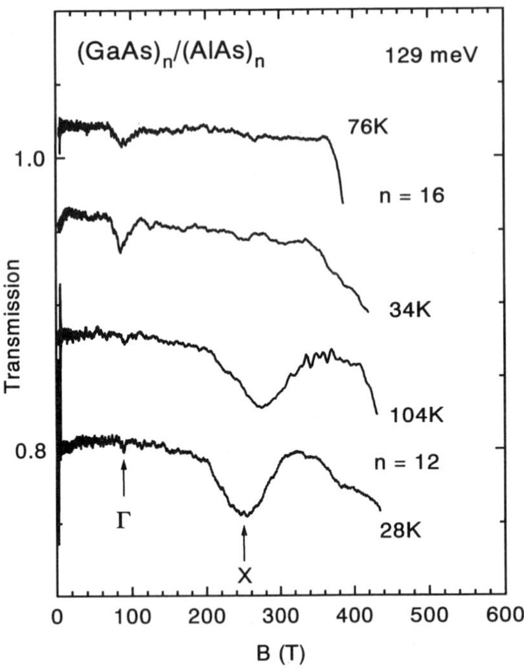

FIG. 4.50. Cyclotron resonance traces for $(GaAs)_n/(AlAs)_n$ short-period superlattices ($n = 12$ and 16) at $\hbar\omega$= 129 meV in a field up to 500 T [372, 373]. $B \parallel <100>$.

and $0.226m$ at 28 K [372]. Considering the energy difference between the Γ and X points, it is expected that the type I to type II transition associated with the Γ-X cross-over should occur at some field, and that we should have observed the X-peak even in the $n = 16$ sample. The absence of the X-peak in the sample of $n = 16$ in such high fields suggests the possibility of either the slow relaxation time between the Γ and X valleys or the transfer of electrons to a deep donor level such as the DX center [372].

4.12.3 *Semimetal-semiconductor transition*

A. Bi

In semimetals, the bottom of the conduction band is situated below the valence band top, so as to have crossed band gap. As magnetic fields are applied, the energy of the conduction band bottom is increased and that of the valence band top is decreased, so that a finite band gap is opened in high enough magnetic fields. Thus a semimetal-semiconductor transition should take place in high magnetic fields.

Bi is a typical semimetal, which has a band overlap of 35 meV. The conduction band minima L_s are at the L-point and the valence band (hole band) tops T_{45}^{-}

are at the Z-point of the Brillouin zone. Just below the L_s band, there is an antisymmetric valence band L_a. In a magnetic field, the band profile shows a complicated field dependence. In the case of $\boldsymbol{B} \parallel$ binary axis, the field dependence of the band extrema is shown in Fig. 4.51. The energy gap $\mathcal{E}_g(0)$ between the L_s and L_a at first decreases because the spin-orbit interaction is very large such that the spin splitting is larger than the Landau level spacing. The energy of the bottom of the T_{45}^- band decreases with increasing field. L_s and L_a come very close to each other around B_c, but due to the interaction, they repel each other. Above the anti-crossing field B_c, the band gap increases with magnetic field. At a magnetic field B_T, the L_s crosses with the T_{45}^- band, and above B_T, the band gap opens up between L_s (electron band) and T_{45}^- (hole band). Thus the semimetal to semiconductor transition takes place at B_T. The band gap $\mathcal{E}_g(0)$ at zero field is only 30 meV. For $\boldsymbol{B} \parallel$ binary, the values have been estimated as $B_c \sim$ 10–20 T and $B_T \sim$ 100 T. Figure 4.52 shows the Landau level energy diagram of Bi when the magnetic field is applied parallel to the binary axis [374].

At $B = B_T$, it is expected that a zero-gap state is realized. The magnetic field-induced zero-gap state is a very interesting regime, because when the band gap is nearly zero, electronic phase transitions may take place in semimetals or semiconductors. The excitonic phase is a state where excitons are condensed when the exciton binding energy becomes larger than the band gap. When the system undergoes a transition to the excitonic phase, it becomes an insulating state, as the excitons have no electric charge [375, 376]. Yoshioka and Nakajima predicted a different type of phase transition, the gas-liquid transition [377, 378]. The excitonic phase has been explored by a group of Brandt *et al.* in Bi-Sb alloys [379]. When Sb is doped into Bi, the band gap $\mathcal{E}_g(0)$ of the alloy $Bi_{1-x}Sb_x$ is decreased with increasing x up to $x \sim 0.06$, and for higher concentration, the alloy becomes semiconductors. Therefore, in Bi-Sb alloys, in which the band gap is smaller than in Bi, the semimetal to semiconductor transition is expected to occur at a lower field. Brandt *et al.* estimated the energy band gap from the temperature dependence of the resistivity, and insisted that they observed an

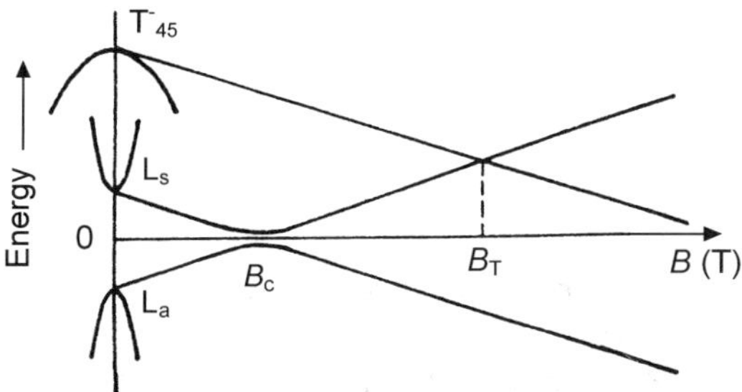

FIG. 4.51. The schematic diagram of the magnetic field dependence of the energy band extrema in Bi when the magnetic field is applied parallel to the binary axis.

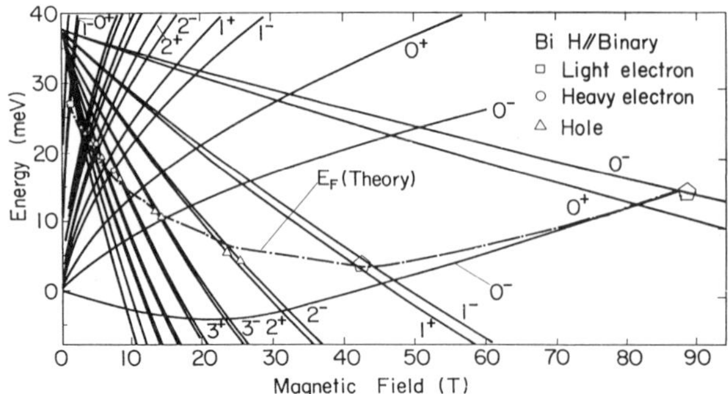

FIG. 4.52. Energies of the Landau levels in the electron, and hole bands and the Fermi level $\mathcal{E}_F$ as a function of magnetic field in Bi [374]. The calculation was made using the band parameters obtained from the analysis of the Shubnikov-de Haas effect. Experimental points of the peaks of the Shubnikov-de Hass effect are shown for light electron, heavy electron and hole bands.

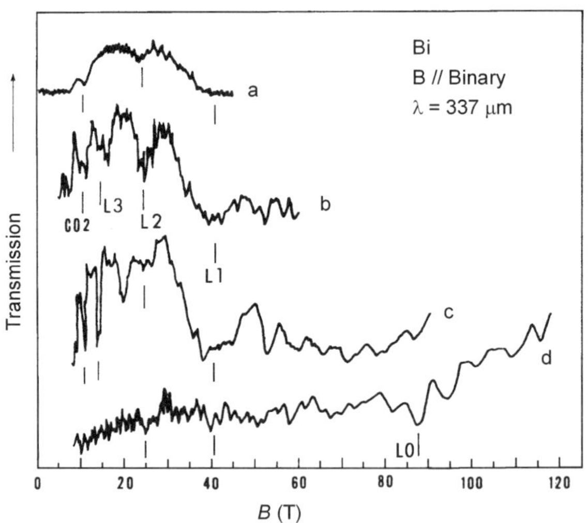

FIG. 4.53. Far-infrared transmission spectra in Bi for $B \parallel$ binary axis [83]. The thickness of the sample and the temperature of the measurement for each figure are (a) 1.61 mm, 6.8 K, (b) 0.60 mm, 4.2 K, (c) 0.40 mm, 8.8 K, (d) 0.62 mm, 4.5 K.

evidence of the excitonic phase. However, no conclusive results have so far been obtained.

Miura *et al.* explored the excitonic phase by applying high magnetic fields exceeding 100 T to Bi, measuring a far-infrared transmission spectra [83]. Figure 4.53 shows examples of the experimental traces of the transmission spectra of the radiation of $\lambda = 337$ μm in Bi when a magnetic field is applied parallel to the binary axis. The transmission was found to increase abruptly at 98 T. This is considered to correspond to the semimetal-semiconductor transition. This transition was further investigated by Shimamoto *et al.* later by the far-infrared strip line technique [330]. The experiment was not easy because of the interference effect of the far-infrared radiation, but evidence of the transition was obtained. Thus the magnetic field-induced semimetal-semiconductor transition has been observed in Bi, but the excitonic phase is still an open question to be investigated in future.

B. InAs/GaSb superlattices

InAs/GaSb superlattices are type II superlattices. At the interface between the InAs layer and the GaSb layer, the valence band top of GaSb is energetically higher than the conduction band bottom of InAs. When the superlattice period is not so short ($d > d_c \sim 15$ nm), electrons in the GaSb layer diffuse into the InAs layer, so that there are always the same number of electrons and holes at the interface, forming a semimetallic state. On the other hand, for $d < d_c$, the band gap opens up and the system becomes a semiconductor [380]. When the magnetic field is applied to the semimetallic samples in the growth direction (perpendicularly to the layers), the energy of the electron subband is increased and that of the hole subband is decreased. Hence, at sufficiently high magnetic

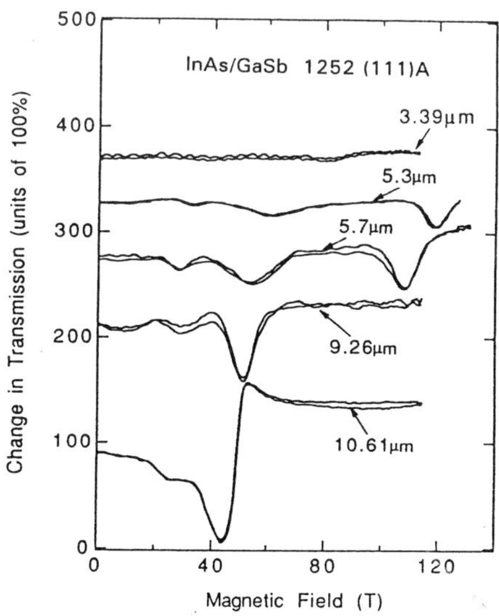

FIG. 4.54. Cyclotron resonance spectra in InAs/GaSb superlattices (grown on a (111)A plane) at different wavelengths [381]. $T \sim 30$ K.

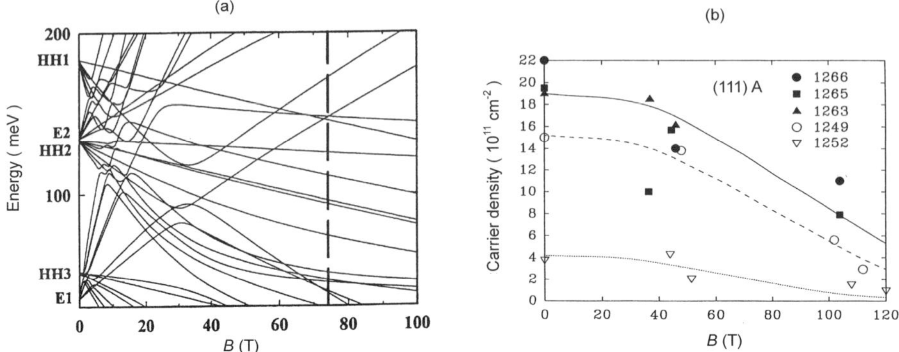

FIG. 4.55. (a) Theoretically calculated Landau level energies for InAs(20 nm)/GaSb(5.0 nm) superlattice [381]. (b) Magnetic field dependence of the carrier density estimated from the absorption intensity.

fields, a semimetal-semiconductor transition is expected to occur as in Bi. Such transition can be monitored well by cyclotron resonance. Figure 4.54 shows experimental traces of cyclotron resonance in InAs/GaSb superlattices at different wavelengths for a sample with (111) plane [381]. As the Landau level structure is very complicated, as shown in Fig. 4.55 (a), several peaks are observed. However, if we are concentrated in the peak corresponding to the effective mass of about $0.05m$, we can see that the absorption intensity is decreased as the photon energy is increased. Actually, Fig. 4.55 (b) shows the integrated intensity of the absorption peak as a function of magnetic field. It is clear that the peak intensity shows a sudden decrease at around 75 T. This is due to the field-induced semimetal-semiconductor transition. Thus the transition field was determined as $\sim$75 T for the (111) sample. For a sample with the (100) plane, the transition field was found to be lower ($\sim$ 40 T) in agreement with the theoretical calculation.

4.13 Magneto-plasma phenomena

4.13.1 *Magneto-plasma*

In conducting substances with many carriers like metals, semimetals and doped semiconductors, electron and hole plasmas play an important role and govern the propagation of electromagnetic radiation. When an electromagnetic wave is incident, the response of matter is determined by the dielectric constant $\epsilon = \kappa\epsilon_0$ (ϵ_0 is the dielectric constant of vacuum). Let the electric and magnetic fields of the radiation be

$$\boldsymbol{E} = \boldsymbol{E}_0 \exp\left[i\omega(t - \frac{\tilde{n}}{c})\hat{\boldsymbol{q}} \cdot \boldsymbol{r}\right], \tag{4.73}$$

$$\boldsymbol{H} = \boldsymbol{H}_0 \exp\left[i\omega(t - \frac{\tilde{n}}{c})\hat{\boldsymbol{q}} \cdot \boldsymbol{r}\right]. \tag{4.74}$$

Then neglecting the magnetization of matter, the Maxwell equations are given by

$$\nabla \times \boldsymbol{E} = -\frac{\partial \boldsymbol{B}}{\partial t} = -i\omega\mu_0 \boldsymbol{H}, \tag{4.75}$$

$$\nabla \times \boldsymbol{H} = -\frac{\partial \boldsymbol{D}}{\partial t} + \boldsymbol{J} = (i\omega\epsilon_0 \boldsymbol{\kappa} + \boldsymbol{\sigma}) \cdot \boldsymbol{E}, \tag{4.76}$$

where we used relations,

$$-eB = \mu_0 \boldsymbol{H}, \quad \boldsymbol{D} = \epsilon_0 \boldsymbol{\kappa} \cdot \boldsymbol{E}, \quad \boldsymbol{J} = \boldsymbol{\sigma} \cdot \boldsymbol{E}. \tag{4.77}$$

Here, $\tilde{n}$ is the complex refractive index

$$\tilde{n} = n - ik, \tag{4.78}$$

and $\hat{\boldsymbol{q}}$ is the unit vector of the propagation direction of the wave. $\boldsymbol{\kappa}$ is regarded as a tensor to include the case where the polarization is nondiagonal. From (4.75) and (4.76), we can derive

$$\nabla \times \nabla \times \boldsymbol{E} = -i\omega\mu_0 (i\omega\boldsymbol{\kappa}\epsilon_0 + \boldsymbol{\sigma}) \cdot \boldsymbol{E}, \tag{4.79}$$

and thus

$$\tilde{n}^2 [\boldsymbol{E} - (\hat{\boldsymbol{q}} \cdot \boldsymbol{E}) \hat{\boldsymbol{q}}] = \boldsymbol{\kappa} \cdot \boldsymbol{E} - i\boldsymbol{\sigma} \cdot \frac{\boldsymbol{E}}{\omega\epsilon_0}. \tag{4.80}$$

In conducting materials containing conduction carriers, application of magnetic fields $\boldsymbol{B}_{ex}$ gives rises to a change in $\boldsymbol{\sigma}$, so that the refractive index $\tilde{n}$ becomes magnetic field dependent. In most cases, the electric vector is perpendicular to the propagation vector $\boldsymbol{q}$, and therefore, (4.80) is reduced to

$$\tilde{n}^2 \boldsymbol{E} = \boldsymbol{\kappa} \cdot \boldsymbol{E} - i\frac{\boldsymbol{\sigma} \cdot \boldsymbol{E}}{\omega\epsilon_0}, \tag{4.81}$$

Let us assume that the external magnetic field $\boldsymbol{B}_{ex}$ is applied in the z-direction, and the electromagnetic wave propagates in the same direction ($\hat{\boldsymbol{q}} \parallel \boldsymbol{B}$). This arrangement is called the Faraday configuration. When the circular radiation with electric field $E_x \pm iE_y$ propagates along the z-direction, the refractive index is obtained from (4.81) and (4.13) as,

$$\hat{n}_\pm^2 = (n_\pm - ik_\pm)^2 = \kappa - i\frac{\sigma_\pm}{\omega\epsilon_0} \tag{4.82}$$

$$= \kappa \left[1 - \frac{\omega_p^2}{\omega(\omega \mp \omega_c - i\frac{i}{\tau})} \right], \tag{4.83}$$

where ω_p is the plasma frequency defined as

$$\omega_p^2 = \frac{ne^2}{m^* \kappa \epsilon_0}. \tag{4.84}$$

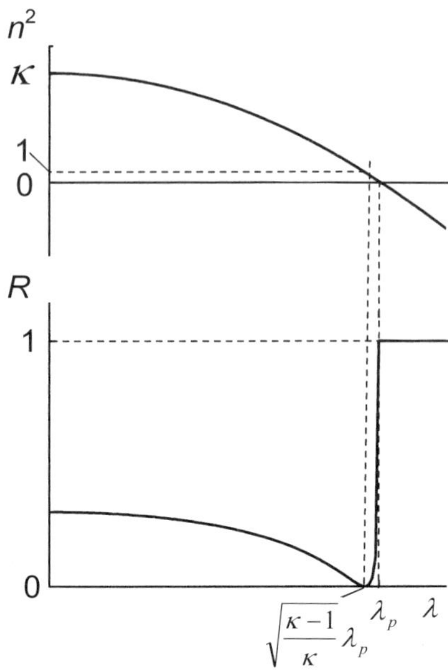

FIG. 4.56. Wavelength dependence of the real part of the dielectric constant n^2 and the reflectivity R.

Let us consider the case when the carrier scattering is negligibly small, and τ can be assumed as ∞. Then (4.83) is reduced to

$$\hat{n}_{\pm}^2 = \kappa\left[1 - \frac{\omega_p^2}{\omega(\omega \mp \omega_c)}\right]. \tag{4.85}$$

Equation (4.85) indicates that in the absence of an external field ($\boldsymbol{B} = 0$), $\omega_c = 0$ and $\hat{n}^2$ is negative for $\omega < \omega_p$, so that $\hat{n}$ is purely imaginary. The reflectivity R for the right incidence is

$$R = \frac{(n-1)^2 + k^2}{(n+1) + k^2}. \tag{4.86}$$

For imaginary $\tilde{n}$, n is zero so that $R = 1$. In other words, the radiation is totally reflected at the incident surface, and the radiation cannot penetrate the medium. For higher frequency (shorter wavelength) than ω, the reflection becomes less than 1, and some part of the radiation can penetrate into the medium through the surface. The wavelength dependence of the refractive index $\tilde{n}$ and the reflectivity is shown in Fig. 4.56. The longest wavelength below which the electromagnetic radiation penetrates the medium is called the plasma edge and is represented by

$$\lambda_p = \frac{2\pi c}{\omega_p} = 2\pi c\sqrt{\frac{m^*\kappa\epsilon_0}{ne^2}}. \tag{4.87}$$

The reflectivity spectrum shows a zero point at $[(\kappa - 1)/\kappa]^{1/2}$. The reflection spectrum as shown in Fig. 4.56 is a characteristic one for the vicinity of the plasma edge in semiconductors, which have high carrier concentration. As τ is finite in reality, the sharp edges at λ_p or $\sqrt{(\kappa-1)/\kappa}$ is broadened. When an external field $\boldsymbol{B}_{ex}$ is applied in the z direction, the plasma edge shifts to shorter and longer wavelength depending on the sense of the circular polarization according to (4.85). In relatively low magnetic fields, when $\hbar\omega_c \ll \hbar\omega_p$ and $\hbar\omega_c\tau \gg 1$, the plasma edge is obtained from the condition that $\tilde{n}_\pm^2$ becomes zero. That is

$$\omega \approx \omega_p \pm \frac{\hbar\omega_c}{2}. \tag{4.88}$$

The above equation implies that the plasma edge shifts to longer or shorter wavelength depending on the polarization and by measuring the shift of the plasma edge with magnetic field, we can obtain $\hbar\omega_c$ and the carrier effective mass [382]. The measurement of the plasma shift is an alternative means to determine effective mass when cyclotron resonance experiment is difficult.

According to (4.85), the magnetic field dependence of the reflectivity at a fixed $\hbar\omega$ should show a dramatic change when the $\hbar\omega_c$ crosses $\hbar\omega$ if $\hbar\omega < \hbar\omega_p$ for right circular polarization (electron-active mode, $\omega_c > 0$). In the range $\omega_c < \omega$, $\hat{n}^2$ is negative, so that the sample is totally reflective. When magnetic field is increased and ω_c exceeds ω, the sample becomes penetrative. In samples with high carrier concentration where the condition $\omega < \omega_p$ holds, such a large change is actually observed. For left circular polarization ($\omega_c < 0$), on the other hand, a plasma edge-like behavior similar to Fig. 4.56 should be observed in the reflection spectra. In usual semiconductor samples with small carrier concentration, $\omega > \omega_p$. Then the zero point of $\tilde{n}^2$ occurs at a positive ω and the total reflection should be observed only in a narrow ω_c range just below ω. In the actual cyclotron resonance experiments for $\omega > \omega_p$, however, this total reflection is obscured because of the existence of the τ term, except in very high mobility samples.

In highly conducting samples, the reflectivity is large and the effect of the magneto-plasma effect as mentioned above appears in the transmission spectra. Furthermore, the radiation travels in the sample back and forth by multiple reflection. In such a case, the expression of the transmission Eq. (4.9) should be replaced by one which takes account of the multiple reflection, and the line shape of the cyclotron resonance is distorted from simple Lorentzian shape.

So far, we have considered the Faraday configuration. In the Voigt configuration ($\hat{\boldsymbol{q}} \perp \boldsymbol{B}$), the characteristic mode is linearly polarized electromagnetic wave, and the corresponding refractive index is obtained as follows. There are two cases depending on the direction of the electric vector of the radiation relative to $\boldsymbol{B}$. For $\boldsymbol{E} \perp \boldsymbol{B}$,

$$\tilde{n}_\perp^2 = \kappa \left[1 - \frac{\omega_p^2(\omega_p^2 - \omega^2 + i\omega/\tau)}{\omega^2\{(\omega_p^2 - \omega^2 + i\omega/\tau)(1 - i\omega/\tau) + \omega_c^2\}} \right]. \tag{4.89}$$

For $\boldsymbol{E} \parallel \boldsymbol{B}$, on the other hand, there is no effect of magnetic field on $\tilde{n}$. Namely,

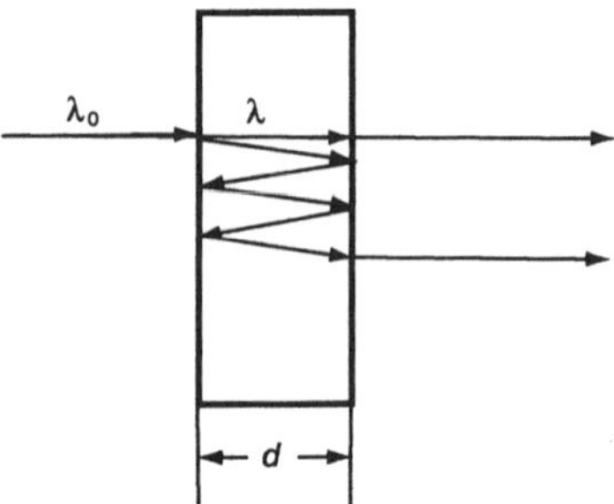

FIG. 4.57. Interference of the Helicon and Alfvén waves.

$$\tilde{n}_{\parallel} = \kappa \left[1 - \frac{\omega_p^2}{\omega(\omega - i/\tau)} \right]. \tag{4.90}$$

The difference in the refractive index for the two different linearly polarized radiation gives rise to the Voigt effect or Cotton-Motton effect.

4.13.2 *Helicon wave and Alfvén wave*

As shown in previous sections, in the case of the Faraday configuration, the right circularly polarized electromagnetic wave can propagate through the sample in the presence of electron plasma if the condition $\omega_c > \omega$ is satisfied, because $\tilde{n}^2 > 0$. This wave is regarded as the electromagnetic wave propagating together with the motion of electron. Such a wave is called a "helicon wave". The velocity of the wave is modified by the presence of carriers through the change in $\tilde{n}$. When the electromagnetic wave with a wavelength of λ_0 for which $\omega_c > \omega$ is incident on the sample, the wavelength is reduced to $\lambda = \lambda_0/n$ when it propagates in the sample as a helicon wave, so that λ varies as a function of magnetic field. The wavelength of the propagating wave is visible by an interference of the wave in the sample with the wave bypassing the sample, or an interference of the wave which suffers multiple reflection within the sample, as shown in Fig. 4.57. If we assume the sample thickness is d, in the former case, the transmitted wave intensity is strengthened by the interference in the resonance condition

$$\frac{2\pi}{\lambda} d - \frac{2\pi}{\lambda_0} d = 2N\pi \qquad (N \text{ is an integer}) \tag{4.91}$$

In the latter case, the resonance condition becomes

$$2d = N\lambda = N\frac{\lambda_0}{n}, \tag{4.92}$$

just like Fabry-Pérot interference. In both cases, therefore, the transmitted radiation intensity for a constant λ_0 shows an oscillation as a function of B.

In semimetals, such as graphite or Bi, which have almost equal number of electrons and holes, a different type of magneto-plasma wave called an Alfvén wave propagates. If we define the plasma frequencies of electrons and holes as ω_{pe} and ω_{ph}, and the cyclotron frequencies of those as ω_{ce} and ω_{ch}, respectively, we obtain

$$\tilde{n}^2 = \kappa\left[1 - \frac{\omega_{pe}^2}{\omega(\omega \mp \omega_{ce})} - \frac{\omega_{ph}^2}{\omega(\omega \pm \omega_{ch})}\right]. \tag{4.93}$$

Here again, τ is assumed to be infinitely large. When the high field conditions, $\omega_{ce} \gg \omega$ and $\omega_{ch} \gg \omega$ hold, $\tilde{n}_{\pm}^2$ becomes

$$\tilde{n}_{\pm}^2 \approx \kappa\left[1 \pm \frac{\omega_{pe}^2}{\omega\omega_{ce}}\left(1 \pm \frac{\omega}{\omega_{ce}} + \cdots\right) \mp \frac{\omega_{ph}^2}{\omega\omega_{ch}}\left(1 \pm \frac{\omega}{\omega_{ch}} + \cdots\right)\right] \tag{4.94}$$

$$\approx \kappa\left[1 \pm \frac{1}{\omega}\left(\frac{\omega_{pe}^2}{\omega_{ce}} - \frac{\omega_{ph}^2}{\omega_{ch}}\right) + \left(\frac{\omega_{pe}^2}{\omega_{ce}^2} + \frac{\omega_{ph}^2}{\omega_{ch}^2}\right) + \cdots\right]. \tag{4.95}$$

As

$$\frac{\omega_{pe}^2}{\omega_{ce}} = \frac{en_e}{\kappa\epsilon_0 B}, \qquad \frac{\omega_{pe}^2}{\omega_{ce}^2} = \frac{n_e m_e^*}{\kappa\epsilon_0 B^2}, \qquad \frac{\omega_{ph}^2}{\omega_{ch}} = \frac{en_h}{\kappa\epsilon_0 B}, \qquad \frac{\omega_{ph}^2}{\omega_{ch}^2} = \frac{n_h m_h^*}{\kappa\epsilon_0 B^2}, \tag{4.96}$$

we obtain

$$\tilde{n}^2 \approx \kappa\left[1 \pm \frac{e}{\kappa\epsilon_0\omega B}(n_e - n_h) + \frac{1}{\kappa\epsilon_0 B^2}(m_e^* n_e + m_h^* n_h)\right]. \tag{4.97}$$

In semimetals, $n_e \approx n_h$, so that

$$\tilde{n}^2 \approx \kappa\left[1 + \frac{\rho}{\kappa\epsilon_0 B^2}\right], \tag{4.98}$$

where ρ is a quantity expressed by

$$\rho = m_e^* n_e + m_h^* n_h, \tag{4.99}$$

and is called "mass density". Similarly to the helicon wave, the change of the wavelength in crystals by magnetic fields is visible in the interference pattern. When the Alfvén wave propagates through a crystal with a thickness of d, maxima are observed when a condition $2d = N\lambda = N\lambda_0/n$ (N is an integer) is fulfilled. In terms of the mass density, this condition is rewritten as

$$\rho_N = \frac{\epsilon_0 B_N^2 c^2}{\omega^2}\left[\left(\frac{N\pi}{d}\right)^2 - \left(\frac{\omega}{c}\right)^2 \kappa\right], \tag{4.100}$$

where ρ_N is the mass density corresponding to the N-th peak of the interference fringe at B_N.

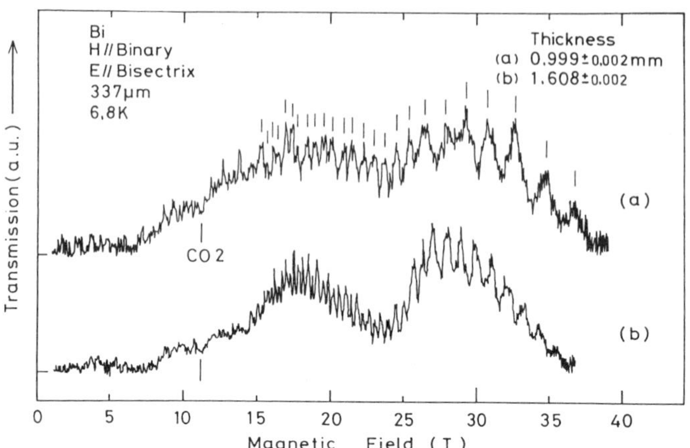

FIG. 4.58. Alfvén wave transmission spectra in Bi with two different thicknesses for magnetic field parallel to binary axis and electric field parallel to bisectrix axis [383]. L denote the Shubnikov-de Haas peaks of electrons with different quantum numbers, and CO denote that of combined resonance of holes. The interference peaks are indicated by vertical bars.

Figure 4.58 shows examples of the experimental traces of the Alfvén wave transmission spectra in Bi for $\boldsymbol{B} \parallel$ binary axis in two samples with different thicknesses [383]. The wavelength of the radiation was 337 μm. The oscillatory pattern with short-periods marked with vertical bars corresponds to the Alfvén wave interference fringes. The oscillation with a much longer period which appears as an envelope is due to the Shubnikov-de Haas effect of electrons ($L = 2$ and $L = 3$), and combined resonance of holes (CO2). When the effect of the carrier scattering is taken into account, $\tilde{n}$ contains a imaginary part which gives rise to the absorption of the radiation. The imaginary part of the refractive index is obtained as

$$k = \frac{2\pi}{\omega c} v_A R(B), \tag{4.101}$$

where v_A is the velocity of the Alfvén wave ($\sim c/\tilde{n}$) and $R(B)$ is the magneto-resistance. We have seen in (4.10) that the transmission of the radiation reflects the magneto-resistance, and the Shubnikov-de Haas oscillation should be observable in the transmission. This is the optically detected Shubnikov-de Haas effect. Combined resonance of holes is also observed through the change of the conductivity.

In a highly anisotropic crystal such as Bi, ρ should be treated as a tensor (ρ_{ij}), for magnetic and electric field polarization directions i and j. As the lattice part of the dielectric constant is also anisotropic, (4.100) should be written as

$$\rho_{ij} = \frac{B_N^2}{4\pi}\left[\left(\frac{N\pi}{\omega d}\right)^2 - \frac{1}{c^2}\epsilon_{jj}\right]. \tag{4.102}$$

For electric field parallel to the binary axis and electric field parallel to the bisectrix axis, the mass density should be

$$\rho_{xy} = N_A m_3 + (N_B + N_C)\frac{(m1 + 3m_2)m_3 - 3m_4^2}{m_1 + 3m_2} + N_h M_3, \tag{4.103}$$

where N_A, N_B, N_C and N_h are the carrier densities of electrons in the A, B and C pockets and holes, M_3 is the effective mass of holes, and m_1, m_2, m_3 and m_4 are the effective mass parameters of electrons. Figure 4.59 shows the mass density obtained from the interference pattern in Fig. 4.58 as a function of B. To plot the experimental points, an assumption was made that the lattice part of dielectric constant $\epsilon_\perp$ is constant. Although the experiment and theory are in good agreement with each other up to 15 T, the discrepancy is significant in higher fields. The discrepancy is considered to be due to the magnetic field dependence of the lattice part of the dielectric constant $\epsilon_\perp$, as $\epsilon_\perp$ excluding the

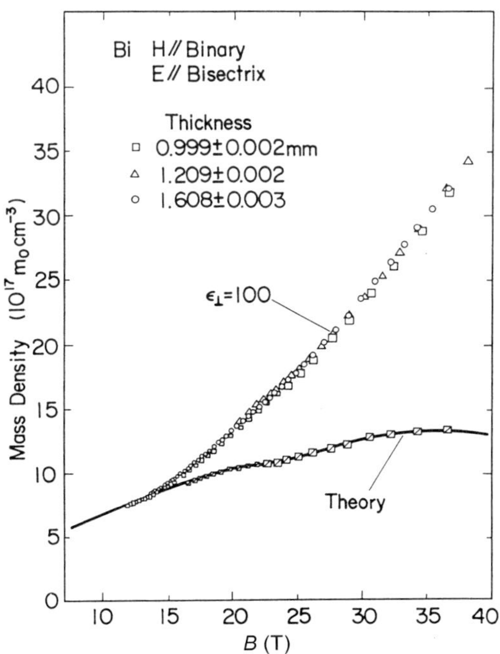

FIG. 4.59. Mass density ρ_{xy} as a function of B [383]. Experimental points were obtained from the interference peaks in Fig. 4.58. The theoretical line was calculated from (4.103).

free carrier contribution consists of the contribution of core electrons and that of the interband transitions. Although the core part should have only a small magnetic field dependence, the interband part would have a considerable magnetic field dependence. Actually, by fitting the experimental points in Fig 4.59, the magnetic field dependence of $\epsilon_\perp$ was obtained [383].

Besides the interference pattern, the plasma effect in the presence of both electrons and holes is also observed as a large structure in the reflection spectra at a field where $\tilde{n}$ crosses zero. The condition is

$$1 - \frac{\omega_{pe}^2}{\omega(\omega \mp \omega_{ce})} - \frac{\omega_{ph}^2}{\omega(\omega \pm \omega_{ch})} = 0. \tag{4.104}$$

This leads to a cubic equation

$$\omega^3 \pm (\omega_{ch} - \omega_{ce}\omega^2 - (\omega_{pe}^2 + \omega_{ph}^2 + \omega_{ce}\omega ch]\omega \pm (\omega_{ph}^2\omega_{ce} - \omega_{pe}^2\omega_{ch}) = 0. \tag{4.105}$$

The last term of the right hand side of (4.105) is usually zero. Then (4.105) is reduced to a quadratic equation. The positive roots of the equation for + and − sign are corresponding to electron-active and hole-active mode. They converge to

$$\omega_p^2 = \omega_{pe}^2 + \omega_{ph}^2 \tag{4.106}$$

at $B = 0$. However, when the effective masses involved in the cyclotron resonance frequency and the plasma frequency are different, namely,

$$m_{pe}^* \neq m_{ce}^*, \qquad m_{ph}^* \neq m_{ch}^*, \tag{4.107}$$

then a new mode appears. The frequency of the new mode becomes zero at $B = 0$, and linearly increases as B is increased. Nakamura, Kido, and Miura found that such a structure actually appears in graphite as shown in Fig. 4.60 [384]. Besides the structures arising from the transitions between the Landau levels, a large structure is observed in the high field range. This is considered to be due to the electron-hole magneto-plasma mentioned above. The field position of the structure is increased with increasing photon energy as expected. However, the photon energy *vs.* magnetic field relation is not linear in contradiction to the classical model. In graphite, four Landau levels remain below the Fermi level in the quantum limit at high magnetic fields as shown in Fig. 4.38 (a). In such a quantum limit, a classical approach is no longer adequate. Nakamura *et al.* explained the non-linear dependence by a quantum mechanical calculation taking account of the transitions between these Landau levels [385].

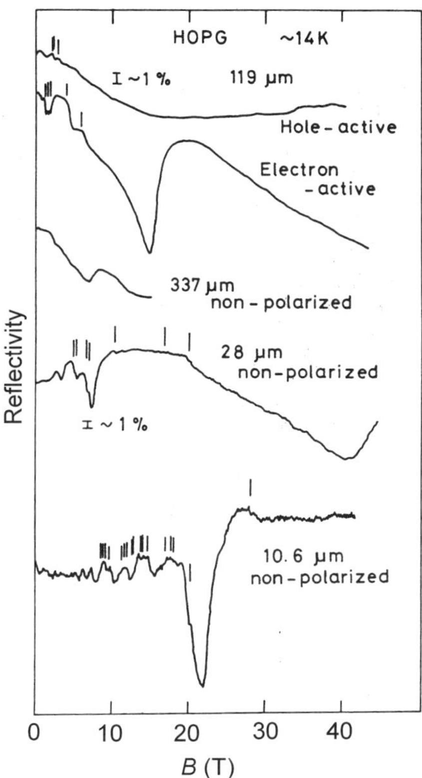

FIG. 4.60. Far-infrared magneto-reflectivity spectra in highly oriented pyrographite (HOPG) at different wavelengths in high magnetic fields [384]. $B \parallel c$. $T = 14$ K. Vertical bars indicate the field positions corresponding to the electronic transition between the Landau levels. A large structure is observed in the higher field side of the Landau level transitions due to the magneto-plasma effect.

5

MAGNETO-OPTICAL SPECTROSCOPY

5.1 Interband magneto-optical transition

5.1.1 *Density of states and the absorption coefficient*

When electromagnetic radiation is incident upon a semiconductor, the radiation is absorbed when the photon energy of the radiation exceeds the band gap, as electrons in the valence band are subjected to a transition to the conduction band. The absorption coefficient is expressed by the same formula as (4.21) based on the transition probability. In the present case, κ and ν denote the valence band and the conduction band, respectively. When the integral $\int u_{\kappa 0}^* \boldsymbol{r} u_{\nu 0} d\boldsymbol{r}$ remains non-zero, only the first term of (4.23) contributes to the absorption, and the transition is called "direct allowed transition". The absorption coefficient can be written as

$$\alpha = \frac{2\hbar\omega\mu^*}{ncE_0^2 V}\frac{2\pi}{\hbar}\sum_{1,2}|M|^2\delta(\hbar\omega - \mathcal{E}_2 + \mathcal{E}_1), \tag{5.1}$$

where M is the matrix element of the perturbation and is expressed as

$$M \approx \frac{eE_0}{m\omega}\frac{1}{\Omega}\int_{cell} u_{10}^* \left(\boldsymbol{p}\cdot\boldsymbol{E}\right) u_{20} d\boldsymbol{r}\int_{crystal} f_1^*(\boldsymbol{r}) f_2(\boldsymbol{r}) d\boldsymbol{r}. \tag{5.2}$$

When $B = 0$, α is obtained as

$$\alpha(B=0) = \frac{2\pi^2\hbar^3}{V}K\sum_{1,2}\delta(\hbar\omega - \mathcal{E}_2 + \mathcal{E}_1), \tag{5.3}$$

with

$$K = \frac{2\mu^* e^2 p_{12}^2}{\pi\hbar^3 ncm^2\omega}.$$

Here p_{12} is the matrix element of the dipole transition between the two bands, suffixes 1 and 2 denote the conduction band and the valence band, respectively, and we assume that the energy dependence of the matrix element is negligibly small. Using the property of the delta function

$$\delta\left(f(x)\right) = \sum_i \frac{1}{|f'(x_i)|}\delta(x - x_i),$$

and the relation

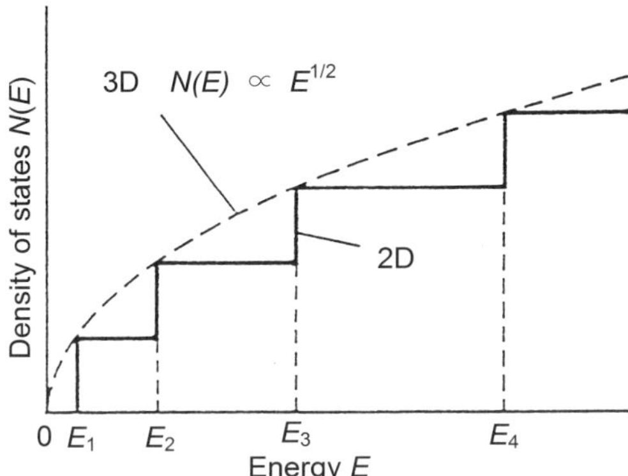

FIG. 5.1. Joint density of states of the three-dimensional (3D) and two-dimensional (2D) electron systems.

$$\hbar\omega - \mathcal{E}_2 + \mathcal{E}_1 = \hbar\omega - \left(\mathcal{E}_g + \frac{\hbar^2 k^2}{2m_1^*} + \frac{\hbar^2 k^2}{2m_2^*}\right),$$

we obtain for three-dimensional systems,

$$\sum_{1,2} \delta(\hbar\omega - \mathcal{E}_2 + \mathcal{E}_1) = \frac{V}{2\pi^2\hbar^3}(2\mu^*)^{3/2}(\hbar\omega - \mathcal{E}_g)^{1/2}. \tag{5.4}$$

Here $\mathcal{E}_g$ is the band gap, and μ^* is called the reduced mass represented by

$$\frac{1}{\mu^*} = \frac{1}{m_1^*} + \frac{1}{m_2^*}. \tag{5.5}$$

The quantity expressed by (5.4) is called the joint density of states. The absorption coefficient is obtained from (4.21) and (5.4);

$$\alpha(B = 0) = K(2\mu^*)^{3/2}(\hbar\omega - \mathcal{E}_g)^{1/2}. \tag{5.6}$$

In two-dimensional systems, the joint density of states has a step-wise form as in individual bands. The positions of the steps correspond to the interval between the subbands in the conduction bands and the valence bands;

$$\mathcal{E}(B = 0) = \mathcal{E}_g + \mathcal{E}_{cn} + \mathcal{E}_{vn'}, \tag{5.7}$$

where $\mathcal{E}_{cn}$ and $\mathcal{E}_{vn'}$ are the n-th and n'-th subband energies of the conduction band and the valence band, respectively. In general, the selection rule is such that $n = n'$ because of the wave function overlap. The joint density of states is depicted in Fig. 5.1 by a solid line. We can see that it is very different from that for the three-dimensional case which is proportional to $\sqrt{\mathcal{E}}$ as represented by a broken line. The absorption coefficient α is proportional to $N(\mathcal{E})$ and also a step-wise function. However, α is accompanied with peaks at each step due to the excitonic effects [386]. In quantum wells, as the well width is increased, the interval of the steps is decreased.

When the direct transition at $\boldsymbol{k}$ =0 is not allowed between the bands u_1 and u_2, weak absorption is still possible between the two bands, because the bands have some components of other bands at $\boldsymbol{k} \neq 0$ due to the band mixing. Such a transition is called "direct forbidden transition". The absorption coefficient in this case is

$$\alpha(B=0) = K'(2\mu^*)^{\frac{5}{2}}(\hbar\omega - \mathcal{E}_g)^{\frac{3}{2}}, \tag{5.8}$$

where

$$K' = \frac{2\mu^* e^2}{\pi n c\omega\hbar^3}|C|^2, \tag{5.9}$$

and C is the matrix element of the second order dipole transition.

When the conduction band minima and the top of the valence band are located at different points in the Brioullin zone with wave vectors $\boldsymbol{K}_c$ and $\boldsymbol{K}_v$, the direct transition between the conduction and the valence bands is not possible conserving the k-vector. The transition occurs with the assistance of absorption or emission of phonons with a wave vector $\boldsymbol{q}$ to satisfy the condition

$$\boldsymbol{K}_c = \boldsymbol{K}_v \pm \boldsymbol{q}. \tag{5.10}$$

The conservation of wave vector and energy at the transition holds involving the phonons. A transition of this type is called an "indirect transition". The absorption coefficient is given by

$$\alpha(B=0) = K_{\pm}(m_1^* m_2^*)^{\frac{3}{2}}(\hbar\omega \mp \hbar\omega_o - \mathcal{E}_g)^2, \tag{5.11}$$

where $K_{\pm}$ is the coefficient corresponding to the emission and absorption of phonons, and $\hbar\omega_o$ is the energy of the involved phonons. The indirect transition has a much smaller absorption intensity as it is the second order transition. Ge and Si are typical examples of the crystals which shows the indirect transition. The weak indirect transition is observed in the low energy tail of the stronger direct transition at the Γ point.

5.1.2 *Magneto-absorption spectra*

A. Direct allowed transition

When we apply magnetic fields, the optical spectra are modified due to the quantization of energy levels. The effect of magnetic fields on the optical spectra is called the magneto-optical effect. The magneto-optical effect is classified into two groups depending on the two different directions of the magnetic field relative to the propagation vector of the electromagnetic wave $\boldsymbol{q}$. When $\boldsymbol{B} \parallel \boldsymbol{q}$, the arrangement of the field is called Faraday configuration (or Faraday geometry). When $\boldsymbol{B} \perp \boldsymbol{q}$, it is the Voigt configuration (or Voigt geometry). In later sections, we consider first the Faraday configuration.

In a magnetic field, the motion of electrons in the plane perpendicular to the magnetic field is quantized to form the Landau levels. $\int_{crystal} f_1^*(\boldsymbol{r}) f_2(\boldsymbol{r}) d\boldsymbol{r}$ in (4.25) does not vanish only when $N = N$. The selection rule in the case of the direct allowed transition is

$$\Delta N = N - N' = 0. \tag{5.12}$$

The sum of the delta function in (5.4) is obtained as

$$\begin{aligned}
\sum_{N,k_y,k_z} \delta\ (\hbar\omega - \mathcal{E}_2(N,k_z) + \mathcal{E}_1(N,k_z)) &= \frac{2L_xL_y}{2\pi l^2} \sum_N \int \frac{L_z}{2\pi} dk_z \delta(\hbar\omega - \mathcal{E}_2 + \mathcal{E}_1) \\
&= \frac{V}{2\pi^2} \frac{1}{l^2} \sum_N \left| \frac{1}{\dfrac{d\left(\hbar\omega - (N+\frac{1}{2})\hbar\tilde{\omega}_c - \frac{\hbar^2 k_z^2}{2\mu^*} - \mathcal{E}_g\right)}{dk_z}} \right|_{\hbar\omega = \mathcal{E}_2 - \mathcal{E}_1} \\
&= \frac{V}{2\pi^2} \frac{eB}{\hbar^2} \frac{\mu^*}{\hbar^2} \sum_N \left| \frac{1}{k_z} \right|_{\hbar\omega = \mathcal{E}_2 - \mathcal{E}_1} \\
&= \frac{VeB}{2\pi^2\hbar^2} (2\mu^*)^{\frac{1}{2}} \sum_N \frac{1}{[\hbar\omega - \mathcal{E}_g - (N+\frac{1}{2})\hbar\omega_c]^{\frac{1}{2}}},
\end{aligned} \tag{5.13}$$

where V is the volume, and

$$\hbar\tilde{\omega}_c = \frac{eB}{\mu^*} = eB\left(\frac{1}{m_1^*} + \frac{1}{m_2^*}\right). \tag{5.14}$$

The absorption coefficient is thus obtained from (5.1),

$$\alpha(B) = K\hbar eB(2\mu^*)^{\frac{1}{2}} \sum_N \frac{1}{\hbar^{\frac{1}{2}}(\omega - \omega_N)^{\frac{1}{2}}}, \tag{5.15}$$

where

$$\hbar\omega_N = \mathcal{E}_g + (N + \frac{1}{2})\hbar\tilde{\omega}_c. \tag{5.16}$$

Equation (5.15) shows that the absorption spectrum becomes an oscillatory function with a diversion at $\omega = \omega_N$. In actual systems the diversion is suppressed because of the broadening of the Landau levels due to the carrier scattering. The broadening is taken into account by replacing the delta function with a Lorentzian function,

$$\int \frac{dk_z}{2\pi} \delta(\hbar\omega - \mathcal{E}_2 + \mathcal{E}_1) \longrightarrow \int \frac{dk_z}{2\pi} \frac{1}{\pi} \frac{\gamma}{(\hbar\omega - \mathcal{E}_2 + \mathcal{E}_1)^2 + \gamma^2}, \tag{5.17}$$

where γ denotes the width of each level. The diverging term in (5.15) is then

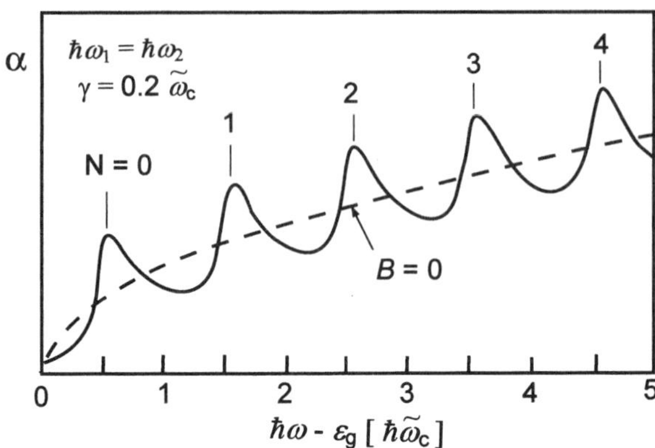

FIG. 5.2. Calculated magneto-absorption spectra for direct allowed transition

$$\frac{1}{(\omega-\omega_N)^{\frac{1}{2}}} \longrightarrow \left\{ \frac{\omega-\omega_N+[(\omega-\omega_N)^2+\gamma^2]^{\frac{1}{2}}}{2[(\omega-\omega_N)^2+\gamma^2]} \right\}^{\frac{1}{2}} . \tag{5.18}$$

This expression was first derived by Roth, Lax and Zwerdling [8]. An example of the calculated absorption spectrum is shown in Fig. 5.2. Reflecting the joint density of states, the spectrum shows an oscillatory behavior. The transition takes place between the Landau levels which have components of the harmonic oscillator function satisfying the selection rule (5.12). In actual semiconductors, the absorption spectrum is more complicated because of the degeneracy of the valence band as described in 2.4.3. Oscillatory magneto-absorption in Ge was studied extensively by Zwerdling *et al.* [387]. As an example of magneto-absorption spectra, we show in Fig. 5.3 (a) the experimental results for InSb obtained by Pidgeon and Brown [13]. We can see oscillatory behavior of the absorption with a number of transmission minima (absorption peaks). Figure 5.3 (b) shows relation between the photon energies of the absorption peaks and the magnetic field. From its appearance, such a diagram is called a Landau fan chart. According to (5.16), the lines in Fig. 5.3 (b) should be straight lines that have equal spacings at constant fields. However, the lines are not straight but bent downwards, and moreover, they are not equally spaced. The bending is due to the non-parabolicity. The non-equal spacing is because the Landau levels in the valence bands have a complicated structure due to the interaction between the heavy hole band and the light hole band.

B. Direct forbidden transition

We briefly consider the magneto-absorption spectra for the direct forbidden case and the indirect case, which are mentioned in the previous section. In the case of direct forbidden case, the dipole transition is not allowed at $\boldsymbol{k} = 0$,

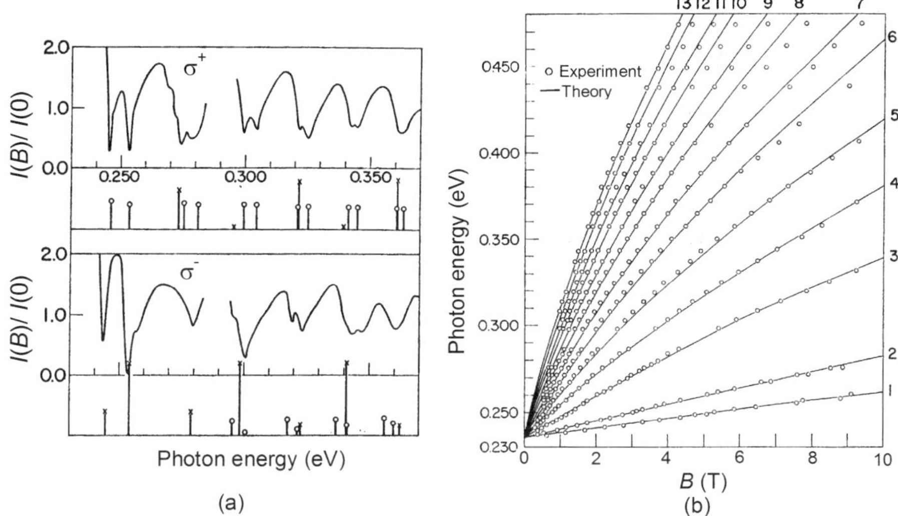

FIG. 5.3. (a) Magneto-absorption spectra in InSb for two different circular polarizations [13]. $I(B)$ denotes the transmitted light intensity through the sample. $B =$ 3.85T. $E \perp B \parallel < 100 >$. The vertical bars beneath the spectra show the theoretical positions and relatice strengths of the transition. (b) Plot of the photon energy of the absorption peak as a function of magnetic field.

but the transition becomes allowed at $\boldsymbol{k} \neq 0$, where the wave function of other bands are mixed together. At $B = 0$ the absorption coefficient is proportional to $(\hbar\omega - \mathcal{E}_g)^{3/2}$. For $B \neq 0$, it can be readily shown that the selection rule for the transition between the Landau levels of the conduction band and the valence band is

$$\Delta N = \pm 1, \tag{5.19}$$

instead of (5.12) for radiation polarized perpendicular to the magnetic field. The absorption coefficient is obtained as

$$\alpha(B) = K' \hbar e B (2\mu^*)^{\frac{3}{2}} \sum_{N,N'} \frac{1}{2}(\hbar\omega_{c1} + \hbar\omega_{c2}) \frac{N\delta_{N',N-1} + (N+1)\delta_{N',N+1}}{(\hbar\omega - \hbar\omega_{N,N'})^{\frac{1}{2}}}, \tag{5.20}$$

where

$$\hbar\omega_{N,N'} = \mathcal{E}_g + (N + \frac{1}{2})\hbar\omega_{c1} + (N + \frac{1}{2})\hbar\omega_{c2}. \tag{5.21}$$

An example of the calculated absorption spectra is shown in Fig. 5.4. As compared with the direct allowed case, the background absorption in the direct forbidden transition is different and the spacing between the peaks is different because of the different selection rules.

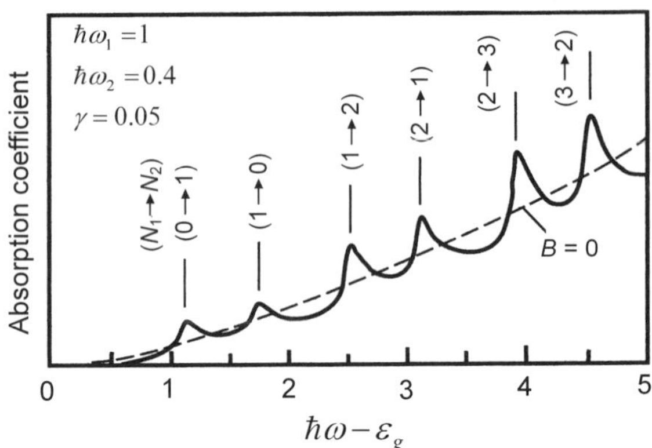

FIG. 5.4. Calculated magneto-absorption spectra for the direct forbidden case. The calculation was made using the parameters shown in the figure.

C. Indirect transition

In the case of indirect transition, the absorption coefficient is represented as

$$\alpha(B) = 2K_{\pm}(eB)^2(m_1^* m_2^*)^{\frac{1}{2}} \sum_{N,N'} F(\hbar\omega - \hbar\omega_{N,N'}), \tag{5.22}$$

where F denotes a unit step function and

$$\hbar\omega_{N,N'} = \mathcal{E}_g \pm \hbar\omega_o + (N + \frac{1}{2})\hbar\omega_{c1} + (N' + \frac{1}{2})\hbar\omega_{c2}. \tag{5.23}$$

The + and – signs in Eq. (5.23) correspond to the emission and absorption of the phonons, respectively. The characteristic point for the indirect transition is that the each transition between Landau levels appear as steps.

D. Inter-valence band transition

Another type of interband transition is the inter-valence band transition. The valence band of Ge, Si, and III-V semiconductors have split-off band Γ_7 as shown in Fig. 2.5. In p-type samples containing holes in the Γ_8 bands, the inter-valence band transition takes place between the Γ_8 bands and Γ_7 band. In Ge, an absorption band due to such a transition was observed [388–390]. The absorption intensity is proportional to the hole density, but it is rather small, because it is a forbidden transition. In Te, the band structure is completely different, but there is also a spin-orbit-split-off band H_5 below the valence band H_4, and there is a strong absorption band arising from the allowed transition between H_5 and H_4 for $E \parallel c$ [391]. Because of the camel's back structure of the valence band as was discussed in Section 3.5, the inter-valence band absorption

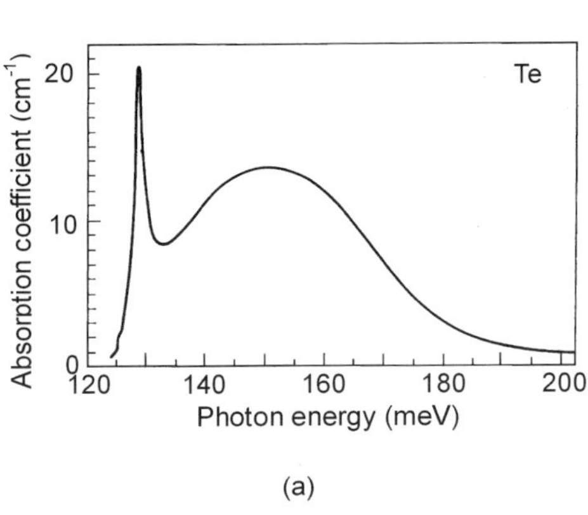

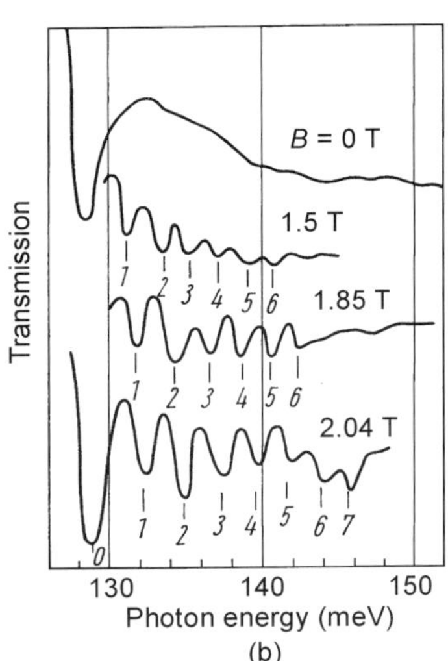

FIG. 5.5. (a) Inter-valence band absorption spectrum in p-type Te [171]. $T = 8$ K. $p = 5.2 \times 10^{15}$ cm^{-3}. (b) Magneto-absorption spectra for $E \parallel$ c at different magnetic fields. $B \parallel$bisectrix. $T = 1.7$ K. $p = 3.5 \times 10^{15}$ cm^{-3}.

shows a complicated shape as shown in Fig. 5.5 (a) [171]. A sharp peak at 128.6 meV is due to the transition at the H point (saddle point) and a broad peak around 150 meV is due to the transition to the valence band top. Figure 5.5 (b) shows the magneto-absorption spectra in various magnetic fields. An oscillatory structure with many peaks was observed. These peak positions were in good agreement with the theory of Nakao *et al.* [168, 392].

E. Two-dimensional systems in quantum wells

In two dimensional systems as in quantum wells, the interband transitions are allowed only between the subbands in the conduction and the valence bands with the same quantum number. When we apply magnetic fields perpendicular to the two-dimensional plane, the Landau levels are formed in each subband. Similarly to the case of three-dimensional systems, the absorption spectra show oscillatory features with the peak positions given by

$$\hbar\omega = \mathcal{E}_g + \mathcal{E}_e(n_s) + \mathcal{E}_h(n_s) + \left(N + \frac{1}{2}\right)\hbar e\frac{B}{\mu^*}, \tag{5.24}$$

where $\mathcal{E}_g$ is the band gap, $\mathcal{E}_e(n_s)$ and $\mathcal{E}_h(n_s)$ denote the n_s-th subband energies of electrons and holes, $N = N_e = N_h$ are the Landau quantum numbers, and μ^* is the reduced mass of electrons and holes. In the two-dimensional systems, the density of states of each Landau level has a delta-function-like shape, because

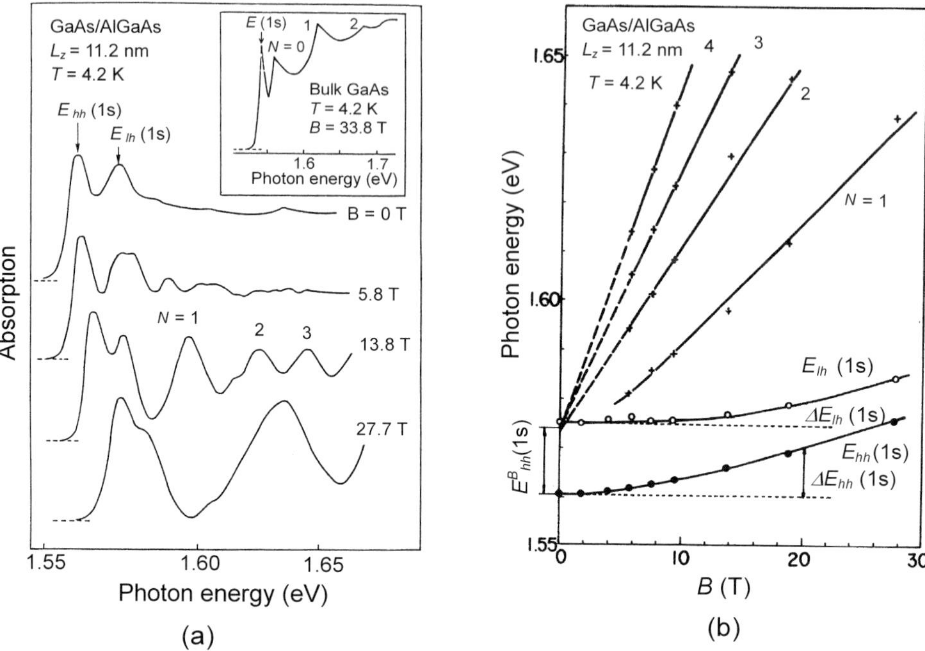

FIG. 5.6. (a) Magneto-absorption spectra in GaAs/AlGaAs quantum wells [393]. The inset shows the absorption spectra in bulk (3D) GaAs. (b) Photon energy in the magneto-absorption spectra in GaAs/AlAs quantum wells as a function of magnetic field.

there is no degree of freedom in the motion along the direction of magnetic field. Therefore, the peaks of the oscillatory absorption are sharper and without the tails in the high energy side.

There have been many studies of the magneto-absorption spectra in two-dimensional GaAs systems. Figure 5.6 (a) shows an example of the magneto-absorption spectra of GaAs/AlGaAs quantum wells observed in high magnetic fields [393]. At zero magnetic fields, there appear two exciton peaks corresponding to the two different exciton states formed with heavy holes ($\mathcal{E}_{hh}(1s)$) and light holes ($\mathcal{E}_{lh}(1s)$). The degeneracy of the subbands of the heavy holes and light holes is lifted by the quantum potential. The binding energy is also different between the heavy hole exciton and the light hole exciton. Therefore, they are observed as two distinct peaks. In the presence of magnetic fields, oscillatory absorption appears in the spectrum, and we can see distinct absorption peaks corresponding to the inter-Landau level transitions ($N = 1, 2, \cdots$ are the quantum numbers of the Landau levels). As the field is increased, the oscillation becomes more and more prominent and the interval between the peaks increases. It should be noted that each peak has a nearly symmetrical form with respect to the center on both sides of the peak. This is due to the fact that the Landau levels of the two-dimensional system have delta function-like joint density of states, in sharp contrast to the three-dimensional case whose absorption spectrum is shown in the inset. That is, in bulk crystals, the peaks show tails in the high field side, reflecting the 3D joint density of states of the Landau levels.

Figure 5.6 (b) shows the photon energy of the absorption peaks $\hbar\omega$ as a function of magnetic field. It is noticeable that the lowest line next to the exciton lines ($N = 1$ line) shows a deviation from a straight line. This is because for the lower lying lines, the excitonic effect is appreciable. For the higher lying lines ($N > 1$), the excitonic effect is not so large, and they are almost straight lines. Extrapolating the lines to the zero field, we find that they nearly converge to a single point. We can see from (5.24) that the photon energy of the converging point should be $\tilde{\mathcal{E}}_g = \mathcal{E}_g + \mathcal{E}_e(n_s) + \mathcal{E}_h(n_s)$, *i.e.* the gap between the subbands of the conduction band and the valence band (in the case of three-dimensional systems, $\tilde{\mathcal{E}}_g$ is just the band gap $\mathcal{E}_g$). In this manner, measurement of magneto-absorption spectra is a powerful means to obtain the band gap $\tilde{\mathcal{E}}_g$ in semiconductors. The difference in energy between $\tilde{\mathcal{E}}_g$ and the exciton peaks at zero field gives the exciton binding energies. Such a measurement can also be exploited to determine the exciton binding energy. We have to keep in mind, however, that the excitons in quantum wells are not exactly two-dimensional because of the finite well width. The well width dependence of the exciton binding energy will be discussed in Section 5.2.3.

5.2 Exciton spectra

5.2.1 *Exciton spectra in magnetic fields*

An exciton is a quasi-particle in which an electron and a hole are bound with each other by the Coulomb interaction. The energy states of excitons resemble those of a hydrogen atom. It is of interest to study excitonic states in magnetic fields in connection with the fundamental problem of a hydrogen atom. We have already seen in Section 2.8 a problem of the energy states of hydrogen atoms or shallow impurities as well as that of excitons in the presence of magnetic field. One difference of the exciton states from those of a hydrogen atom is that the positively charged particles (holes) have a comparable mass with that of the negatively charged particles (electrons), and they move around together in the crystal. For the magnetic field effects on exciton states, we can use arguments similar to those for impurity states as in Chapter 2, if we replace m^* with μ^*. As the dielectric constant κ is larger than 1 and the reduced mass μ^* is smaller than m, the Bohr radius is much larger than $a_B = 0.0529$ nm and the binding energy is much smaller than $Ry = 13.61$ eV in comparison to those of an hydrogen atom. For example, in bulk GaAs, they are $a_B^* = 10.2$ nm and $Ry^* = 4.2$ meV. Therefore, we can easily realize a condition that γ is larger than 1. In this section,

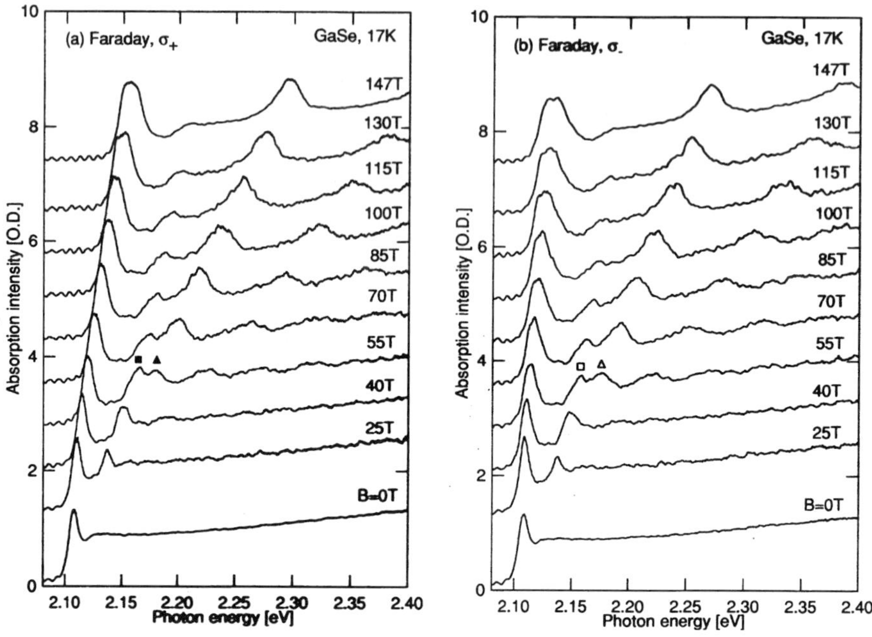

FIG. 5.7. Magneto-absorption spectra in GaSe [399]. (a) For polarization σ^+. (b) For polarization σ^-. Square and triangle points indicate the $2s$ and $3d_0$ states, respectively.

we will see various aspects of high magnetic fields on excitonic states.

In some crystals whose band gap is larger than ordinary III-V compounds, exciton effects are more prominently observed as a large structure due to a large binding energy. As we saw in Section 2.8, the exciton line should show the diamagnetic shift proportional to B^2 at relatively low magnetic fields for $\gamma < 1$. It tends to be proportional to B and parallel to the Landau levels for $\gamma > 1$. Such a transition of the feature of the diamagnetic shift in high magnetic fields sweeping B across the region $\gamma \approx 1$.

As a typical example of the magneto-absorption spectra of excitons over a wide range of γ, we show here experimental results of GaSe. GaSe has a layered crystal structure and is known to show a distinct exciton peak. As the triplet exciton is weakly allowed for $\boldsymbol{E} \perp$ c, the exciton peak is observed easily in the absorption spectra for a relatively thick sample. This is a convenient point in comparison to other substances where the absorption coefficient is very large so that the absorption spectra can be obtained only for a very thin sample. Therefore, it has been regarded as one of the standard materials in which we can study the problem of excitons in high magnetic fields. Many studies have

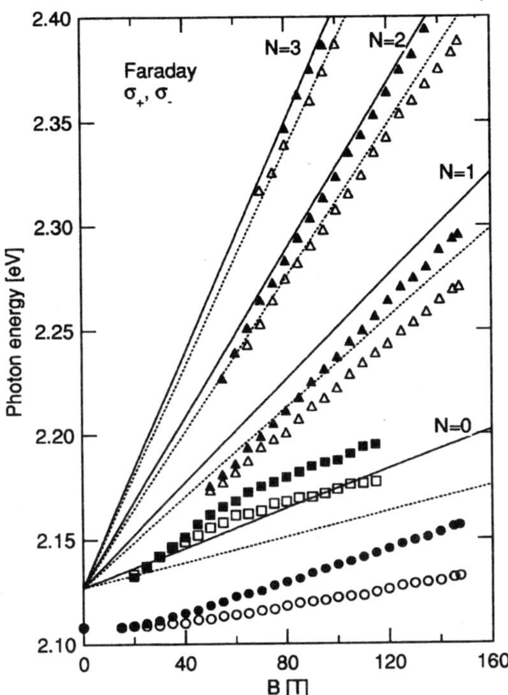

FIG. 5.8. Photon energy of the absorption peak as a function of magnetic field in GaSe [399]. Solid and open points denote the σ^+ and σ^- polarizations, respectively, and solid and dotted lines denote the Landau level transitions for σ^+ and σ^- polarizations, respectively, for the two different spin states.

been made on the magneto-absorption spectra of excitons in GaSe [394–398]. The condition $\gamma \sim 1$ is realized at $B \sim 50$ T. Figure 5.7 shows the magneto-absorption spectra in GaSe for two different circular polarizations [399]. We can see the 1s and 2s exciton lines at $B = 0$ T. In high field range, the $3d_0$ line is also visible, in addition to the higher lying levels. Figure 5.8 shows the photon energy of the absorption peaks as function of magnetic fields. The 1s line shows the B^2 dependence below $\sim$ 50 T, but it tends to show B dependence in higher fields. For $N \geq 2$, the peak positions almost coincide with the theoretically calculated Landau level transitions, but for $N = 0$ and 1, the experimental points are below the theoretical lines due to the large exciton effects. From the Zeeman splitting (difference between the 2 circular polarizations), the g-factor can be obtained. We will come back to the details of the excitons of GaSe in Section 5.6.1.

5.2.2 *Dielectric constant*

As shown in (2.207) and (2.216), the exciton parameters such as the binding energy and the diamagnetic shift coefficient σ are dependent on the dielectric constant $\epsilon = \kappa\epsilon_0$. However, the dielectric constant is very frequency dependent. In the low frequency limit, it is a static dielectric constant $\epsilon(0) = \kappa_0\epsilon_0$, to which both the core electron and the lattice deformation contribute. In the high frequency limit, the lattice part cannot contribute to the polarization, and it becomes the high frequency dielectric constant $\epsilon(\infty) = \kappa_\infty\epsilon_0$, which is smaller than $\epsilon(0)$. The change from $\epsilon(0)$ to $\epsilon(\infty)$ occurs at around the optical phonon frequency. Therefore, we have to be careful regarding the problem which dielectric constant should be responsible for the exciton binding. A detailed investigation of the dielectric constant to be used for the exciton binding was made by Haken [400, 77]. Generally, the frequency of the internal motion of the exciton is roughly

$$\omega_{ex} \sim \frac{\hbar}{\mu^* a_B^{*2}}, \tag{5.25}$$

where μ^* is the reduced mass of the exciton. If ω_{ex} is smaller than the optical phonon frequency ω_o, the static frequency is responsible for the exciton parameters. This condition is

$$a_B^* > \sqrt{\hbar/\mu^*\omega_o}. \tag{5.26}$$

For a smaller exciton radius, the dielectric constant $\epsilon(\infty)$ is more likely to be responsible for determining the exciton parameters. Actually, in typical Wannier-Mott excitons, which have large Bohr radii, we should use $\epsilon(0)$. However, in many cases for smaller radii, the diamagnetic shift is in between those expected from the two different dielectric constants.

Figure 5.9 shows the diamagnetic shifts of the ground state excitons in CdS [395] and HgI_2 [401] in high magnetic fields.

It should be noted that the experimentally observed diamagnetic shift is in between the theoretically expected curves assuming $\epsilon(0)$ and $\epsilon(\infty)$.

5.2.3 *Quasi-two-dimensional excitons in quantum wells*

Excitons in two-dimensional space provide an interesting playground for studying a fundamental problem of a hypothetical hydrogen atom in two-dimensional space. In two-dimensional space, the Bohr radius of the ground state is half of that in the three-dimensional case due to the difference in the degree of freedom of the electron motion. Therefore, the exciton binding energy is four times larger than in three-dimensional space. Namely,

$$\begin{aligned} a_B^*(2D) &= \frac{1}{2}\, a_B^*(3D), \\ \mathcal{E}_0(2D) &= 4\, \mathcal{E}_0(3D). \end{aligned} \tag{5.27}$$

It means that the two-dimensional excitons are more stable than the three-dimensional ones which are easily separated to electrons and holes at high temperatures. It can be expected that the two-dimensional exciton lines are visible up to room temperature, which is very rare in the three-dimensional case. In fact,

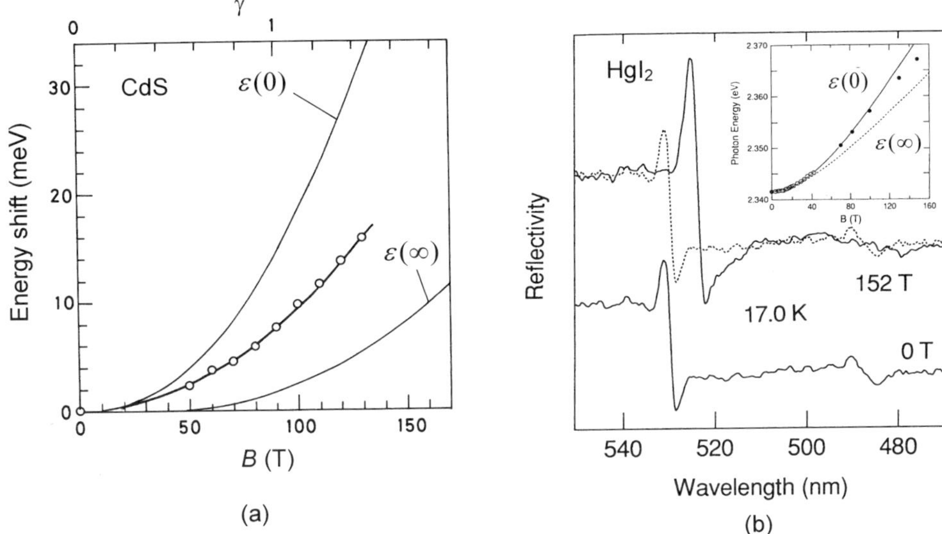

FIG. 5.9. Diamagnetic shifts of the ground state excitons. (a) Magneto-absorption data of A exciton in CdS [395]. $B \perp$ c, $E \perp$ c. The thin solid lines indicate the theoretical lines assuming $\epsilon(0)$ and $\epsilon(\infty)$. (b) Magneto-reflection data of HgI_2 [401]. The dotted line in the upper trace shows the zero field data (the same as the lower trace) for comparison. The inset depicts the calculated diamagnetic shift as compared with the experimental data.

the quasi-two-dimensional excitons in quantum wells persist at a room temperature. This is a desirable characteristic for fabricating electro-optical devices such as semiconductor lasers utilizing sharp exciton lines.

The problem of the magnetic field effect on two-dimensional excitons was first studied by Akimoto and Hasegawa [78] as discussed in Section 2.8.2. Although the calculation by Akimoto and Hasegawa was devoted to the analysis of excitons in GaSe, it is in fact more or less three-dimensional, and there has been no real system where the excitons have exactly two-dimensional character.

Excitons in quantum wells are confined in a quantum potential. The wave function of excitons is extended in the z-direction (perpendicular to the layers) within the well width, so that they are not two-dimensional excitons in a rigorous sense. However, we find many features of two-dimensional excitons for quantum wells with small well widths. As the well width L_w is increased, excitons tend to three-dimensional ones. Therefore we can see the cross-over between the two-dimensional to three-dimensional excitons by varying L_w. The cross-over should occur when L_w is close to the Bohr radius. The Bohr radius and the binding energy of the excitons in quantum wells are not so simple as (2.243) or (2.244), and should be calculated by taking account of the quantum potential. Putting

$\rho^2 = x^2 + y^2$ (x and y are the relative coordinates of electrons and holes in the quantum well layers), the Hamiltonian can written as

$$\mathcal{H} = -\frac{\hbar^2}{2\mu_\pm^*}\left(\frac{1}{\rho}\frac{\partial}{\partial\rho}\rho\frac{\partial}{\partial\rho} + \frac{1}{\rho^2}\frac{\partial^2}{\partial\psi^2}\right) - \frac{\hbar^2}{2m_e^*}\frac{\partial^2}{\partial z_e^2} - \frac{\hbar^2}{2m_\pm^*}\frac{\partial^2}{\partial z_h^2} - \frac{e^2}{4\pi\kappa\epsilon_0|\boldsymbol{r}_e - \boldsymbol{r}_h|} + V_z(z_e) + V_h(z_h), \tag{5.28}$$

where $V_z(z_e)$ and $V_h(z_h)$ are quantum well potentials for electrons and holes,

$$V_e(z_e) = \begin{cases} 0 & |z_e| < \frac{L}{2} \\ V_e & |z_e| > \frac{L}{2} \end{cases}$$
$$V_h(z_h) = \begin{cases} 0 & |z_h| < \frac{L}{2} \\ V_h & |z_h| > \frac{L}{2}, \end{cases} \tag{5.29}$$

and $\mu_\pm^*$ and $m_\pm^*$ are the effective masses of holes in the well layer ($x-y$ direction) and perpendicular direction (z-direction), given by

$$\frac{1}{\mu_\pm^*} = \frac{1}{m_e^*} + \frac{1}{m}(\gamma_1 \pm \gamma_2),$$
$$\frac{1}{m}(\gamma_1 \mp 2\gamma_2). \tag{5.30}$$

The + and − signs designate heavy and light holes, respectively. It should be noted that the heavy holes have larger mass than the light holes in the z-direction but smaller mass in the well layer. This is the characteristic point for the valence bands in zinc-blende crystals.

To obtain the eigenvalues and the eigenfunctions of the Hamiltonian (5.28), the variational method is often employed. The variational function is represented by

$$\Psi_n = f_e(z_e)f_h(z_h)g_n(\rho, z, \phi), \tag{5.31}$$

where f_e and f_h are the envelope functions in the z-direction, with z denoting the relative coordinate of electrons and holes perpendicular to the well layer. They satisfy the following equations.

$$\left[\frac{p_e^2}{2m_e} + V_e(z_e)\right] f_e(z_e) = \mathcal{E}_e f_e(z_e),$$
$$\left[\frac{p_h^2}{2m_h} + V_h(z_h)\right] f_h(z_h) = \mathcal{E}_h f_h(z_h), \tag{5.32}$$

and have the forms,

$$f_{e,h}(z_{e,h}) = \begin{cases} \cos k_{e,h} z_{e,h} & |z_e| < \dfrac{L}{2} \\ B_{e,h} \exp(-k_{e,h}|z_{e,h}|) & |z_e| > \dfrac{L}{2}. \end{cases} \tag{5.33}$$

Green and Bajaj took a variational function

$$g_{1s}(\rho, z, \phi) = (1 + \alpha z^2) \exp(-\delta^2 + z^2)^{1/2}, \tag{5.34}$$

and minimized the variational energy by choosing the optimum α and δ as variational parameters [402]. Shinozuka and Matsuura took a slightly different functional form [403]. Both calculations assume the hydrogen-atom-like function, with some anisotropy. The calculated results showed that the binding energy increases as the well width is reduced. For very small values of L_z, the binding energy decreases again with decreasing L_z. This is because the wave function

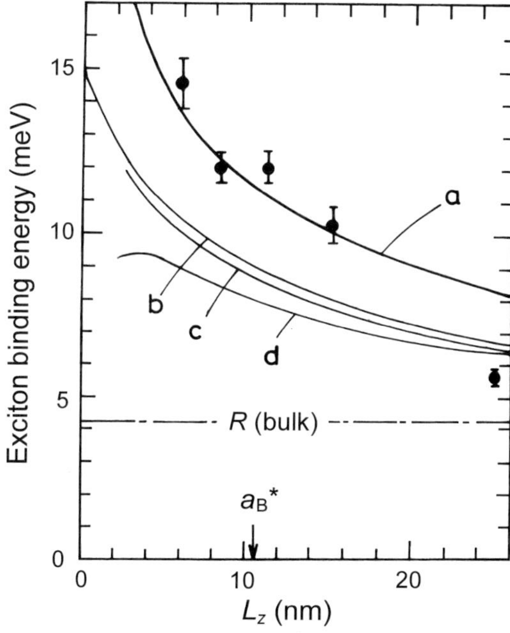

FIG. 5.10. Binding energy of the ground state (1s) of heavy hole excitons in GaAs/AlAs multiple quantum wells as a function of well width [393, 404]. Theoretical lines a–d are shown for comparison: (a). assuming an infinite quantum potential barrier height and the reduced mass $\mu^*_{hh} = 0.062m$, (b). assuming infinite barrier height and $\mu^*_{hh} = 0.040m$ [405], (c). calculated by Green and Bajaj assuming an infinite barrier height and $\mu^*_{hh} = 0.040m$, (d). calculated by Green and Bajaj assuming a finite barrier height for $Ga_{0.7}Al_{0.3}As$ and $\mu^*_{hh} = 0.040m$ [402]. a^*_B and R(bulk) are the effective Bohr radius and the binding energy of the three-dimensional excitons in bulk GaAs.

penetrates into the barrier layers and this effect becomes significant for very narrow quantum wells.

Figure 5.10 shows the experimentally obtained binding energy $\mathcal{E}_0$ of the heavy hole excitons in GaAs/AlAs quantum wells as a function of well width L_z. The exciton energies were obtained by a measurement as shown in Fig. 5.6 [393,404]. It can be seen that $\mathcal{E}_0$ is increased from the 3D value ($\sim$ 4meV) of GaAs for large L_z to the 2D value (4 times larger) as L_z is decreased. Theoretical curves calculated by Green and Bajaj [402], are also shown in the figure for comparison. Although the general tendency is in agreement with the experiment, there is a significant discrepancy between the experiment and the theory. This is probably due to the misestimation of the hole effective mass in the theory. Actually, the experimental data are better explained by a simple calculation assuming an infinite barrier height and the empirical reduced mass $\mu^*_{hh} = 0.062m$. Although the valence bands have complicated Landau level structure, the theory takes the mean mass of the heavy hole band.

The diamagnetic shift of the heavy hole exciton ΔE_d is shown in Fig. 5.11, for various well widths L_w. It is clearly seen that for $L_w = 5.8$ nm, ΔE_d almost follows the theoretical curve for the exactly two-dimensional exciton, but as L_w is increased it approaches the three-dimensional curve.

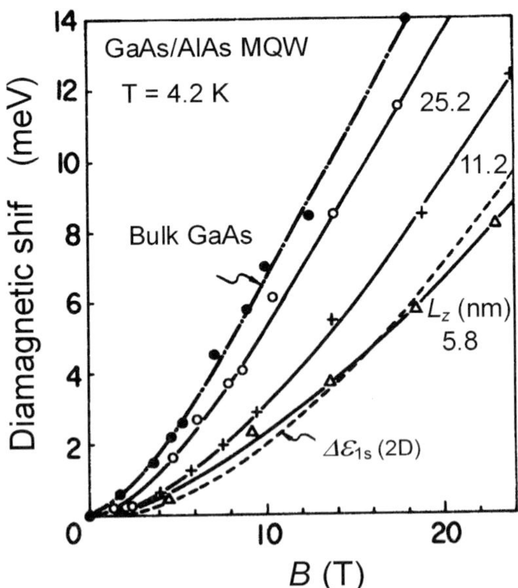

FIG. 5.11. Diamagnetic shift of the ground state (1s) of heavy hole excitons in GaAs/AlAs multiple quantum wells for different well widths L_z [393, 404]. Solid lines show guides for the eyes. The broken line stands for the theoretically expected line for a 2D exciton.

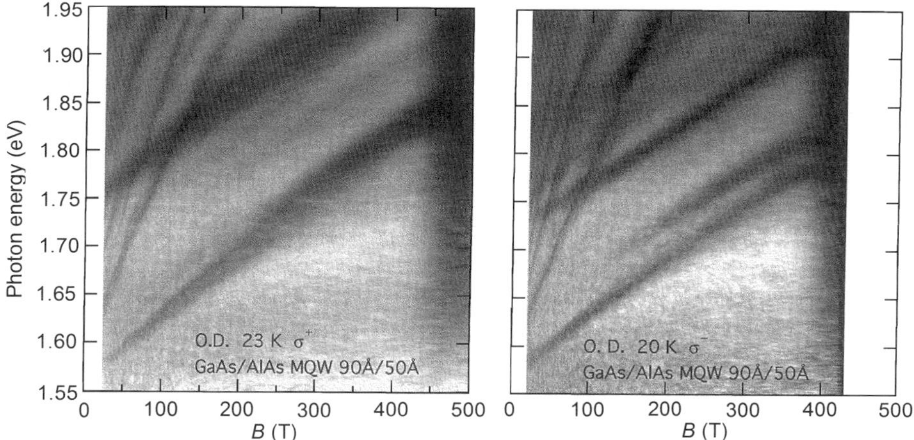

FIG. 5.12. Streak spectra of the magneto-absorption in a GaAs/AlAs quantum well for (a) σ^+ and (b) σ^- circular polarization in very high magnetic fields generated by electromagnetic flux compression [406]. The thickness of the GaAs and AlAs layers were 9.0 nm and 5.0 nm, respectively.

In the example described above, the effect of magnetic fields on 2D excitons is still in the weak field regime, even though a high magnetic field was necessary to resolve the Landau quantization. In higher magnetic fields, the field effects in the high field regime should be observed. Figure 5.12 shows streak spectra of the magneto-absorption in a GaAs/AlAs quantum well in very high magnetic fields up to 450 T generated by electromagnetic flux compression [406]. We can see complicated magnetic field dependence of the absorption line of the ground state exciton as well as the Landau level transitions, and those in the higher subbands. Some lines increased their intensity and width. There are also some lines which appeared only in the high field range above some field. Figure 5.13 shows the photon energy of the absorption peaks in Fig. 5.12 as a function of magnetic field. Theoretically calculated lines based on Ando's theory [24,25] are also shown in the figure. A good agreement between theory and experiment was obtained for low lying lines, apart from the exciton effect.

An interesting feature is the ground state exciton line. For the σ^+ radiation, it shows an anomaly shifting to a higher energy at around 250 T. This may be related to the 2D to 3D transition of excitons in high magnetic fields. As discussed in Section 2.8, in very high magnetic fields, the wave function extension of the ground state of electrons bound in a Coulomb potential (see Fig. 2.17). The effect is very prominent for $a_\perp$ in the direction parallel to the magnetic field. However, $a_\parallel$ in the plane perpendicular to the magnetic field is also decreased. The field dependence is $\sim \log(B/B_c^*)$ ($B_c^* = B_c^* = 2.35\times10^5$ T for a hydrogen atom and $B_c^* \approx 7.5$ T for excitons in GaAs) and much smaller than for $a_\perp$, but the decrease of $a_\parallel$ should become visible in very high magnetic fields. If $a_\parallel$

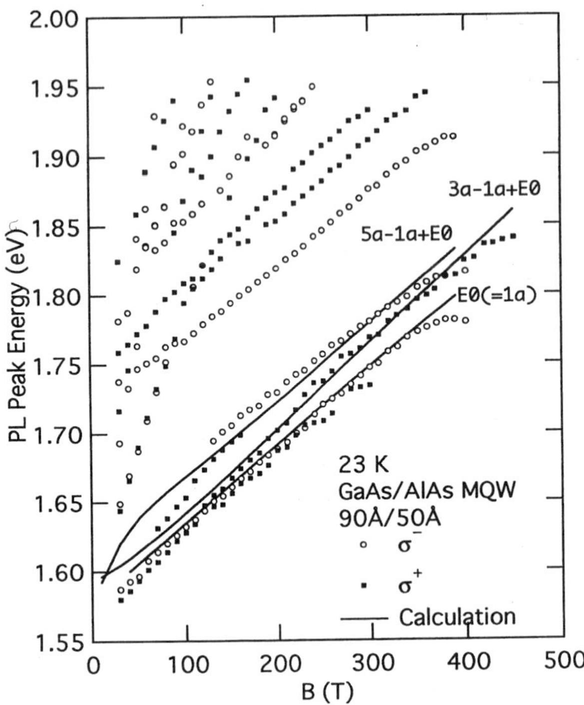

FIG. 5.13. Magnetic field dependence of the photon energy of the absorption peaks in Fig. 5.12 [406]. The solid lines are the theoretically calculated lines. e denotes the lowest Landau level of the conduction band and $1a$, $3a$, and $5a$ denote the Landau levels in the valence band.

becomes smaller than the quantum well width L_z, the exciton states would have a transition from the quasi-2D state to 3D state, as confinement effect by the quantum potential would diminish. The magnetic field at which the $a_{\|}$ becomes equal to the well width of 9.0 nm is exactly 250 T. At higher magnetic fields, the excitons would behave as 3D excitons. Recalling that 2D excitons have a larger binding energy and smaller diamagnetic shift than 3D excitons, we can expect a shift of the line to the higher energy and the increase of the diamagnetic shift. This is consistent with the experimental observation. The field-induced shrinkage of the wave function extension along the magnetic field direction is seldom discernible, but the result above might be a clear demonstration of such an effect.

The problem of the wave function shrinkage of excitons by extremely high magnetic fields is of interest in connection with the problem of the ground state of an hydrogen (H) atom in ultra-high magnetic fields [407–409]. As shown in 2.8, in the high field limit the wave function extension of an H atom becomes smaller and smaller as the field is increased, and in such a situation, H atoms form a solid crystal. We need an extremely high magnetic field to realize such a state. Some celestial objects possess huge magnetic fields. In neutron stars, it is considered that magnetic fields of 10^{8-11} T exist due to the extremely high density, while white dwarfs are considered to have magnetic fields of the order of 100 T. In neutron stars, hydrogen atoms are believed to exist over a surface layer a few centimeters thick. It has been predicted that hydrogen atoms subjected to such very high magnetic fields form a molecular chain H_n along the magnetic field, and in a higher field, these one-dimensional molecules further form a bcc crystal [407–409]. Although it is almost impossible to realize such a condition with respect to magnetic fields in hydrogen atoms, an equivalent condition may be possible to realize for excitons. Exploration of the magnetic field-induced solid crystal excitons is likely to become a future interesting target of the use of very high magnetic fields.

5.2.4 *Magneto-photoluminescence*

When we illuminate samples by radiation from a laser *etc.* and excite electrons from the valence band to the conduction band, light emission is observed as the electrons and holes recombine. Such a process is called radiative recombination, and the radiation created by such a process is called photoluminescence (PL). The electrons and holes relax to low energy states close to the ground state before the recombination. Therefore, the luminescence spectra provide information about these states. We can observe a variety of states, such as energy band extrema, free excitons, bound excitons associated with impurities and defects, and charged excitons. When the light intensity is intense, exciton molecules are also observed. To analyze the photoluminescence (PL) spectra, we have to take care to identify the origin of the light emission. The PL measurement is generally easier than the absorption measurement, because we need not worry about the shape of the samples or transparency of the substrates. When the substrate is opaque, we have to remove it by etching for absorption measurements. It is especially useful for measurement in quantum wires and dots. In PL spectra, peaks are observed at photon energies corresponding to the low lying energy states. Even for free excitons, peaks usually appear at a lower energy than the absorption peak. This is because the electrons and holes fall to the low energy states if there is any inhomogeneity in the samples. The difference in energy between the PL and the absorption is called Stokes shift. The amount of the Stokes shift, as well as the line-width is a measure of the quality of the crystal.

For measuring such PL spectra, we fix the photon energy of the excitation (usually at an energy higher than the energy range of interest) and measure the luminescence spectra. There is also a technique for measuring the intensity of the luminescence at a fixed photon energy by varying the photon energy of the excitation. The spectra are obtained as a function of the excitation photon energy. Such spectra are called PL excitation spectra. As the luminescence intensity is nearly proportional to the absorption of the incident radiation, the excitation spectra are usually very close to the absorption spectra. Therefore, PL excitation spectra are measured when the direct absorption measurement is difficult. This technique is often employed in steady fields. However, it is not easy to measure the excitation spectra in pulsed fields, because the time is too short to sweep the exciton wavelength.

5.3 Magneto-optics of excitons in doped quantum wells

5.3.1 *Magneto-optical spectra in doped quantum wells*

In modulation doped quantum wells, where shallow impurities are doped only in barrier layers, high mobility is realized because there is no impurity scattering within the well layers. The carriers are supplied from the impurities doped in the barrier layers, so that we can create an ideal two-dimensional metallic state. The optical process in metallic systems is different from that in ordinary semiconductors. In n-type quantum wells, electrons fully occupy states up to the Fermi energy $\mathcal{E}_F$, as shown in Fig. 5.14. Electronic transition by absorbing photons is not allowed to the conduction band states below $\mathcal{E}_F$. Therefore, the transition occurs only for photon energies larger than $\mathcal{E}_g + \mathcal{E}_F$. It looks as if the band gap shifts to higher energy by $\mathcal{E}_F$ due to the presence of carriers. This shift is called the Burstein-Moss shift [410]. Furthermore, the carrier-carrier interaction decreases the self energy of carriers, so that the band gap in doped samples

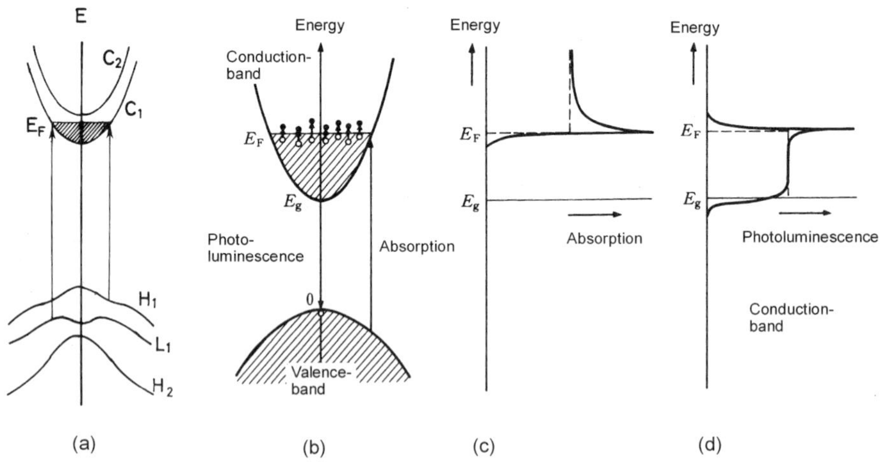

FIG. 5.14. (a) Transition of electrons between the conduction band and the valence band absorbing photons in doped quantum wells. (b) Diagram showing the perturbation of the surface of the Fermi sea by the excitation of an electron. (c) two-dimensional density of states in a doped quantum well. The absorption is allowed only to the states above the Fermi level. (d) The density of states considering the band gap renormalization, the absorption spectra and the photoluminescence spectra.

is smaller than in undoped samples. This effect is called band gap renormalization [411]. As regards excitons, the carriers screen the Coulomb interaction between electrons and holes, so that the binding energy becomes smaller as the carrier density increases [412].

Mahan predicted the existence of a kind of excitonic state even in a metallic state [413]. The excitonic state in metals is called the Mahan exciton. As shown in Fig. 5.14 (b), when a hole is created in the valence band by absorbing a photon which excites an electron to the conduction band, electrons in the vicinity of the Fermi level are perturbed by the Coulomb potential of the hole, and many small energy excitations are caused in the Fermi sea. Such a many body interaction gives rise to a sharp peak in the absorption spectra or the PL spectra at a photon energy corresponding to the Fermi energy as shown in Fig. 5.14 (d). This anomalous peak is called the Fermi edge singularity [413–416]. Such anomalies have been observed in the optical spectra of some metals in the X-ray range. Doped quantum wells are an interesting playground to study the Fermi edge singularity. As the Fermi energy in semiconductor quantum wells is much smaller than in normal metals, we can expect to observe a prominent magnetic field effect.

Optical spectra in p-type modulation doped quantum wells have been studied by Miller and Kleinman [417]. In the absorption spectra of p-type GaAs/AlGaAs quantum wells at low temperatures, the heavy hole exciton peak diminishes and only the light hole exciton is visible. This is because the heavy hole subband edge is occupied by holes and the optical transition from the band edge is not possible. This effect is called the phase space occupation effect. As the temperature is increased, the heavy hole exciton peak appears, because some electrons are thermally excited and occupy the band edge state.

Figure 5.15 shows the spectra in p-type GaAs/AlGaAs quantum wells measured by Iwasa, Lee and Miura [418]. We can see in Fig. 5.15 (a) that in comparison to an undoped sample, which shows both the heavy hole exciton and light hole exciton peaks, the heavy hole exciton is missing in the doped sample at zero field. However, as the magnetic field is increased, it appears and grows. This is because the density of states of the Landau levels is increased with increasing field, creating more electrons as well as holes in the lowest Landau level. In addition, the exciton binding energy is increased with fields, which also contributes to the intensity of the heavy hole exciton. Figure 5.15 (b) shows the temperature dependence of the spectra at zero field. The heavy hole exciton peak appears and grows as the temperature is increased.

Concerning the field-induced appearance of the heavy hole exciton peak, Iwasa *et al.* found that the effect of the density of states with the field gives the dominant contribution. The intensity of the heavy hole exciton is shown in Fig. 5.16 as a function of magnetic field [418]. Above a field at which the system enters the quantum limit B_{QL}, some electrons are populated in the lowest Landau level and we can expect the appearance of the exciton peak. It was found, however, that the field where the heavy hole exciton peak starts showing up

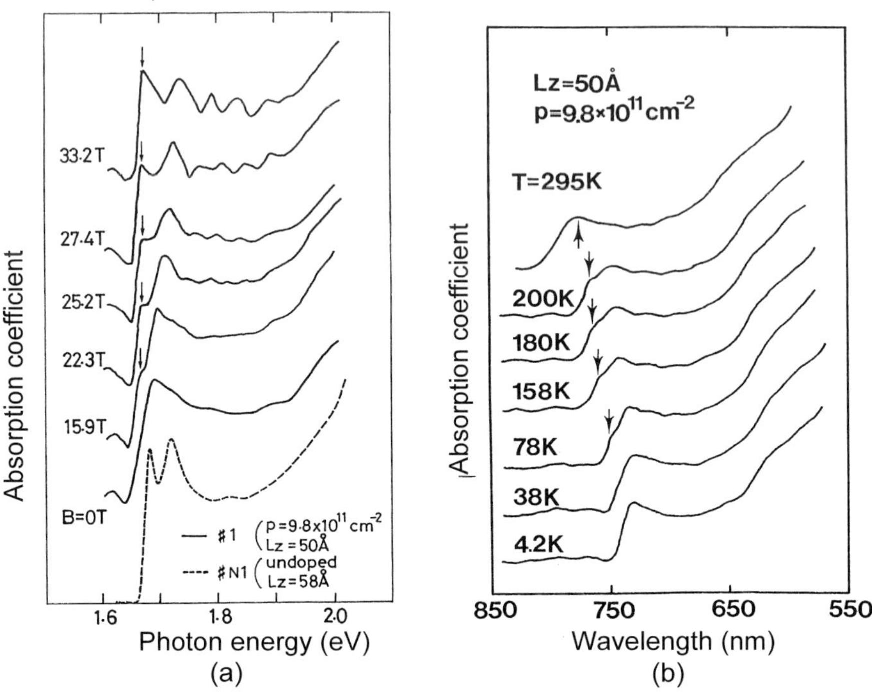

FIG. 5.15. (a) Magneto-absorption spectra in p-doped GaAs/AlGaAs quantum wells. $T = 4.2$ K. The bottom graph is for an undoped sample. (b) Exciton absorption spectra in the same sample at different temperatures.

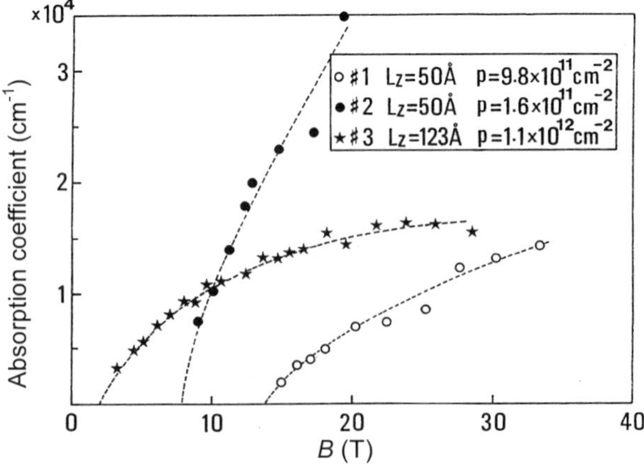

FIG. 5.16. Absorption intensity of the heavy hole exciton line as a function of magnetic field for various p-doped GaAs/AlGaAs quantum wells [418].

B_{th} is lower than B_{QL}. For example, in sample 1, B_{QL} is estimated to be 40 T, whereas B_{th} is about 14 T as seen in the figure. Furthermore, contrary to a naive expectation, B_{th} is higher for the relatively small L_z, where the excitons are considered to be more stable than in samples with larger L_z. These phenomena probably arise from the nature of the excitons in high magnetic fields; the excitons are formed not only from the lowest subband, but also from higher lying subband including light hole ones, and the interaction with the light hole subbands may play an important role.

5.3.2 *Fermi edge singularity*

For n-type samples, Pinczuk *et al.* found that the onset of the absorption spectrum in GaAs/AlGaAs heterostructures was at an energy higher than the photoluminescence (PL) peak by $\mathcal{E}_F(1+m_e^*/m_h^*)$, corresponding to the Moss-Burstein shift. Furthermore, a sharp peak was observed at the onset of the absorption at low temperatures. The peak is considered to be due to the Fermi edge singularity (FES) [419]. Skolnick *et al.* found a prominent structure typical to the FES in the PL spectrum of InGaAs/InP heterostructures [420]. For the GaAs/AlGaAs system, Lee, Iwasa, and Miura observed the FES effects in both the absorption and the PL spectra as shown in Fig. 5.17 [421]. The spectra show theoretically expected features as shown in Fig. 5.14. The PL spectrum consists of a nearly rectangular shape, reflecting the joint density of states, and the width of the luminescence band is approximately equal to the Fermi energy $\mathcal{E}_F$. A distinct enhancement of the luminescence intensity was observed at the high energy edge. The absorption spectrum exhibits a sharp peak at nearly the same energy. As the energy corresponds to the transition at the Fermi level, the peaks are considered to be due to the Fermi edge singularity. The peaks disappear quickly as the temperature is increased, in contrast to the exciton peak which persists even

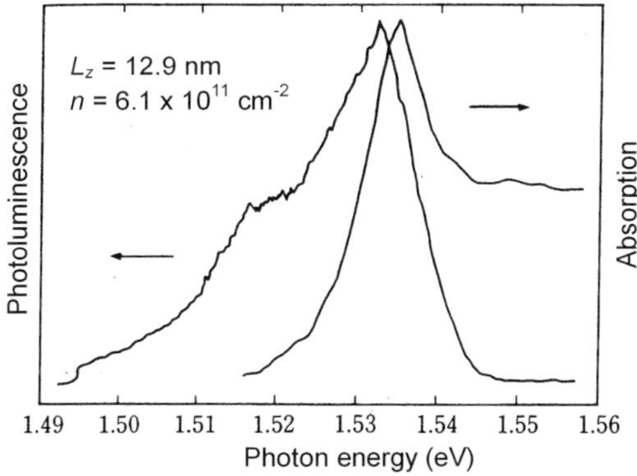

FIG. 5.17. Absorption and photoluminescence spectra in an n-type modulation-doped quantum well of GaAs/AlGaAs [421]. $T = 4.2$ K.

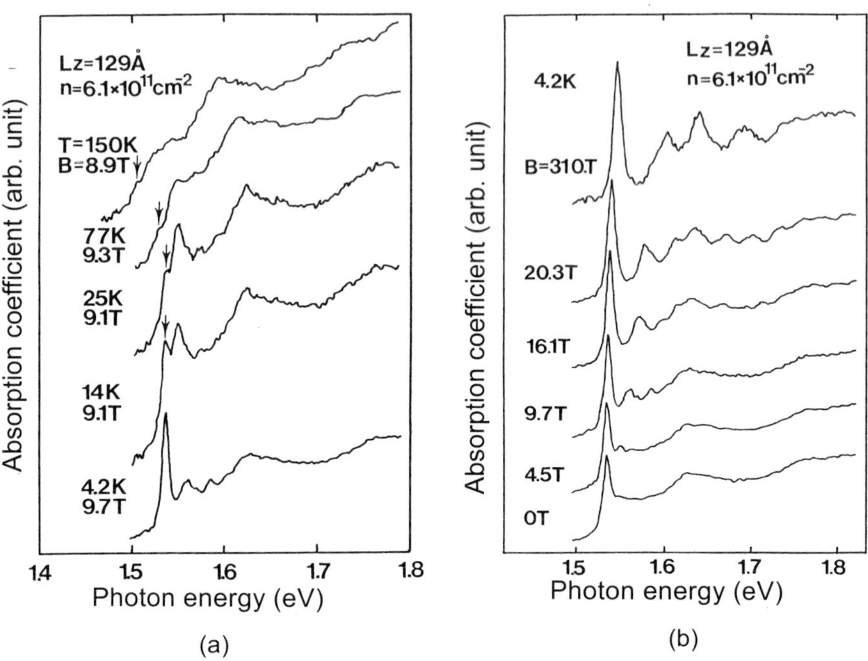

FIG. 5.18. Magneto-absorption spectra in n-doped quantum well of GaAs/AlGaAs for magnetic fields applied perpendicular to the layers [422]. $L_z = 12.9$ nm, $n = 6.1\times10^{11}$ cm^{-2}. (a) Magneto-absorption spectra at around $B \sim 9$ T at different temperatures. The sharp peak at $T = 4.2$ K and the peak denoted by arrow for high temperatures is the Fermi edge singularity. (b) Magneto-absorption spectra at different magnetic fields.

at rather high temperatures.

Properties of the FES in high magnetic fields are extensively investigated by Lee, Miura, and Ando [422]. Figure 5.18 shows the magneto-absorption spectra involving the FES in high magnetic fields. At low temperatures, the spectra show a sharp peak due to the FES as well as the inter-subband transitions at higher energies. When a magnetic field is applied perpendicularly to the layers, oscillatory absorption peaks are observed due to the inter-Landau level transitions from the levels lying below the Fermi level. As the temperature is increased at $B \sim 9$ T, the FES peak diminishes as expected, although the ordinary exciton peak in un-doped samples can usually be observed at these high temperatures. When we apply higher magnetic fields, however, a peak reappears even at a high temperature of 77 K. The peak at high temperatures and high magnetic fields is considered to be the exciton line which is stabilized by magnetic field. As the magnetic field is increased at a low temperature, the FES peak grows as shown

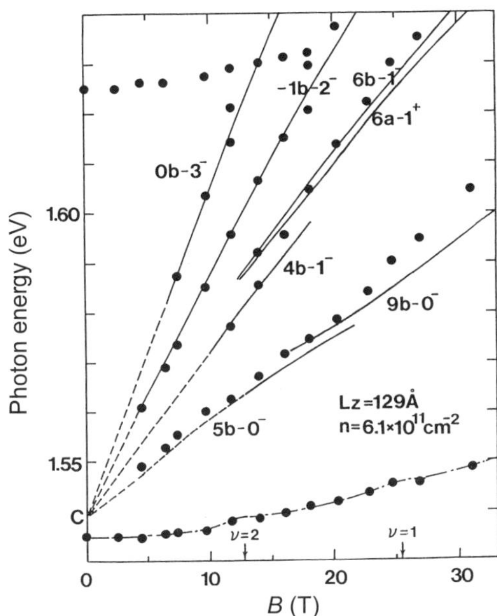

FIG. 5.19. Photon energies of the absorption peaks as a function of magnetic field in n-doped quantum well of GaAs/AlGaAs [422]. The sample is the same as for Fig. 5.18. Solid lines are the calculated lines by the theory of Ando for the inter-Landau level transitions.

in Fig. 5.18 (b). In high field limit, it is regarded as the exciton line. The FES peak is considered to change to the exciton line continuously.

Figure 5.19 shows the photon energy of the peak positions as a function of magnetic field for the same quantum wells. The peak of the Fermi edge singularity shows a diamagnetic shift similar to that of excitons. The lines corresponding to the inter-Landau level transitions converge to a single point C at zero field. This point is about 4 meV above the FES peak. It is an interesting point that there seems to be a binding energy for the FES state. Moreover, the FES level shows a B^2-like diamagnetic shift in the low field region just like ordinary excitons. It is also noticeable that the FES line exhibits a structure at integer fillings $\nu = 1$ and 2. A similar structure at integer fillings is also reported by Perry *et al.* [423]. As for the inter-Landau level transition lines, the experimental points are well explained by the theory of Ando, as shown in the solid lines.

5.3.3 *Anomalies at integer and fractional filling of the Landau levels*

Optical study of the integer and fractional quantum Hall effect in doped quantum wells or heterostructures has long been of great interest to many authors [424–427]. Buhmann *et al.* found in GaAs/AlGaAs heterojunctions that the magnetic field dependence of the photoluminescence peak energy shows a step-wise vari-

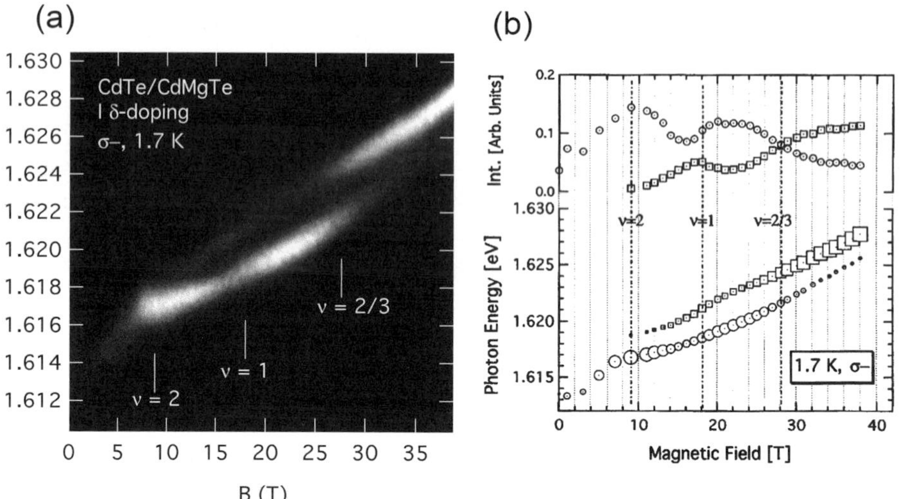

FIG. 5.20. (a) Magneto-photoluminescence spectra of excitons in CdTe/$Cd_{1-x}Mg_xTe$ ($x = 0.135$) [428, 429]. The sample is an n-doped quantum well with carrier concentration of 4.4–4.8 $\times$ 10^{11} cm^{-2}, mobility of 82,000 cm^2/Vs and well width of 10 nm. (b) The peak shift and the intensity variation of the upper (square points) and the lower (circle points) PL peaks for σ^- polarization. The size of the photon energy points indicates the intensity of the line.

ation at the filling $\nu = 2/3, 1/3, 4/5, 3/5, 2/5, 1/5, 1/7, 1/9$ [426]. The same authors also found that a new line grows up in the lower photon energy (by ~1.4 meV) and the intensity of the new line relative to the original line shows minima at $\nu = 1/5, 1/7, 1/9$ [427]. They associated the phenomenon with a pinned Wigner solid.

Here, we show a result obtained for a high mobility II-VI quantum well sample. Figure 5.20 shows photoluminescence spectra from CdTe/CdMgTe quantum wells in pulsed high fields up to 40 T [428, 429]. The photoluminescence peak exhibits a splitting of about 2.4 meV at $B > 15$ T. As shown in Fig. 5.20 (b), the photon energy of the two peaks has some structures and their intensity shows abrupt changes at integer ($\nu = 1, 2$) and fractional ($\nu = 2/3$) fillings. At higher fields for $\nu < 2/3$ the higher field peak becomes predominant. Similar spectral anomalies have observed in GaAs/AlGaAs quantum wells by Heiman *et al.* [430]. The relatice intensity of the doublet structure showed an activation type temperature dependence above $T > 8$ K.

5.4 Magneto-optical spectra of short-period superlattices

5.4.1 $\Gamma - X$ *cross-over for* $\boldsymbol{B} \perp$ *layers*

When the period of the multiple-quantum wells is sufficiently short, wave functions penetrating to the barrier layers are connected with each other and a mini-band is formed with a dispersion perpendicular to the well layers (in the k_z-direction). Therefore, conduction across the barrier layer becomes possible. Such short-period superlattices are regarded as a new kind of anisotropic three-dimensional system rather than a two-dimensional system. The band structure of short-period superlattices is complicated due to band folding in the Brillouin zone. When the magnetic field is applied in the perpendicular direction to the layers, Landau levels are formed for motion perpendicular to the layers.

As discussed in Section 4.12.2, in GaAs/AlAs short-period superlattiecs, the conduction band minima at the X points in the AlAs barrier layer are relatively low. As the well width becomes small, the conduction band minimum at the Γ point is raised, so that in sufficiently short-period superlattices, the lowest conduction minima will be located at the X point in the AlAs layer. As the valence band maximum is at the Γ point, such a band alignment results in the indirect band gap and the type II configuration as shown in Fig. 4.48 (b). It is contrary to the case of long period superlattices where the band line-up is type I with a direct gap at the Γ point in the GaAs layer. The type I to type II transition takes place at the monolayer number of about 8 of the GaAs layer. In the type II regime, the optical transition probability is much smaller than the type I case because the electrons and holes are located mainly in the GaAs layer and the AlAs layer, respectively, so that the transition is indirect both in the real space and the $\boldsymbol{k}$ -space. Therefore the absorption and photoluminescence (PL) should be much smaller.

There are two kinds of X points, X_{xy} and X_z: X_z is located in the direction of the accumulation of the layers and X_{xy} is in the perpendicular direction. The X_z band is lower in energy when the well width L_z is large, since the effective mass in the z-direction (accumulation direction) is larger than in the X_{xy} band. However, as L_z is decreased, the band widths of the both bands are increased, and the lowest energy of the X_{xy} band may become lower than that of the X_z band. The position of the lowest conduction band is thus a subtle problem. As the X_z band is folded back in the Brioullin zone, the band structure of the short-period superlattices is very complicated.

Magneto-optical measurements are useful for investigating the character of the mini-bands. Figure 5.21 shows the magneto-photoluminescence spectra in $(GaAs)_5(AlAs)_6$ for magnetic fields parallel ($\boldsymbol{B}_{\parallel}$) and perpendicular ($\boldsymbol{B}_{\perp}$) to the quantum well layers [431]. The PL peak due to the exciton transition shows a large diamagnetic shift for $\boldsymbol{B}_{\perp}$ because of the relatively small effective mass in the plane perpendicular to the field, but the observed peak position does not show any discernible change for $\boldsymbol{B}_{\parallel}$. The observed large anisotropy of the diamagnetic shift clearly indicates that the conduction band minimum is at the X point and the system is type II. Furthermore, from the direction of the anisotropy, we can

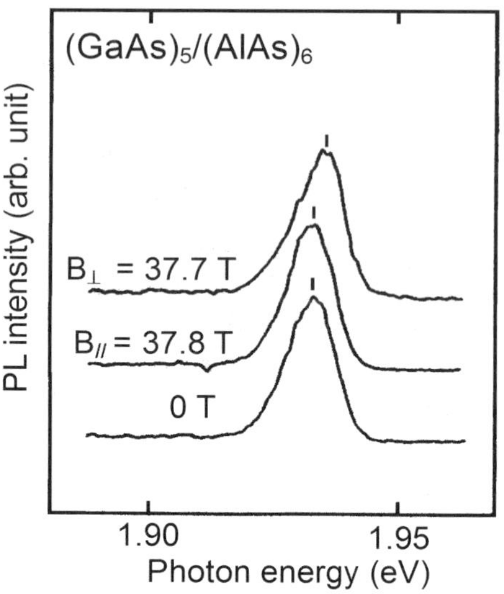

FIG. 5.21. Magneto-photoluminescence spectra of a $(GaAs)_5(AlAs)_6$ short-period superlattice sample for magnetic fields parallel ($B_{\|}$) and perpendicular ($B_{\perp}$) to the quantum well layers [431].

conclude that the minimum is at the X_z point rather than the X_{xy}-point.

As the pressure coefficients of the band edge are different between the Γ point and the X point, the type I to type II transition is induced by applying hydrostatic pressure to a type I sample at some critical pressure p_c, as shown in Fig. 5.22. Such a transition has been observed and reported by many authors [432–434]. This is called pressure-induced type I-type II transition. A similar type I-type

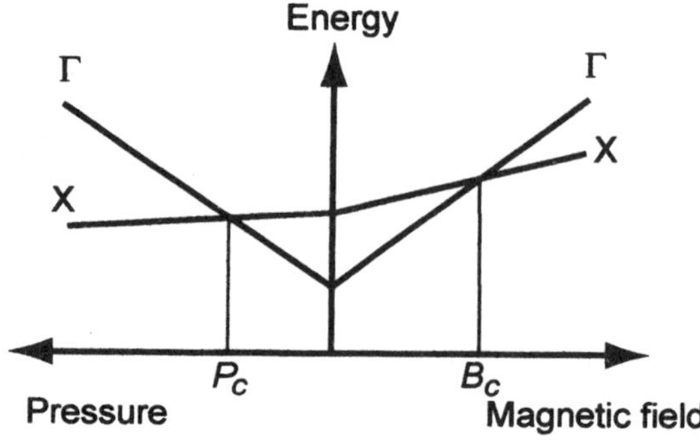

FIG. 5.22. Schematic diagram of the conduction band minima at the Γ and X points as a function of hydrostatic pressure (left scale) and magnetic field (right scale). At some critical value of pressure p_c and B_c, the type I-type II transition takes place.

II transition can also be induced by applying magnetic fields, since the effective mass is different between the Γ point and the X point. In other words, the increase of the band edge at the Γ point with field is much larger than that of the X point due to the small effective mass as shown on the right-hand scale of Fig. 5.22.

Such magnetic field-induced type I-type II transitions have been extensively studied by Sasaki *et al.* using very high magnetic fields generated by the single turn coil technique [370, 435]. Figure 5.23 shows the exciton PL spectra in a $(GaAs)_{15}(AlAs)_{13}$ superlattice. At zero field, this sample has character of type I, where the conduction band at the Γ point is located at 26.5 meV lower than the X point and a distinct direct exciton peak was observed. We can see an exciton peak which shows diamagnetic shift as we apply magnetic field. As is evident in the figure, however, the exciton peak diminishes suddenly above a certain field. The intensity recovers as the field is lowered again in the down sweep. This is

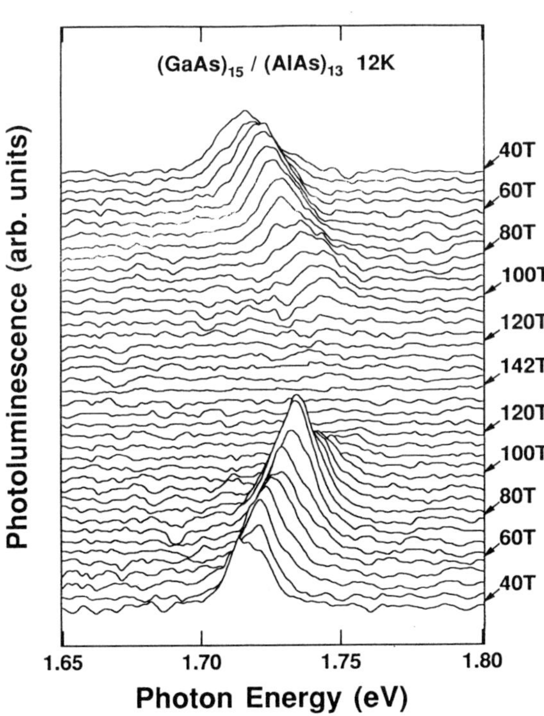

FIG. 5.23. Magneto-photoluminescence spectra in a $(GaAs)_{15}(AlAs)_{13}$ superlattice observed at different magnetic fields up to 142 T generated by the single turn coil technique [370]. The time evolves from top to bottom. The magnetic field height is shown on the right for each graph.

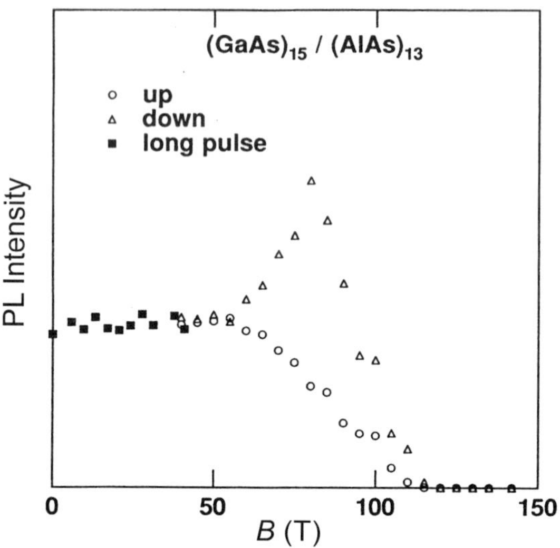

FIG. 5.24. Dependence of the integrated photoluminescence intensity on magnetic field in $(GaAs)_{15}(AlAs)_{13}$ observed in a short pulse field [370].

due to the magnetic field-induced type I-type II transition. In other words, as soon as the sample undergoes the transition to type II, the luminescence intensity decreases because of the indirect nature of the ground state exciton. Figure 5.24 shows the magnetic field dependence of the PL intensity of the exciton peak. It is clear that the intensity diminishes at high enough magnetic fields. However, a remarkable point is that the intensity exhibits a prominent peak in the falling sweep of the magnetic field and there is large hysteresis between the up-sweep and the down-sweep. The hysteresis is brought about because the magnetic field generated by the single turn coil technique is swept in a short time ($\sim$7 μs) which is comparable with the relaxation time for electrons at the X point in the type II regime. Actually, the relaxation time of electrons at the X point is estimated to be of the order of microseconds by the time resolved PL experiments [436]. In fields higher than the cross-over field (in this case $B_c = 55$ T), electrons are accumulated in the X point minima due to the slow relaxation time and a part of these electrons are thermally excited to the Γ point and contribute to the photoluminescence to some extent in the down sweep. It was shown that the magnitude of the hysteresis is well reproduced by the solution of rate equations. Such a measurement of B_c of the type I-type II transition gives an approximate effective mass of AlAs. It was found that the effective mass obtained from the type I-type II transition is in good agreement with the result of the cyclotron resonance [312]. This is another example of the usefulness of the short pulse field

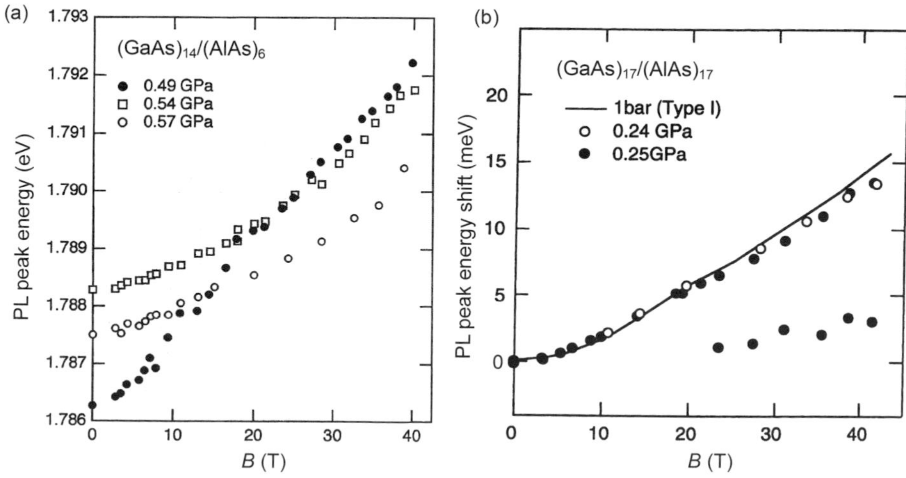

FIG. 5.25. Two types of the type I to type II transitions observed in the diamagnetic shift of exciton ground state in short-period superlattices. (a) diamagnetic shift in $(GaAs)_{14}(AlAs)_6$ short-period superlattice [437]. (b) Diamagnetic shift in $(GaAs)_{17}(AlAs)_{17}$ [406]. Hydrostatic pressure was applied to the sample in order to observe the transition within the field range of the experiment.

for studying the relaxation phenomena as we discussed in Section 4.3.1.

The experiment to observe the magnetic field-induced type I-type II transition requires generally very high magnetic fields unless the energy difference between the Γ and X points $\Delta_{\Gamma-X}$ has a suitably small value which is within a range attainable by the applied magnetic field. If we apply hydrostatic pressure in addition to magnetic field, however, we can control $\Delta_{\Gamma-X}$ to a suitable value, since $\Delta_{\Gamma-X}$ is very sensitively changed by hydrostatic pressure. Then we can use steady fields or long pulse fields to observe the type I-type II transition, and in this case the PL from the X point has been also observed after the transition [406, 437].

In Fig. 5.25, we plot the diamagnetic shift of exciton ground state observed in photoluminescence in two different short-period superlattices. Depending on the superlattice period, different features were observed at the type I-type II transition. In a $(m, n) = (14,6)$ sample (Fig. 5.25 (a)), the system is in the type I regime at a pressure of $p = 0.49$ GPa, showing a relatively large diamagnetic shift, while at pressures higher than $p_c = 0.52$ GPa it shows type II behavior so that the diamagnetic shift is smaller (see the graph for $p = 0.54$ GPa and 0.57GPa). In a magnetic field of 20 T and at a pressure of $p = 0.49$ GPa, the type I to type II transition occurs and the slope of the diamagnetic shift $\Delta\mathcal{E}_{1s}(B)$ decreases towards that of the type II. Thus the transition is continuous, showing only the change of the slope of the $\Delta\mathcal{E}_{1s}(B)$. The intensity of the PL at the peak does not show a large change at the transition.

In a $(m, n) = (17,17)$ sample (Fig. 5.25 (b)), on the other hand, the system is type I below $p_c = 0.30$ GPa. From the energy difference $\Delta_{\Gamma-X}$ and the difference in the slope of the diamagnetic shift of excitons, the transition field is estimated to be about 20 T at $p = 0.24$ GPa, and about 30 T at $p = 0.25$ GPa. At these fields, does indeed the PL intensity of the exciton peak corresponding to the type I transition start decreasing, and gradually diminishes in the higher field. In addition, at around the transition field, a weak new peak emerges below the type I peak. In other words, the PL peak from both the type I transition and the type II transition are observed simultaneously. In samples with $(m, n) = (15,15)$ and (19,19), similar behavior was observed at the transition. The difference in the PL features at the transition between (14,6) and other samples can be ascribed to the difference in the thickness of the AlAs barrier layers (L_b). In the former type of samples with small L_b, the exciton state in the type II regime still keeps partly the type I character due to the larger coupling, whereas in other samples with larger L_b, the transition is observed more distinctly [406].

5.4.2 *Photoluminescence in type II short-period superlattices and quantum wells*

Generally, type II superlattices or quantum wells show much weaker photoluminescence (PL) than type I superlattices. However, there are some exceptions such as GaP/AlP superlattices and Si/SiGe superlattices [438]. In Si/SiGe, the effect of the strain in the heterostructure is responsible for the intense PL [439]. In GaP/AlP short-period superlattices or quantum wells, anomalously intense PL of exciton lines was found although both constituents are indirect gap semiconductors [440]. In addition, the band line-up is such that the valence band top is at the Γ-like band of the GaP layer and the conduction band minima are at the X- or Z-like band of the AlP layer. Therefore, the system is indirect both in the real space and the $\boldsymbol{k}$-space. As regards the position of the conduction band minima, there has been a controversy concerning whether it is X_z, X_{xy}, or Z point. Concerning the origin of the intense PL in GaP/AlP short-period superlattices, it has been argued that the effect of the folding of the energy band in the Brillouin zone might be the origin of the large intensity, since the folded X_z point comes to the Γ point and has the quasi-Γ character [441–444]. However, as the intense PL is also observed in just a single quantum well with a neighboring confinement structure [445], the zone-folding effect is not responsible for the large PL intensity. The group of Kamimura attributed the strong intensity of the PL to the disorder of the interface between the GaP and AlP layers [446, 447].

Magneto-photoluminescence in GaP/AlP short-period superlattices has been extensively studied by Uchida *et al.* [448–450]. Figure 5.26 shows the magneto-photoluminescence spectra for $(\mathrm{GaP})_m(\mathrm{AlP})_n$ short-period superlattices of (m, n) = (4,4), (5,5), (6,6), and (7,7) for a magnetic field perpendicular to the layers [449]. A sharp intense peak originating from the ground state exciton is observed for each sample. A striking feature of the spectra is that the intensity of the exciton peak is decreased with increasing magnetic field. Furthermore, the

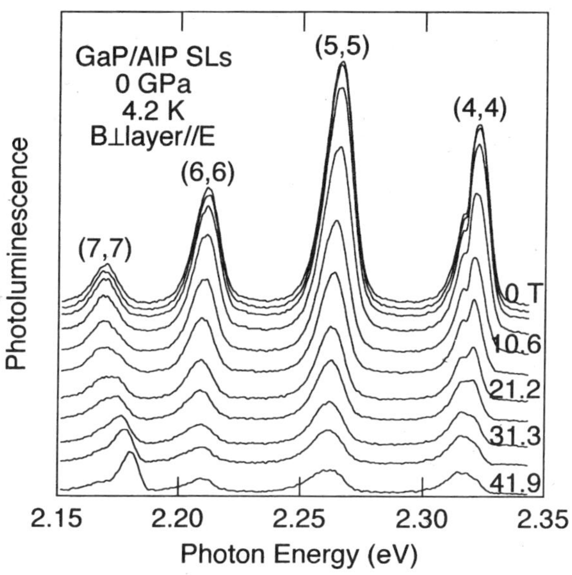

FIG. 5.26. Photoluminescence spectra for $(GaP)_m(AlP)_n$ short-period superlattices of (m, n) = (4,4), (5,5), (6,6), and (7,7) at different magnetic fields [449]. The four samples were accumulated on the same substrate. The magnetic field was applied perpendicular to the layers.

exciton peak shows a prominent red shift with field. These are completely opposite to the ordinary magnetic field dependences of the exciton peak. For ordinary Wannier excitons, the exciton peak shows a diamagnetic shift (blue shift) and its intensity increases with increasing magnetic field, due to the shrinkage of the wave function with field. Figure 5.27 shows the magnetic field dependence of the integrated intensity of the PL peak and the photon energy of the peak for each sample when the magnetic field is perpendicular to the superlattice layers ($B_\perp$). Figure 5.28 shows the dependences for magnetic field parallel to the layers ($B_\parallel$). It is clearly seen that for $B_\perp$ both the intensity and the photon energy show prominent decrease with increasing field except for the (7,7) sample. In the (7,7) sample, the intensity decreases up to 25 T, but in higher magnetic fields it starts increasing. The photon energy of the peak shows a blue shift above about 10 T. In other samples with shorter period, the intensity drop with field is quite dramatic. For $B_\perp$, on the other hand, the field dependence of the PL intensity is very small and the red shift was much less. So, the remarkable PL intensity decrease and the red shift seem to be the characteristic properties of the indirect excitons for $B_\perp$.

It was found that these anomalous features of the excitons in short-period superlattices arise from the localization of electrons and holes in different quan-

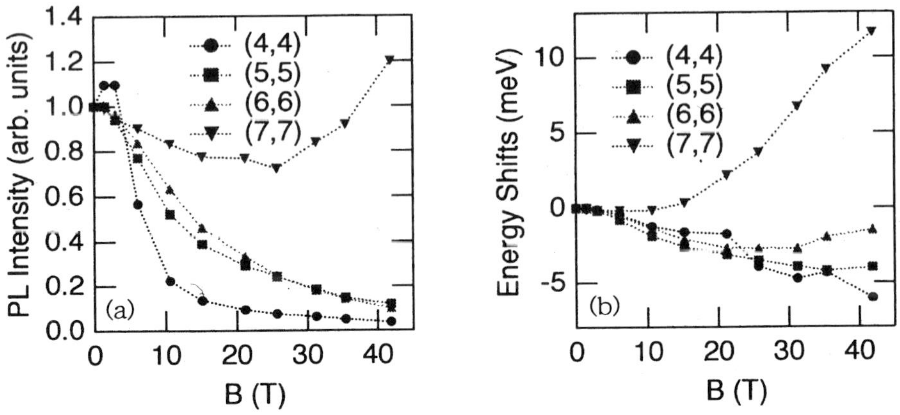

FIG. 5.27. Magnetic field dependence of the intensity of the PL (a) and the photon energy of the ground state exciton peak (b). The magnetic field was applied perpendicular to the layers.

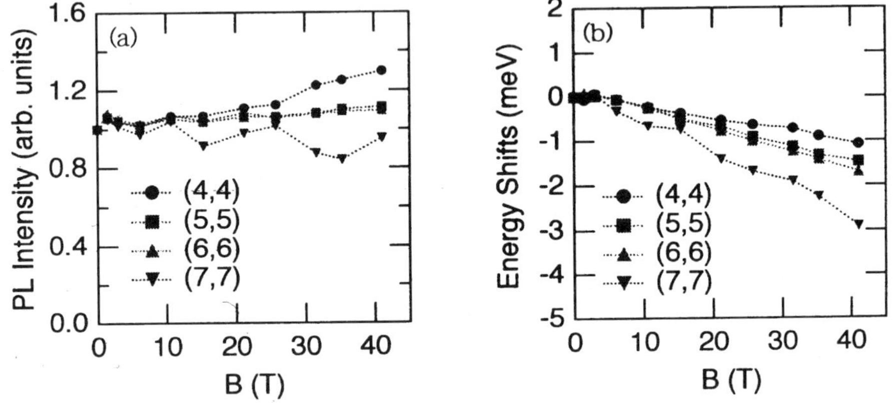

FIG. 5.28. The same dependences as in Fig. 5.27, but the magnetic field was applied parallel to the layers.

tum wells [448, 449]. In other words, electrons are mainly populated in the AlP layers and holes are in the GaP layers in GaP/AlP superlattices. Because of the non-uniform quantum well thicknesses, electrons and holes would be partly localized in each layer. If such localized wave functions of electrons and hole are overlapped, it would be helpful to increase the PL intensity. When a high magnetic field is applied, wave functions of both electrons and holes would show shrinkage and the overlap would be decreased. This effect would cause the decrease of the PL intensity. The red shift of the exciton peak can also be explained by the same mechanism. If the PL life time of the excitons are prolonged by the wave function shrinkage, excitons are more relaxed to areas with lower energies if the recombination time of excitons becomes longer than the relaxation time of excitons to lower energies. After the relaxation to the lower energy areas, the exciton would show a red shift. Schubert and Tsang derived an expression of the spectra involving such a relaxation process and showed that the peak should shift to the low energy side and become asymmetric with the high energy side being suppressed [451]. The asymmetric peak shape has been observed in GaP/AlP short-period superlattices.

The decrease of the intensity and the red shift of the exciton peak seem to be intrinsic properties of the indirect excitons in which electrons and holes are located in separate places both in the real space and $\boldsymbol{k}$ space. In fact, similar behavior of the indirect excitons has been observed in small CdSe/ZnSe quantum dots, as will be described in Section 5.5.1.

5.4.3 *Magneto-optics for $\boldsymbol{B} \parallel$ layers*

When we apply magnetic fields in the direction parallel to the quantum well layers of the short-period superlattice, the Landau levels are broadened as discussed in Section 4.11.1, if the energy of the cyclotron motion becomes comparable with the mini-band width. The broadening of the Landau levels in the conduction band has been seen in Fig. 4.40. This effect of the broadening appears in magneto-absorption as well. Figure 5.29 shows the magneto-absorption spectra in a short-period superlattice of $(GaAs)_m(Al_{0.47}Ga_{0.53}As)_n$ $((m, n) = (7, 5))$ parallel and perpendicular to the superlattice layers at nearly 38 T. A large difference is seen between the spectra for $\boldsymbol{B} \parallel$ layers ($\boldsymbol{B}_{\parallel}$) and $\boldsymbol{B} \perp$ layers ($\boldsymbol{B}_{\perp}$). That is to say, for $\boldsymbol{B}_{\perp}$, the oscillatory structure is visible up to the high energy range in the absorption spectrum, whilst for $\boldsymbol{B}_{\parallel}$, the oscillation is indiscernible in the energy range higher than at a threshold corresponding to the mini-band edge. This is because the joint density of states of the Landau levels is broadened at energies higher than the mini-band edge. This effect was first found by Belle *et al.* in the PLE spectra of $(GaAl)_5(Al_{0.5}Ga_{0.5}As)_6$ [452].

For analyzing interband magneto-absorption spectra, we have to take account of the complicated Landau level structure of the valence band. In contrast to the Landau levels of the conduction band, which can be calculated by solving a simple Schrödinger equation Eq. (3.113), it is necessary in the case of the valence band to solve the Schrödinger equation for a Hamiltonian of a 4×4 matrix form because of the four-fold degeneracy of the valence band [339, 453]. Setting the $\boldsymbol{k}$ -vector component in the field direction k_z to be zero, the 4×4 matrix can be decoupled into two 2×2 matrices,

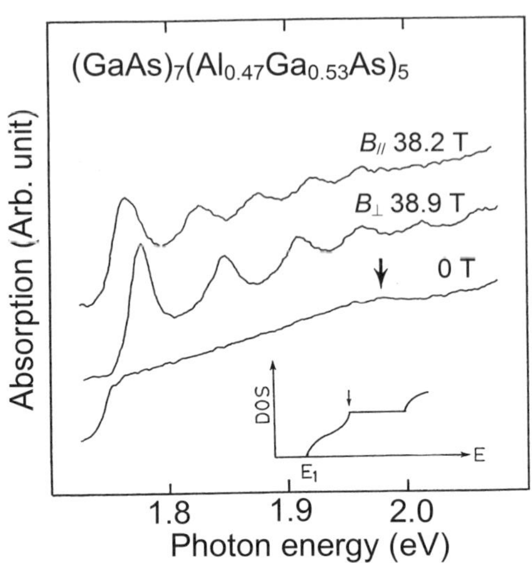

FIG. 5.29. Magneto-absorption spectra for $(GaAs)_7(Al_{0.47}Ga_{0.53}As)_5$. Nearly the same magnetic fields of nearly 38 T were applied parallel ($B_{\|}$) and perpendicular ($B_{\perp}$) to the superlattice layers. $T = 4.2$ K.

$$\mathcal{H} = \frac{\hbar^2}{m}\begin{vmatrix} \begin{array}{l} \frac{\gamma_1+\gamma_2}{2}\left[-\frac{\partial^2}{\partial x^2}+\frac{(x-X)^2}{l^4}\right] \\ \pm\frac{1}{l^2}\left(\frac{3}{2}\kappa+\frac{27}{8}q\right)+\frac{mU(x)}{\hbar^2} \end{array} & \begin{array}{l} \frac{\sqrt{3}}{2}\gamma_2\left[-\frac{\partial^2}{\partial x^2}-\frac{(x-X)^2}{l^4}\right] \\ \mp\frac{\sqrt{3}\gamma_3}{l^2}\left[(x-X)\frac{\partial}{\partial x}+\frac{1}{2}\right] \end{array} \\ \begin{array}{l} \frac{\sqrt{3}}{2}\gamma_2\left[-\frac{\partial^2}{\partial x^2}-\frac{(x-X)^2}{l^4}\right] \\ \pm\frac{\sqrt{3}\gamma_3}{l^2}\left[(x-X)\frac{\partial}{\partial x}+\frac{1}{2}\right] \end{array} & \begin{array}{l} \frac{\gamma_1-\gamma_2}{2}\left[-\frac{\partial^2}{\partial x^2}+\frac{(x-X)^2}{l^4}\right] \\ \mp\frac{1}{l^2}\left(\frac{1}{2}\kappa+\frac{1}{8}q\right)+\frac{mU(x)}{\hbar^2} \end{array} \end{vmatrix}, \quad (5.35)$$

where $(\gamma_1, \gamma_2, \gamma_3, \kappa, q)$ are the Luttinger parameters as discussed in Section 2.4.3, X is the center coordinate of the cyclotron motion and $U(x)$ is the quantum well potential. The upper sign corresponds to $J_z = 3/2$ and $-1/2$ states and the lower sign to $J_z = -3/2$ and $1/2$ states. Equation (5.35) can solved numerically as a function of X.

Figure 5.30 shows the calculated energy of the Landau levels in the valence band for $(GaAs)_7(Al_{0.47}Ga_{0.53}As)_5$. We can see that the Landau levels are broadened for the hole energies higher than the mini-band edge. The broadening is obtained from the energy dispersion of the states over the entire range of X as shown in the inset of Fig. 5.31. We can see that the Landau levels are broadened for hole energies higher than the mini-band edge. The broadening is obtained from the energy dispersion of the states over the entire range of X as shown in the inset of Fig. 5.31. The absorption coefficient can also be calculated from the

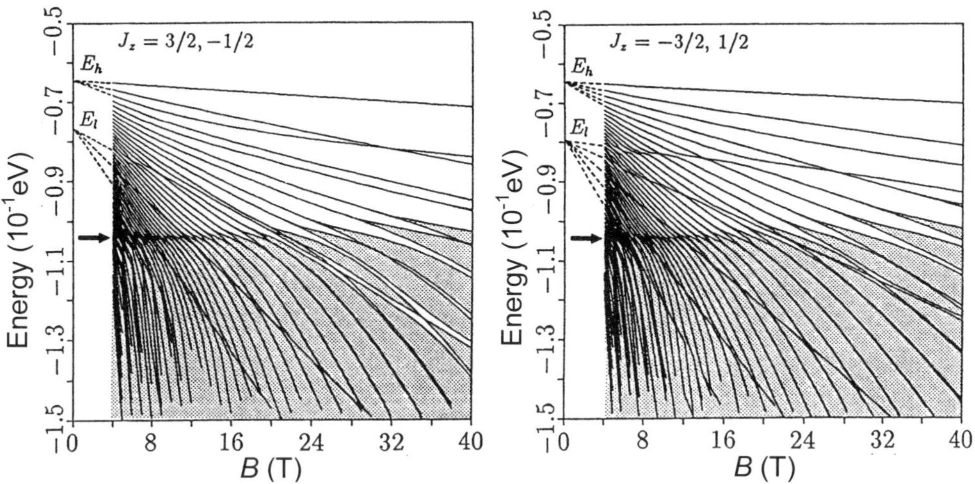

FIG. 5.30. Calculated Landau level energies in the valence band for $(GaAs)_7(Al_{0.47}Ga_{0.53}As)_5$ for $J_z = 3/2$ and $-1/2$ states (left) and $J_z = -3/2$ and $1/2$ states (right) as a function of magnetic field [339]. The arrow shows the mini-band edge.

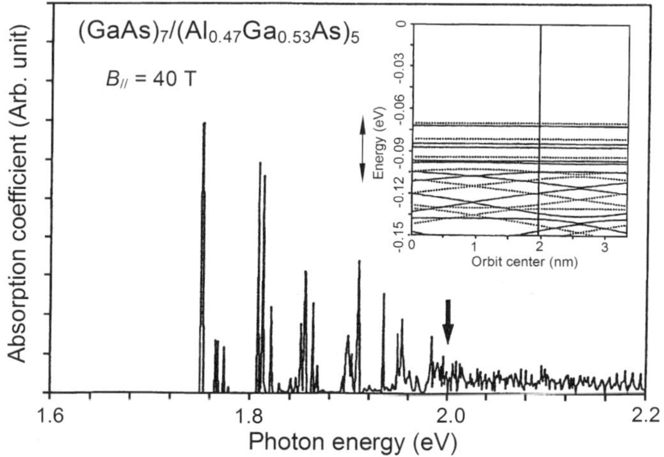

FIG. 5.31. Calculated absorption spectra for $(GaAs)_7(Al_{0.47}Ga_{0.53}As)_5$ when the magnetic field of 40 T is applied parallel to the superlattice layers [339]. The inset shows the calculated valence band energies as a function of X. The arrows indicate the mini-band edge.

wave functions calculated from (5.35). Figure 5.31 shows the calculated absorption coefficient taking account of all the allowed transitions including $\Delta n \neq 0$, It can be seen that although bunches of sharp peaks are seen up to the energy close to the mini-band edge, no distinct peak is seen in the higher fields. The centers of the bunches of the peaks are in good agreement with the experimental observation.

In very short-period superlattices in which the superlattice period is not much larger than the crystal period, it is questionable whether the effective mass approximation is in principle valid. However, the above results for the GaAs/AlGaAs short-period superlattices imply that the experimental results for the position of the absorption peaks are in good agreement with the theoretical calculation and that the effective mass theory is still valid even beyond the applicable limit.

5.5 Magneto-optics of quantum dots and quantum wires

Excitons confined in quantum wires and quantum dots show combined effects of magnetic fields and quantum potentials in high magnetic fields. Similarly to the case of quantum wells where quasi-two-dimensional excitons are observed, nearly one-dimensional or nearly zero-dimensional excitons are observed, respectively in quantum wires and quantum dots.

5.5.1 *Quantum dots*

In quantum dots, electrons and holes are confined by the quantum potentials in all three dimensions. When we apply magnetic fields in the perpendicular direction to the substrate layers, we can observe an effect arising from the competition between the quantum potential and the magnetic field potential in the lateral direction. Nowadays, self-organized quantum dots with a very small dimension can be grown by the Stranski-Krastanov (SK) mode utilizing the lattice mismatch between the substrate and the dot materials, as described in Section 2.5.4. The non-uniformity of the dot size gives rise to broadening of the spectra, but high quality samples have reasonably uniform size distribution, so that we can study the properties of quantum dots using these samples. The diamagnetic shift of excitons in quantum dots is significantly smaller than in the bulk because of the confinement. Therefore, we need high magnetic fields to investigate the magnetic field effect. The magnetic energy levels in quantum wells can be described by Darwin-Fock energy levels as shown in Fig. 2.13 if we assume a circular quantum potential. The exciton levels are associated with these magnetic energy levels.

In the early days of research on self-organized quantum dots, Wang *et al.* measured the diamagnetic shift in InAlAs/AlGaAs quantum dots which appears in the magneto-photoluminescence spectra using pulsed high magnetic fields up to 40 T [454]. They found that the diamagnetic shift of the ground state excitons in quantum dots is reduced significantly in comparison to the bulk material for both directions of the magnetic field parallel and perpendicular to the growth direction. In a similar manner to the discussion in Section 2.8.1, the diamagnetic shift of the exciton ground state in the low field limit is derived as [455, 456]

$$\Delta \mathcal{E}_{1s} = \frac{e^2 < \rho^2 >}{8\mu^*}, \tag{5.36}$$

where ρ is the extension of the exciton wave function perpendicular to the magnetic field and μ^* is the reduced mass of the exciton. For quantum dots in which the exciton wave function is squeezed, the low field approximation can still be applicable even in high magnetic fields. Thus from the diamagnetic shift, the effective exciton radius $\sqrt{< \rho^2 >}$ was estimated to be 5 nm, which is smaller than the quantum dot radius of ~9 nm [454]. This value is also smaller than the 2D exciton radius of 6.4 nm estimated from the 3D exciton radius of 10.4 meV of InAlAs, clearly showing the confinement of the exciton wave function by the quantum potential. As shown in this example, the diamagnetic shift of excitons is a useful means to estimate the extension of the exciton wave function in the plane perpendicular to the magnetic field.

As an example of the diamagnetic shift of excitons in quantum dots, Fig. 5.32 shows the photoluminescence spectra of InAs/GaAs and InAs/InGaAs quantum dots measured by Hayden *et al.* [457, 458]. They used two different samples. One is InAs quantum dots grown on a (100) surface of GaAs. The second one is InAs quantum dots grown on a (311) surface of InGaAs. The PL peaks from the ground state exciton states are clearly seen in both samples. Figure 5.33 shows the diamagnetic shift of the excitons as a function of magnetic field. Reflecting the larger confinement effect, the diamagnetic shift is always smaller for the magnetic field perpendicular to the growth direction (parallel to the layers) than in the parallel direction. A remarkable result is that the diamagnetic shift is considerably larger for the quantum dots on the (311)A surface than those on (100) surface. For the quantum dots on the (311)A surface, the exciton wave function size perpendicular to the growth direction (parallel to the layer) is estimated from the diamagnetic shift to be 13±2 nm which is even larger than the average quantum dot diameter of 5–8 nm. This implies that in the case of the (311)A quantum dots, the electrons and holes are less strongly confined by the quantum potential. On the other hand, for the (100) quantum dots, the wave function extension is estimated to be 5.5 nm. As the average dot size is ~10 nm, the electrons and holes are well confined in the quantum potential.

In self-organized InP quantum dots embedded in an InGaAs/GaAs quantum well, magnetic field dependence of the energy levels characteristic of the quantum dots represented as a Darwin-Fock diagram was observed in magneto-photoluminescence in high magnetic fields [460]. The lines of the Darwin-Fock states tend to the Landau level-like states, with a large reduction of the slope from the calculation assuming the single-particle model. From the amount of the reduction, the electron-hole correlation effect was estimated.

In comparison to quantum dots of III-V compounds, there have been much less studies of those of II-VI compounds. Magneto-photoluminescence spectra

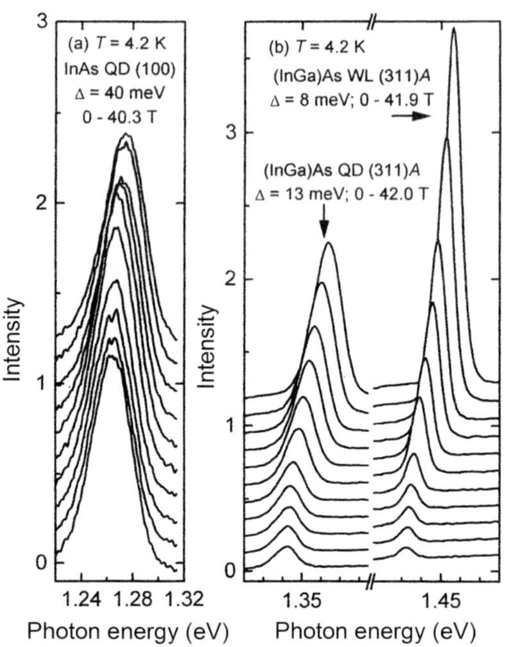

FIG. 5.32. Magneto-photoluminescence spectra of (a) InAs quantum dots on a (100) surface of GaAs, and (b) InGaAs quantum dots on a (311)A surface of GaAs [457]. In the latter figure, the PL peak from the InGaAs wetting layer (WL) is also shown.

of CdSe/ZnSe quantum dots were studied by Yasuhira *et al.* [461, 462]. The samples with different dot sizes were grown by the MBE technique on GaAs substrates with a ZnSe buffer layers. In relatively small quantum dots with a diameter of about 5 nm, peculiar magnetic field effects on the exciton line were observed. Figure 5.34 shows the magneto-photoluminescence spectra of excitons in high magnetic fields. An astonishing result is that the energy shift of the exciton peak by magnetic field is in the negative direction. Moreover, the PL intensity also decreases with increasing field. It is completely opposite to the normal field dependence of the energy shift and the intensity of the exciton peak, and similar to the case of GaP/AlP short-period superlattices. In a larger dot with a diameter of 10 nm, the energy shift is just the normal diamagnetic shift with a B^2 dependence. and the intensity was increased as the field was increased. Therefore, the anomalous energy shift and the intensity decrease are considered to be characteristic of excitons in small quantum dots of II-VI compounds. The similarity of the magnetic field dependence of the excitons in the small quantum dots to the GaP/AlP quantum well reminds us of the indirect transition, and the electrons and holes spatially localized in a different sites.

Figure 5.35 shows a schematic energy diagram for a large dot and a small

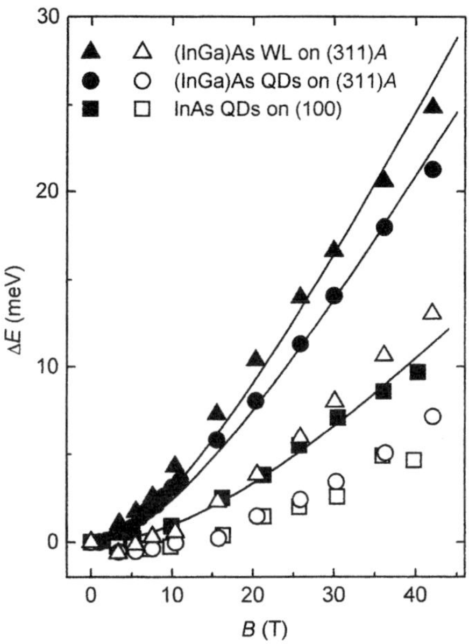

FIG. 5.33. Diamagnetic shift of excitons as a function of magnetic field observed in the magneto-photoluminescence spectra shown in Fig. 5.32 [457]. The closed points indicate the data for B parallel to the growth direction and the open points indicated the data for B perpendicular to the growth direction. The solid lines were calculated by the method of MacDonald and Ritchie [459].

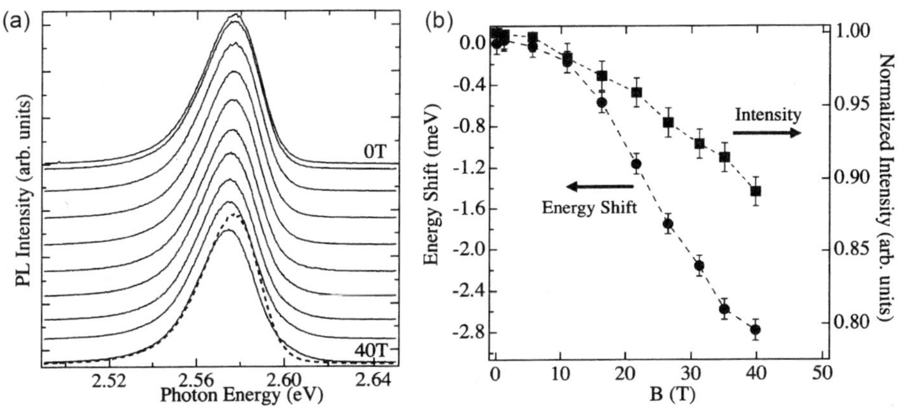

FIG. 5.34. Magneto-photoluminescence spectra in a relatively small dot of CdSe/ZnSe [461, 462]. The dot size is about 5 nm in diameter.

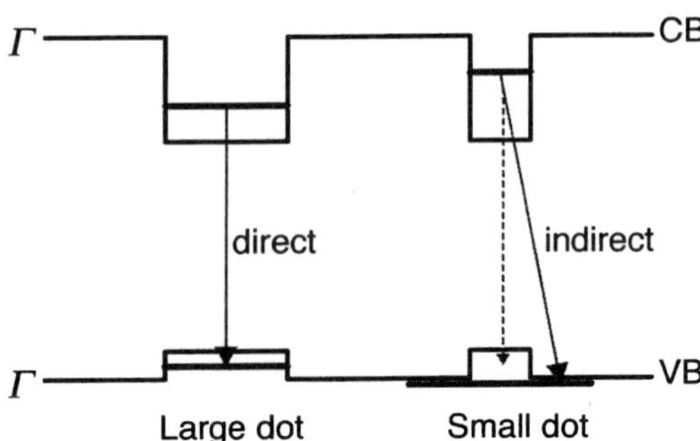

FIG. 5.35. Model of the band off-set and the transition in a large dot and small dot of II-VI compounds.

dot. In a large dot, the optical transition is direct. It is known, however, that the valence band off-set of the CdSe/ZnSe interface is very small, so that the hole confinement would be small. In a small dot, the strain potential due the piezoelectric effect should be fairly large, so that the holes would be trapped in potential minima around the dot, as shown in the right-hand side of the figure. Then the transition is indirect in a real space. Thus the intensity of the PL would be decreased with increasing magnetic field by the shrinkage of the wave function. The negative shift of the PL peak is also explained by the same model as we considered in the case of GaP/AlP system.

5.5.2 *Quantum wires*

Next, let us see an example of the magneto-optical spectra of quantum wires. Nagamune *et al.* succeeded in measuring the diamagnetic shift of excitons in GaAs quantum wires fabricated in a saw-tooth shaped V-groove at the bottoms of valleys in GaAs/AlAs subsequent layers. The wire has the shape of a triangular bar, as shown in the inset of Fig. 5.36 [463]. When we apply magnetic fields, their effect is expected to be different for different magnetic fields. Figure 5.36 shows the diamagnetic shift of excitons in the photoluminescence spectra for different direction of magnetic fields. The diamagnetic shift of excitons in bulk GaAs substrate layer is also plotted for comparison. It is seen in the figure that the diamagnetic shift of the quantum wire is smaller than that of the bulk due to the confinement. In addition, the magnitude of the diamagnetic shift shows a large anisotropy, reflecting the quantum confinement different among the directions.

In the low magnetic field range, the diamagnetic shift is proportional to B^2, so that from (2.216), the reduced mass of excitons μ_i^* $(i = x, y, z)$ was obtained. The reduced masses for all the three directions thus obtained are $\mu_x^* = 0.293m$, $\mu_y^* = 0.126m$, $\mu_z^* = 0.081m$. These values are considerably larger than that of the bulk sample, $\mu^* = 0.059m$ due to the confinement effect in the quantum wire.

In the high magnetic field range, the diamagnetic shift is proportional to B. In this regime, if we assume that the quantum potential form for the confinement is approximately represented as

$$V(x,y) = \frac{1}{2}m^*\omega_{0x}^2 x^2 + \frac{1}{2}m^*\omega_{0y}^2 y^2, \tag{5.37}$$

the energy levels for magnetic fields along the x, y, z axes are obtained as follows,

$$\mathcal{E} = \begin{cases} (N+\frac{1}{2})\hbar\omega_{0x} + (M+\frac{1}{2})\hbar(\omega_{cx}^2 + \omega_{0y})^{\frac{1}{2}}, & (B \parallel x), \\ (N+\frac{1}{2})\hbar(\omega_{cy}^2 + \omega_{0x}^2)^{\frac{1}{2}} + (M+\frac{1}{2})\hbar\omega_{0y}, & (B \parallel y), \\ (N+M+1)\frac{1}{2}\hbar(\omega_{cz}^2 + 4\omega_0^2)^{\frac{1}{2}} + (N-M)\hbar\omega_{cz}, & (B \parallel z). \end{cases} \tag{5.38}$$

Here, M and N are positive integers. From these equations, the size of the quantum confinement potential can be obtained. The height and the width were estimated to be 9.5 nm and 17.7 nm, respectively. These values were very close to the geometrical size of the wire $h = 10$ nmA$w = 20$ nm observed by transmission electron microscopy (TEM) [463].

Magneto-photoluminescence spectra in a little different type of quantum wires are shown in Fig. 5.37. Quantum wires with rhombic cross section were grown by molecular beam epitaxy (MBE) at giant steps on a vicinal GaAs (110) surface slightly tilted towards the $< 110 >$ direction [31]. As we can stack wires in

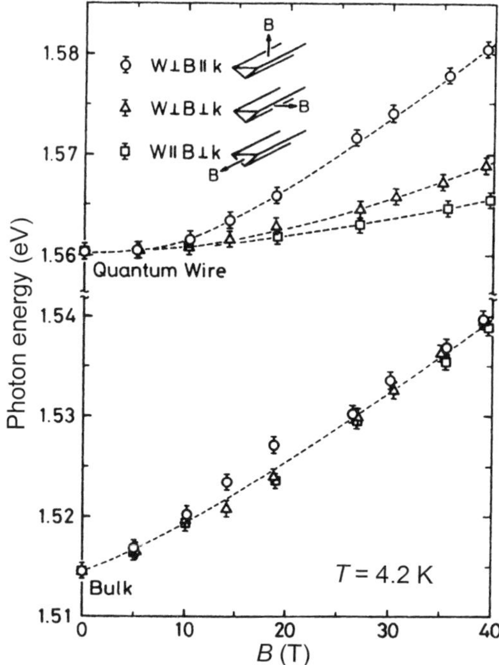

FIG. 5.36. Magneto-photoluminescence spectra in GaAs/AlGaAs quantum wires in high magnetic fields [463].

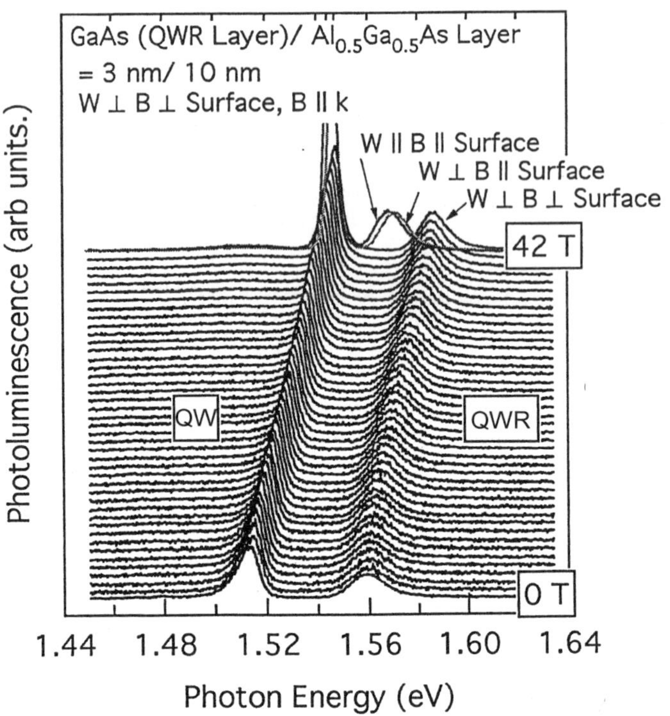

FIG. 5.37. Magneto-photoluminescence spectra in GaAs/AlGaAs quantum wires grown on a vicinal GaAs surface in high magnetic fields applied to different directions relative to the wires [406]. The peaks in the high energy side (QWR) and in the low energy side (QW) are the exciton line from the quantum wire and that from the substrate quantum well, respectively. W denotes the direction of the quantum wires. The spectra for $W \parallel B\perp$ surface are shown in different magnetic fields from 0 T to 42 T are shown (from bottom to top). The spectrum is compared with those for $W \parallel B \parallel$ surface and $W\perp B \parallel$ surface at 42 T.

accumulated layers and the density of the wires can be very large, we can expect some effects of the coupling between the quantum wires. Figure 5.38 shows the diamagnetic shift of the exciton line as a function of magnetic field in different directions. Similarly to the case shown in Fig. 5.36, the diamagnetic shift is the smallest when the magnetic field is parallel to the quantum wire axis. When the magnetic field is perpendicular to the wire axis, the diamagnetic shift is smaller for the field direction parallel to the stacking layers than the perpendicular direction. It is reasonable if we consider the area of the surface of the quantum wire perpendicular to the field.

Another interesting feature of this type of wire is the coupling between the wires. It was found that as the AlGaAs barrier layer thickness was decreased from 3 nm to 1 nm, the diamagnetic shift systematically increased, indicating

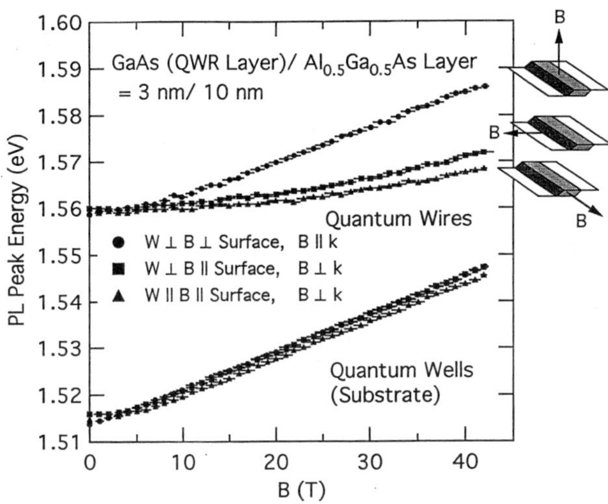

FIG. 5.38. Diamagnetic shift of the exciton line for GaAs/AlGaAs quantum wires and quantum wells (substrate) [406].

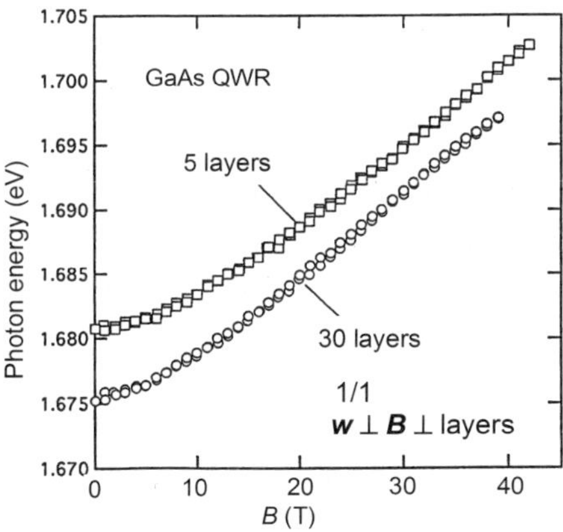

FIG. 5.39. Comparison of the diamagnetic shift of the exciton peak in GaAs/AlGaAs quantum wires between the different numbers of stacked layers, 5 layers and 30 layers [406]. The GaAs and AlGaAs layer thicknesses were both 1 nm.

the effect of the coupling between the quantum wires. Figure 5.39 shows the comparison of the diamagnetic shift between quantum wires with different number of stacking layers [406]. Although the cross section of the quantum wires in the two samples has the identical shape and size, it was found that the diamagnetic shift is significantly larger in the sample with 30 stacking layers than with 5 layers. This result clearly shows that there is a coupling between the quantum wires and formation of a mini-band through the AlGaAs barriers.

5.6 Magneto-optics of layered semiconductors

Some crystals, such as GaSe, BiI_3, PbI_2, HgI_2, CdI_2, ZnP_2, *etc.*, which have layered crystal structures show prominent exciton peaks. This is because the exciton binding energy is relatively large, reflecting large effective masses of electrons and holes in comparison to ordinary II-VI or III-V semiconductors. In these crystals, the exciton wave function is anisotropic, reflecting the crystal structure. Another characteristic feature of the excitons in this class of crystals is a relatively small Bohr radius due to the large binding energy. Table 5.1 lists the binding energy and the Bohr radius of excitons in layered crystals in comparison to other typical semiconductors.

These materials provide a rich source of interesting subjects to be studied concerning excitons, such as anisotropy, magnetic field effect on small radius excitons, and excitons arising from the stacking faults in the crystals. Many investigations have been made on the magneto-optics of excitons in layered crystals. In this section, we show some typical examples of interesting features concerning excitons in these materials.

Table 5.1 Binding energy, Bohr radius and reduced mass of excitons in different crystals

Material	Band gap (eV)	Binding energy (meV)	Bohr radius (nm)	Reduced mass μ^*_{ex} (m_0)	Reference
GaAs	1.520	4.2	14	0.048	
Cu_2O[(1)]	2.173	96.8	6.6	0.38	[464]
CdS	2.582	28	2.7	0.18	[465, 395]
GaSe	2.131	22	3.5	0.14	[399]
PbI_2	2.55	63	1.9	0.17	[466]
ZnP_2	1.605	46	1.6	0.45(x) 0.39(y) 0.15(z)	[467–469]
BiI_3	2.25	180	0.6	0.5	[470, 471]
$C6PbI_4$[(2)]	2.727	386	0.51		[472]

(1) Yellow series

(2) $(C_6H_{13}NH_3)_2PbI_4$

5.6.1 *GaSe*

As already shown in Section 5.2.1, GaSe shows sharp and clear exciton absorption peaks. As the absorption intensity of the exciton lines is not too large, the

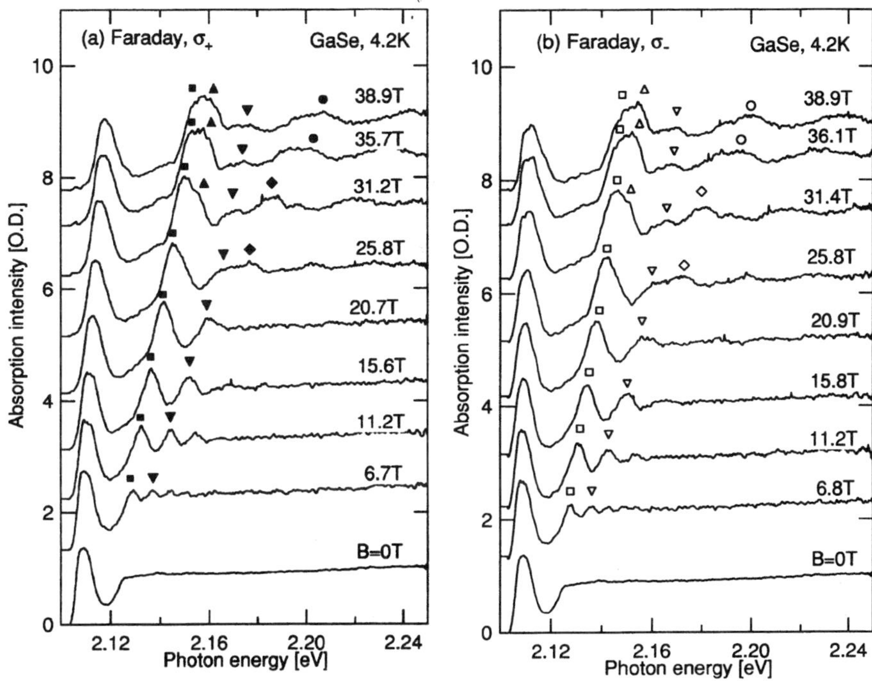

FIG. 5.40. Magneto-absorption spectra in GaSe for the Faraday configuration with (a) σ^+ and (b) σ^- circular polarizations [399].

lines can be easily observed in cleaved samples, so that the exciton in GaSe has attracted much interest. In early days, it was expected that nearly two-dimensional excitons might be observed due to a large anisotropy arising from the layered structure [78, 394]. Later it was revealed that the band structure is not so anisotropic in this material, so that the excitons have nearly isotropic character [395–398]. The conjunction problem between the hydrogen-like states and the Landau-like states has been discussed for the region $\gamma \sim 1$ [396].

The magneto-absorption spectra were observed in GaSe up to 150 T using the single turn coil and up to 40 T using non-destructive long pulse fields by Watanabe, Uchida and Miura [399]. The data up to 150 T have already been shown in Fig. 5.7 and Fig. 5.8. Figure 5.40 shows the magneto-absorption spectra in GaSe up to 40 T for the Faraday configuration with two different circular polarizations. Detailed structures in the low field range are seen. Figure 5.41 shows the photon energies of the magneto-absorption peaks as a function of magnetic field for both the Faraday configuration and the Voigt configuration [399]. For the Voigt configuration, two different linear polarizations π and σ were employed. According to theory of Schlüter, the band edge direct exciton states of GaSe consist of the Γ_4^s singlet state and the $\Gamma_3^t + \Gamma_6^t$ state [473]. The former is

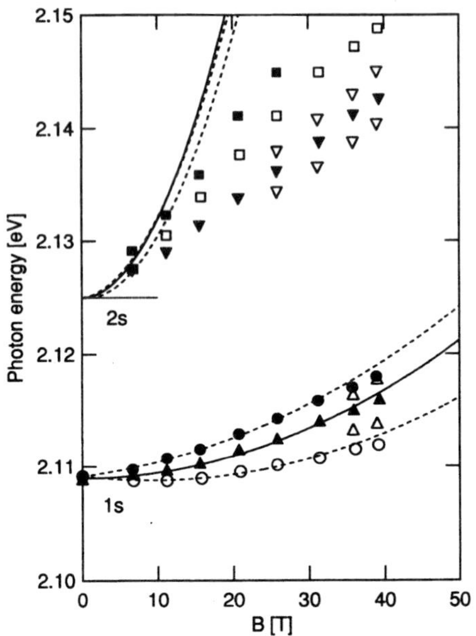

FIG. 5.41. Magnetic field dependence of the photon energies of the absorption peaks for the Faraday and the Voigt configuration [399]. Solid circles and squares denote the data for σ^+ and open circles and squares denote the data for σ^-. Open triangles and inverse triangles denote the data for the π polarization, and solid triangles and inverse triangles denote the data for the σ polarization in the Voigt configuration. The broken lines show the calculated result for excitons in the Faraday configuration (σ^+ and σ^-), and the solid lines that in the Voigt configuration.

allowed for $E \parallel$c, whereas the triplet Γ_6^t is weakly allowed for $E \perp$c due to the spin orbit interaction in the zero field. The triplet exciton line consists of three degenerate states and split into three components in a magnetic field. Two of the three components should be observed in the Faraday configuration for σ^+ and σ^-, respectively. For the Voigt configuration, two of the three components should be observed for the π polarization and the other line should be observed for the σ polarization in the center. This was actually the case in the observed spectra.

From the Zeeman splitting (difference between the two circular polarizations), the g-factor was obtained as $g_\parallel = 2.8$ and as $g_\perp = 2.1$, in good agreement with the previous data [474]. The $2s$ line shows the diamagnetic shift with a B^2 dependence only up to about 10 T and deviates downwards in higher fields. The intensity becomes weaker in higher fields. From the analysis of the diamagnetic shift of the $1s$ state, the anisotropy factor of the exciton reduced mass was estimated as $\mu_\parallel^*/\mu_\perp^* = 0.9$.

The $2s$ level becomes higher than the $N = 0$ Landau level line in the high

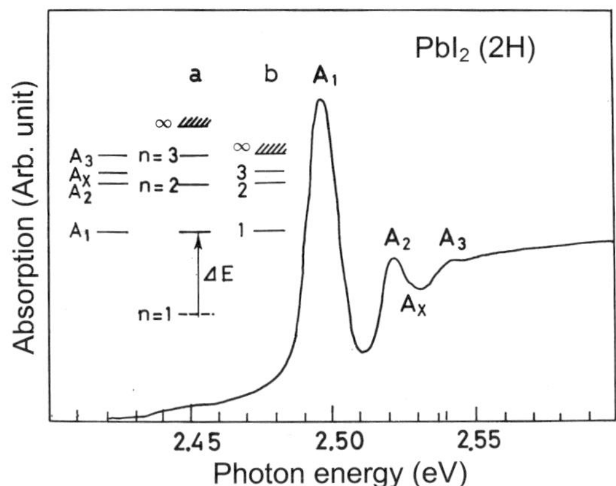

FIG. 5.42. Exciton spectra in PbI_2 [466]. Inset shows different models of the exciton series.

field range as shown in Fig. 5.8. This apparently contradicts the prediction of the non-crossing rule, but the reason is not understood.

5.6.2 *PbI_2*

A. Excitons in PbI_2

PbI_2 shows particularly sharp exciton peaks among layered crystals. A series of exciton absorption peaks are observed, but there has been much controversy concerning the assignment of the peaks. There are several polytypes of PbI_2. 2H, 4H, 6H, 12R, $\cdots$, which show slightly different spectra from each other. Therefore, in order to discuss the spectra exactly, we need to use a single polytype crystal. In high quality single crystals, we can usually observe four peaks, A_1, A_2, A_x and A_3, as shown in Fig. 5.42. These peaks do not fit to the hydrogen-like series, and different models were proposed regarding the assignment of the peaks. Model a assigns the peaks A_1, A_2, A_3 as $n = 1$, 2, 3 of the hydrogen series [475]. From the distance between A_2 and A_3, the binding energy of 142 meV was obtained. However, the observed position of the A_1 peak is 82 meV higher than expected from the hydrogen series. Nikitine and Perry attributed the significant discrepancy to the central cell correction. Actually, it is generally observed in other materials that only $n = 1$ exciton deviates from the hydrogen series, since the wave function of the $n = 1$ is smaller than the higher states. However, the deviation is particularly large in PbI_2 based on this model. Harbeke and Tosatti ascribed this large deviation to the effectively repulsive electron-hole interaction originating from the cationic character of the excitons [476]. Baldini and Franchi considered that two series of excitons are overlapped in the spectra, so that the A_3 is really the $n = 2$ peak in one of the series, and the A_2 is the ground state of another series (the model is not shown in the figure) [477]. According to their assignment the binding energy should be 60.5 meV. Thanh *et al.*, on the other hand, proposed a Model b that is based on the assumption that there is an additional peak A_x which is assigned as $n = 3$ [478]. Based on this model, the binding energy should be 30 meV. Thus there was a serious controversy concerning the assignment of the exciton series. The use of high magnetic field is very useful for solving such problems as shown below. Excitons in PbI_2 exhibit a variety of properties. So we will look at some details below.

B. $\boldsymbol{B} \parallel$ *c-axis*

Figure 5.43 shows magneto-absorption spectra of excitons in 2H-PbI_2 [466]. We can see clearly diamagnetic shift of the exciton lines A_1, A_2, and A_3. Figure 5.44 shows the diamagnetic shift of excitons in high magnetic fields for two different circular polarizations σ^+ and σ^-. In PbI_2, the condition $\gamma = 1$ is satisfied at around $B \approx 200$ T. Therefore, even in high magnetic fields around 150 T, approximation for the low field condition still holds, so that the diamagnetic shift is proportional to B^2. It also indicates that although the exciton radius is only 1.9 nm, only about four times as large as the in-plane lattice constant, the Wannier exciton model still holds in the case of $\boldsymbol{B} \parallel$ c-axis. It was found that the difference in the photon energy between the peaks for σ^+ and σ^-, ΔE_s is a linear function of the magnetic field and expressed by

$$\Delta E_s = g_{\perp} \mu_B B, \tag{5.39}$$

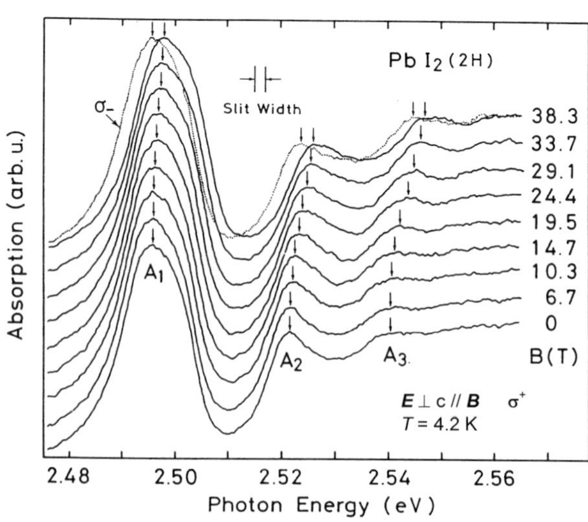

FIG. 5.43. Magneto-absorption spectra in 2H-PbI_2 for a circular polarization σ^+ [466]. Only for the highest field of 38.3 T, the spectrum for σ^- is shown by a dotted line for comparison.

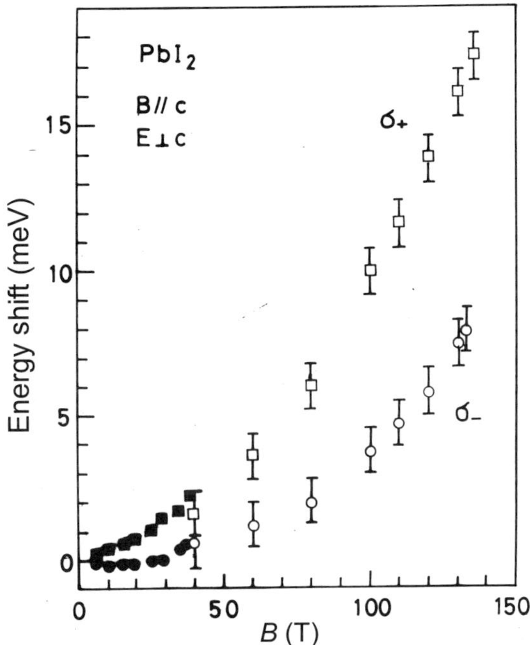

FIG. 5.44. Diamagnetic shift in PbI_2 [466]. The data up to 40 T were obtained for a 2H crystal and those up to 150 T were for a 4H crystal.

and from the coefficient, we can obtain the effective g-factor for $\boldsymbol{B}$ $\|$c as $g_{\perp c}$ $= 0.89 \pm 0.09$. By taking the average of the data for σ^+ and σ^-, the linear Zeeman term can be eliminated and the diamagnetic shift coefficient σ (see Eq. (2.216)) was obtained as $\sigma = (9.68 \pm 0.39)\times 10^{-4}$ meV/T^2. These data were for 2H-crystals in the measurements below 40 T. The data for 4H-crystals up to 150 T, on the other hand, gave slightly different values, $g_{\perp c} = 1.1$ and $\sigma = 6.88\times 10^{-4}$ meV/T^2.

To analyze the exciton spectra in anisotropic substances as layered crystals, we have to take account of the anisotropy in the dielectric constant $\kappa\epsilon$ and the reduced mass of excitons μ^*. According to Gerlach and Pollman, the anisotropy of the exciton spectra is characterized by an anisotropy factor α [479]

$$\alpha = 1 - \frac{\epsilon_\perp \mu_\perp^*}{\epsilon_\| \mu_\|^*}, \tag{5.40}$$

and the binding energy Ry^* (2.207) and the diamagnetic shift coefficient σ^* (2.218) is expressed using α as

$$Ry^* = \frac{\mu^* e^4}{32\pi^2\hbar^2 \kappa_\| \kappa_\perp \epsilon_0^2}|Z(\alpha)|^2, \tag{5.41}$$

$$\sigma^* = \frac{4\pi^2 \kappa_{\parallel} \kappa_{\perp} \epsilon_0^2 \hbar^2}{e^2 \mu_{\perp}^{*3}} \frac{n^2(5n^2+1)}{6|Z(\alpha)|^2}, \tag{5.42}$$

where

$$Z(\alpha) = \begin{cases} \frac{1}{\sqrt{\alpha}} \arcsin \sqrt{\alpha} & (\alpha > 0) \\ 1 & (\alpha = 0) \\ \frac{1}{\sqrt{|\alpha|}} \text{arcsinh} \sqrt{|\alpha|} & (\alpha > 0) \end{cases} . \tag{5.43}$$

In (5.42), we considered the case for $\boldsymbol{B}$ $\|$c. In the range 0$\leq$ α $\leq$0.6, we can set $Z(\alpha) \approx 1$. From the anisotropy of the dielectric constant, we can approximate $Z(\alpha)$ to be 1 unless there is a very large anisotropy in the reduced mass. As seen in Section 5.2.2, it is not always straightforward what kind of dielectric constant should be used to calculate the exciton parameters. The dielectric constants of PbI_2 are reported from infrared measurements to be: $\kappa_{\parallel}(0) = 9.3$, $\kappa_{\parallel}(\infty) = 5.9$, $\kappa_{\perp}(0) = 26.4$, $\kappa_{\perp}(\infty) = 6.1$ [480]. For the ground state of PbI_2, the binding energy is certainly larger than the LO phonon frequency, so that we should use a value closer to the high frequency dielectric constant κ_∞, rather than the static dielectric constant κ_0. Using $\kappa(\infty)$ and σ, we obtain $\mu^* = 0.168m$ and $Ry^* =$ 61.4 meV.

The coefficient of the diamagnetic shift of the A_2 line is about five times larger than that of the A_1 line, so that it is unlikely that the A_2 line is the ground state of another series as Baldini and Franchi insisted. Also, because of the large polaron effect expected in PbI_2, the binding energy determines from the $n = 2$ and $n = 3$ lines should be smaller than the binding energy determined for the ground state (61.4 meV). Therefore, model a is ruled out. From these considerations, we can deduce that model b in Fig. 5.42 is more plausible. The discrepancy between the ground state binding energy expected from model b (30 meV) and the experimentally determined value 61.4 meV is ascribed to the polaron effect.

C. $\boldsymbol{B} \perp$ *c-axis*

Although an analysis of the exciton spectra based on the Wannier exciton model was applicable in the case of $\boldsymbol{B} \parallel$ c, a different feature was observed for $\boldsymbol{B} \perp$ c-axis. Figure 5.45 shows the magneto-absorption spectra for the Voigt configuration, as well as the zero-field spectra and the graphs for the Faraday configuration for comparison [481]. For the Voigt configuration with $\boldsymbol{B} \perp$ c, the peaks A_1 and A_2 split into two peaks for the linear polarization and the split peaks are different depending on the direction of the polarizations, σ ($\boldsymbol{E} \perp \boldsymbol{B}$) and π ($\boldsymbol{E} \parallel \boldsymbol{B}$). Figures 5.46 (a) and (b) show the magnetic field dependence of the photon energies of the absorption peaks for the Faraday and the Voigt configurations, respectively. In both figures (a) and (b), the lines for the A_1 and A_2 peaks are shown. First, we notice the qualitatively similar features between

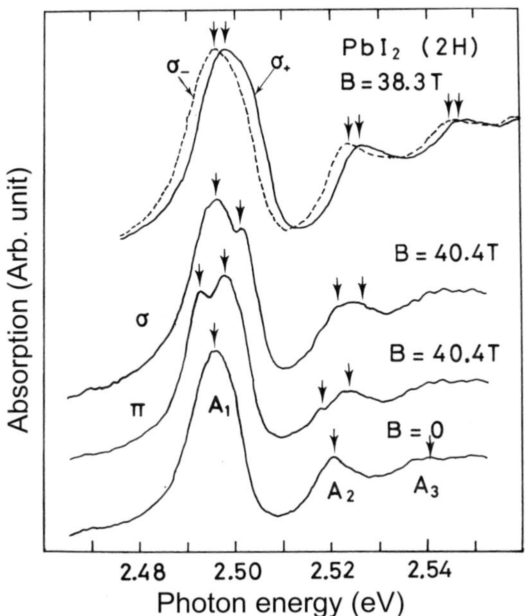

FIG. 5.45. Exciton spectra in 2H-PbI_2 for the Voigt configuration and the Faraday configuration [481]. π and σ denote the Voigt configuration with $E \parallel B$ and $E \perp B$, respectively, and σ^+ and σ^- denote the Faraday configuration with a right- and left-circular polarization, respectively.

A_1 and A_2, as mentioned above. Next, it is seen that in the Voigt configuration, the lines for both σ and π polarizations appear to have level repulsions.

The properties in the Voigt configuration can be well explained by the cationic exciton model that assumes that the exciton is completely localized in the cation site (Pb site in PbI_2). In fact, according to the band calculation, wave functions of both the lowest conduction band and the uppermost valence band are mostly concentrated around the Pb site [482], and the cationic character of the excitons was revealed in the study of the spectra in $Pb_{1-x}Cd_xI_2$ alloys [483]. In the cationic exciton model [482], the valence band and the conduction band of PbI_2 are considered to consist of the 6s and 6p atomic orbitals of Pb, respectively, so that they are expressed as

Conduction band: $P_-\alpha, P_-\beta, P_0\alpha, P_0\beta, P_+\alpha, P_+\beta$,

Valence band: $S\alpha, S\beta$,

where S and P_i ($i = -, 0, +$) are the atomic wave functions of Pb with 6s and 6p orbitals, and α and β are the spin functions. By the crystal field potential and the spin-orbit interaction, the p-orbitals of the conduction band split into three levels, A, B, and C, with wave functions ϕ_A, ϕ_B, ϕ_C, and the exciton states are associated with the valence band with wave function ϕ_0 to the lowest conduction

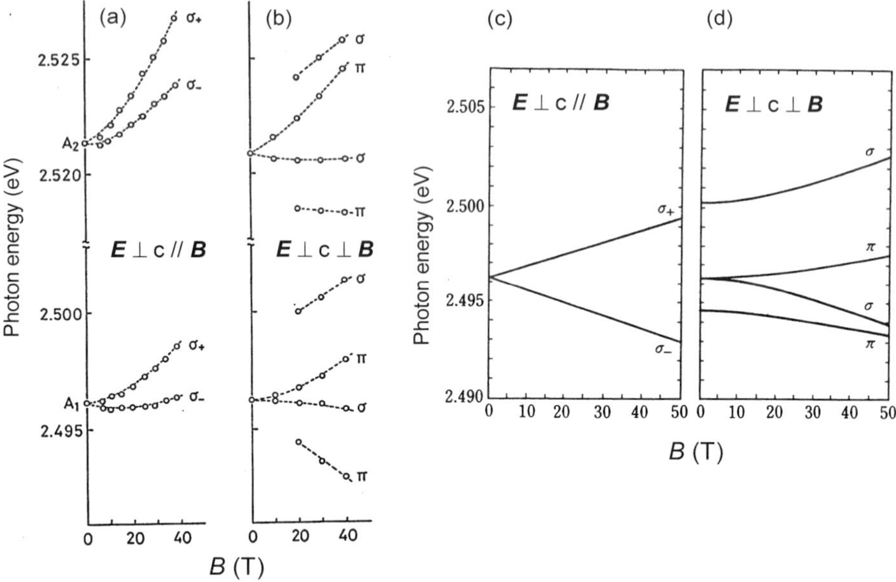

FIG. 5.46. (a) Experimental points of the photon energies of the absorption peaks for the Faraday configuration in 2H-PbI_2 [481]. $B \parallel$ c. (b) Experimental points of the photon energies of the absorption peaks for the Voigt configuration. $B\perp$ c. (c) Calculated lines of the spin-Zeeman energy for the Faraday configuration. (d) Calculated lines of the spin-Zeeman energy for the Voigt configuration.

band A. The exciton states are split into four levels Γ_1, Γ_2 and Γ_3 (two-fold degenerate at zero field), with wave functions, ϕ^1, ϕ^2, ϕ^3_-, ϕ^3_+, as schematically shown in Fig. 5.47 [481]. The selection rule of the transition is as shown in the figure. When an external magnetic field is applied, the spin-Zeeman energy is represented by a Hamiltonian

$$\mathcal{H}_B = -\frac{\mu_B}{\hbar}(l_e + g_e s_e)B + \frac{\mu_B}{\hbar}(l_h + g_h s_h)B, \tag{5.44}$$

where g_e and g_h are the g-factors of electrons and holes. The matrix elements for the Hamiltonian (5.44) among the (ϕ^1, ϕ^2, ϕ^3_-, ϕ^3_+) states are then given by

$$\begin{pmatrix} 0 & KB & 0 & 0 \\ K^*B & 0 & 0 & 0 \\ 0 & 0 & LB & 0 \\ 0 & 0 & 0 & -LB \end{pmatrix}, \tag{5.45}$$

for $\boldsymbol{B} \parallel z$, and

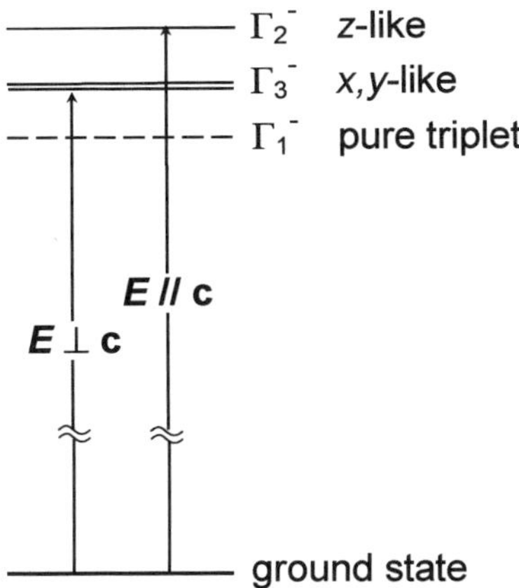

FIG. 5.47. Band edge exciton levels and the allowed transitions in PbI_2 [481]

$$\begin{pmatrix} 0 & 0 & MB & -MB \\ 0 & 0 & NB & NB \\ M^*B & N^*B & 0 & 0 \\ -M^*B & N^*B & 0 & 0 \end{pmatrix}, \tag{5.46}$$

for $\boldsymbol{B} \parallel x$, where K, L, M, and N are constant terms comprising the matrix elements and the potentials. In the case of $\boldsymbol{B} \parallel z$, $\boldsymbol{E} \parallel x, y$, the states ϕ_-^3 and ϕ_-^3 are observed. The Zeeman term is diagonal in (5.45) so that the Zeeman energy is

$$\mathcal{E}_z = \mathcal{E}_3 \pm LB, \tag{5.47}$$

as shown in Fig. 5.46 (c). Hence we can expect a linear Zeeman splitting. This was actually the splitting which was observed for $\boldsymbol{B} \parallel$ c, apart from the B^2-like diamagnetic shift. For $\boldsymbol{B} \parallel x$, the spin-Zeeman energy is obtained by diagonalizing (5.46) as

$$\begin{aligned} \mathcal{E}_x^1 &= \tfrac{1}{2}\left[(\mathcal{E}_1 + \mathcal{E}_3) \pm \sqrt{(\mathcal{E}_1 - \mathcal{E}_3)^2 + 8M^2B^2}\right], \\ \mathcal{E}_x^2 &= \tfrac{1}{2}\left[(\mathcal{E}_2 + \mathcal{E}_3) \pm \sqrt{(\mathcal{E}_2 - \mathcal{E}_3)^2 + 8N^2B^2}\right]. \end{aligned} \tag{5.48}$$

where $\mathcal{E}_x^1$ is allowed for $\boldsymbol{E} \parallel \boldsymbol{B}$ (π polarization) and $\mathcal{E}_x^1$ is for $\boldsymbol{E} \perp \boldsymbol{B}$ (σ polarization). The energy levels are calculated as in Fig. 5.46 (d) and shows semi-quantitatively good agreement with the experiment. The calculation above does not take into account the exciton wave-function extension, and any diamagnetic shift, which are necessary in order to obtain better agreement. However, we can see that the cationic exciton model is well applicable to the analysis of the Zeeman splitting of the excitons in layered crystals such as PbI_2 [481].

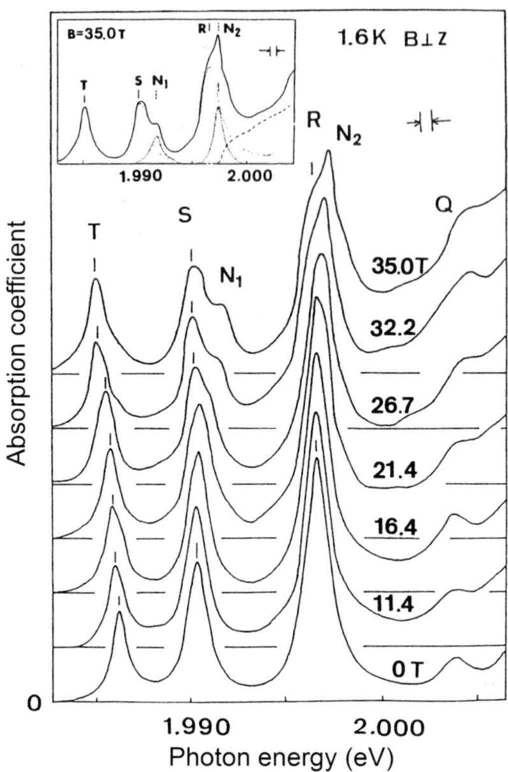

FIG. 5.48. Magneto-absorption spectra in BiI_3 [470]. $B \parallel c$, $k \perp c$,

5.6.3 BiI_3

BiI_3 shows indirect absorption bands followed by the direct absorption band in the higher energy side. Below the indirect absorption edge, a series of very narrow absorption lines are observed. The lines are named Q, R, S, T from the high energy side. As some of these lines fit well with the inverse hydrogen-series, Gross, Perel, and Shekhmametev attributed this series to the formation of a bielectron (or biholes) [484]. In later studies, however, it was found that the absorption lines are very sample dependent, and that they originate from the excitonic transitions perturbed by a stacking fault formed between the layers. Thus the lines are denoted as stacking fault excitons [485,486]. The nature of the stacking fault excitons are clarified in detail by applying a high magnetic field parallel to the layers [470].

Figure 5.48 shows the magneto-absorption spectra for the magnetic field parallel to the layers ($\boldsymbol{B} \parallel c$) [470]. The direction of light propagation is perpendicular to the c-axis ($\boldsymbol{k} \perp c$, Voigt configuration). New lines N_1 and N_2 appear as the field is increased. The R, S, and T lines shift to the low energy side (red shift)

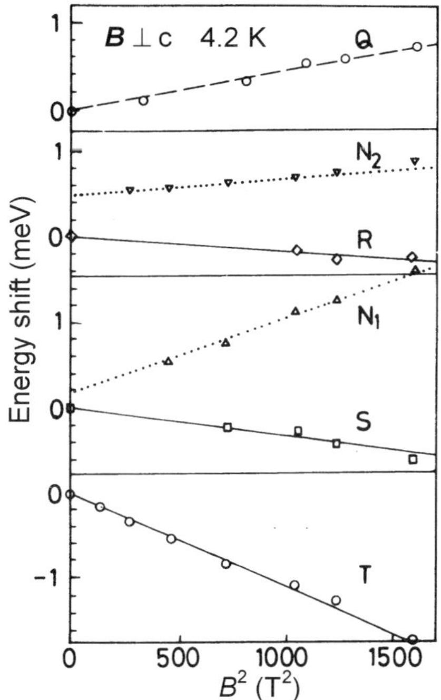

FIG. 5.49. Shift of the photon energies of the absorption peaks Q, R, S, T, and N as a function of B^2 [470]. B ||c, k ⊥c.

as the field is increased, while the other lines show a blue shift. It was found that the magnetic field dependence of all the lines is linear to B^2 as shown in Fig. 5.49. The absorption lines and their magnetic field dependence are now well explained on the basis of the stacking fault excitons and the cationic exciton model [470, 471]. As in the case of PbI_2, the bulk exciton transition is predominantly from s to p states of the cations in the crystal (Bi). When a stacking fault is introduced in the layered crystal, an interchange of positions of the cations and the anions takes place through the plane of the fault. As the exciton wave function is confined in nearly one unit cell, such a stacking fault greatly modifies the ground state of the bulk exciton and produces new exciton lines. Based on the cationic exciton model, eight states are expected to be observed theoretically. Experimentally, six states — Q, R, S, T lines and N_1 and N_2 lines which grow in magnetic fields — are observed. These lines were successfully elucidated as the stacking fault excitons [470, 471].

5.6.4 $(C_6H_{13}NH_3)_2PbI_4$

As another unique example of layered structure, we introduce here perovskite type crystal $(C_6H_{13}NH_3)_2PbI_4$. This material belongs to a family of crystals

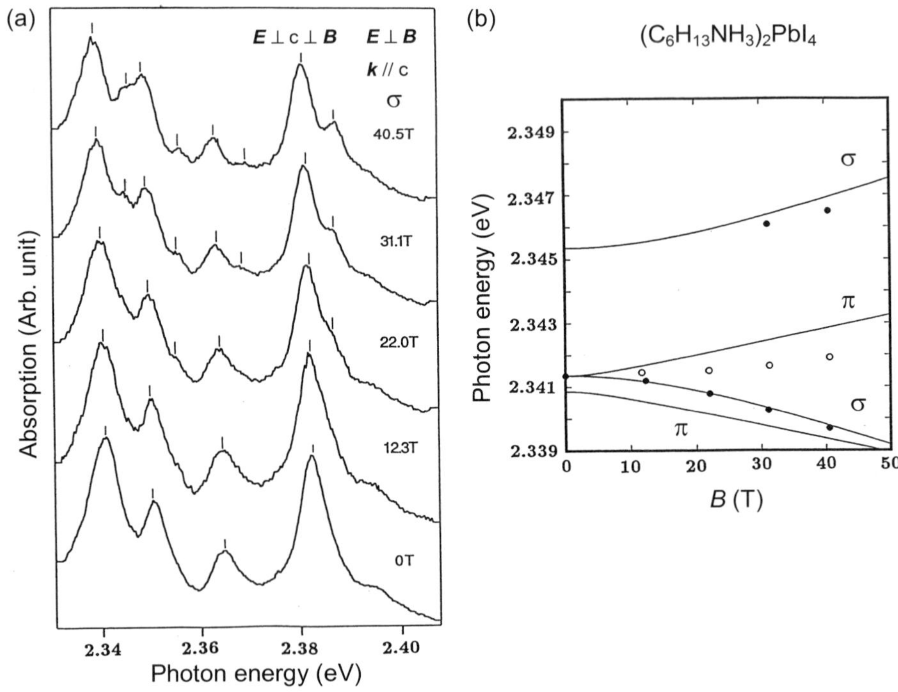

FIG. 5.50. (a) Magneto-absorption spectra of $(C_6H_{13}NH_3)_2PbI_4$ in the Voigt configuration [472]. ($B\perp$ c) and $E\perp B$ (σ polarization). (b) Photon energies of the absorption peaks of the first peak series for the Voigt configuration. Solid lines are theoretical lines calculated based on the cationic exciton model.

$(C_nH_{2n+1}NH_3)_2PbI_4$ which are natural quantum well crystals consisting of PbI_4 monomolecular layers sandwiched between organic barrier layers formed by chains represented as $C_6H_{13}NH_3$ [487]. It is of interest because the charge carriers are tightly confined in monolayers of PbI_4, and moreover the confinement is by a low dielectric organic material, so that the dielectric confinement may be observed [488].

Figure 5.50 (a) shows the magneto-absorption spectra of $(C_6H_{13}NH_3)_2PbI_4$ at different magnetic fields for the Voigt configuration. In this case the magnetic field is perpendicular to the c-axis ($\boldsymbol{B}\perp$ c) and $\boldsymbol{E}\perp\boldsymbol{B}$ (σ polarization) [472]. At zero field, four peaks are observed as well as a small peak in the higher energy range. Each of the four peaks showed a splitting and energy shift in magnetic fields. The four peaks show qualitatively almost similar behavior in a magnetic field. Therefore, the origin of the four peaks is probably the phonon side bands, rather than splitting due to the crystal fields or spin-orbit interaction, although the details are unclear. Splitting and shift of the peaks were also observed for the σ polarization as well as the π polarization. Figure 5.50 (b) shows the photon energies of the absorption peaks as a function of magnetic field as compared with theoretical calculations based on the cationic exciton model. Thus, as in the case of PbI_2, the magnetic field dependence of the photon energies of the split peaks for the σ and π polarizations is well explained by the cationic exciton model [481].

For the Faraday configuration with $\boldsymbol{B} \parallel$ c, on the other hand, a diamagnetic shift proportional to B^2 and a linear Zeeman effect were observed. The g-factor and the diamagnetic shift coefficient of the four absorption lines are listed in Table 5.2. Both values are not very different among the four lines. It should be noted that the g-factor is much larger than in PbI_2, but σ is much smaller. From these values, we can obtain the binding energy of the exciton and the Bohr radius employing the Wannier exciton model, although it is not clear whether the Wannier exciton model is applicable for such a small radius exciton.

Table 5.2 Values of the g factor and the diamagnetic shift coefficient σ of the four absorption lines in $C_6H_{13}NH_3)_2PbI_4$ [472]

	line 1	line 2	line 3	line 4	PbI_2
g	1.80	1.81	1.70	1.61	0.89
σ (10^7eV/T^2)	3.53	3.12	2.70	2.16	9.7

5.7 Faraday rotation

5.7.1 *Principle of Faraday rotation*

Faraday rotation is a fundamental optical phenomenon related to the polarization of the electromagnetic wave in a magnetic field discovered by M. Faraday in 1845. When linearly polarized light propagates through a sample in the parallel direction to the applied magnetic field, the polarization plane is rotated by an angle θ in proportion to magnetic field B and the sample thickness d. The relation is expressed in the normal case as

$$\theta_F = VBd. \tag{5.49}$$

The coefficient V is called the Verdet constant. As the angle is proportional to the field, the Faraday rotation is useful as a supplementary tool for calibrating the magnetic field accurately. The Faraday rotation is usually measured with a set up as shown in Fig. 5.51. The sample is sandwiched by two linear polarizers. The first polarizer is called simply "polarizer"and the second one is "analyzer". The transmitted light intensity through the polarizer, sample and the analyzer is

$$I = I_0 \cos^2(\theta_F + \alpha_0), \tag{5.50}$$

where I_0 is the intensity of the incident light and α_0 is difference in the polarization angle between the two polarizers. An example of the raw data of the

Faraday rotation for GaP in very high magnetic fields is shown in Fig. 5.52. It is seen that the signal has a cosine function as shown in (5.50) and the two signals for $\alpha_0 = +45°$ and -45° are symmetric.

The normal modes of a radiation propagating along the magnetic field (z-direction) are the circularly polarizations whose electric vector are expressed by complex quantities,

$$\hat{\sigma}^+ = \frac{1}{\sqrt{2}}(E_x + iE_y), \tag{5.51}$$

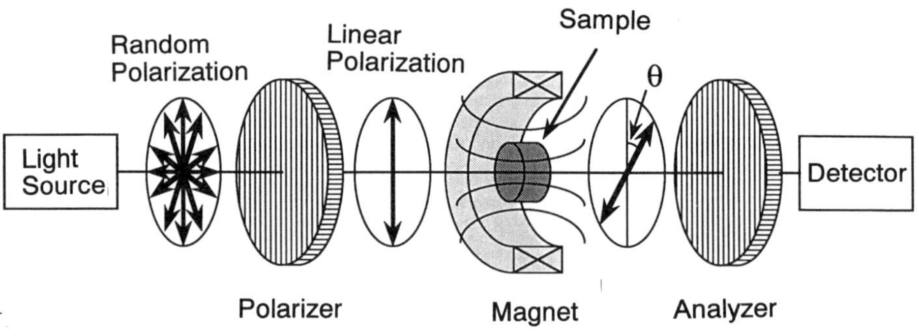

FIG. 5.51. Experimental set up for observing Faraday rotation.

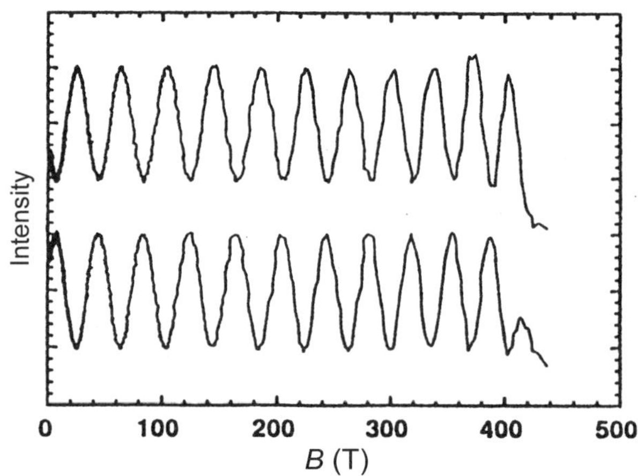

FIG. 5.52. Faraday rotation signal observed for GaP, measured in very high magnetic fields up to 420 T generated by electromagnetic flux compression. The upper and the lower trace are for the angles $\alpha_0 = +45°$ and –45°, respectively. The wavelength of the radiation was 632.8 nm and the thickness of the sample 0.45 mm.

$$\hat{\sigma}^{-} = \frac{1}{\sqrt{2}}(E_x - iE_y). \tag{5.52}$$

$$\tag{5.53}$$

The dielectric constant of an isotropic material in a magnetic field is written as

$$\boldsymbol{\epsilon} = \begin{pmatrix} \epsilon_{xx} & \epsilon_{xy} & 0 \\ -\epsilon_{xy} & \epsilon_{yy} & 0 \\ 0 & 0 & \epsilon_{zz} \end{pmatrix}. \tag{5.54}$$

The off-diagonal component ϵ_{xy} is induced by the magnetic field, and usually we can put $\epsilon_{xy} \propto B$. The dielectric constant is represented by the real and imaginary parts of the refraction index as

$$\epsilon = (n - ik)^2. \tag{5.55}$$

If we define the dielectric constant and the refractive index for the circular polarizations as

$$\epsilon_{\pm}(\omega) = \epsilon_{xx} \mp i\epsilon_{xy} = (n_{\pm} - ik_{\pm})^2, \tag{5.56}$$

These quantities can be used as the scalar coefficients for the circularly polarized field vector. The electric vector of a radiation linearly polarized along the x-direction can be written as a combination of these two normal modes,

$$E_x = E_{\sigma^+} + E_{\sigma^-}. \tag{5.57}$$

Faraday rotation arises if there is a difference in the refractive index between the two normal modes of the circularly polarized radiations (n_+ and n_-), because the two components propagate with different velocities through the sample and form a linearly polarized radiation at the exit with a different phase from that at the entrance. The dielectric constant consists of the real part $\epsilon^{(1)}$ and the imaginary part $\epsilon^{(2)}$ and is given by

$$\epsilon_{\pm} = \epsilon_{\pm}^{(1)} + i\epsilon_{\pm}^{(2)}. \tag{5.58}$$

The absorption coefficient is given by

$$\alpha_{\pm} = -\frac{\omega\epsilon_{\pm}^{(2)}}{cn_{\pm}}. \tag{5.59}$$

The rotation angle is expressed as

$$\begin{aligned} \theta_F &= \frac{\omega d}{2c}(n_+ - n_-) \\ &= \frac{\omega d(n_-^2 - n_+^2)}{4nc} \\ &= \frac{\omega d(\epsilon_-^{(1)} - \epsilon_+^{(1)})}{4nc}. \end{aligned} \tag{5.60}$$

(5.61)

In semiconductors, the dielectric constant and thus the refractive index are predominantly determined by the virtual optical transition between the valence band and the conduction band, and represented as

$$\epsilon_{\pm}(\omega) \cong n_{\pm}^2 \cong \sum_{ij} \frac{f_{ij}}{\mathcal{E}_{ij}^2 - (\hbar\omega)^2}, \tag{5.62}$$

where $\mathcal{E}_{ij}$ is the difference energy between the initial and the final states of the virtual transition and the f_{ij} is the matrix element of the transition.

The difference in the refractive index is written as

$$n_+ - n_- = \frac{n}{\mathcal{E}} \Delta\mathcal{E}_{ij}, \tag{5.63}$$

where $\Delta\mathcal{E}_{ij}$ is the transition energy difference between the right circularly polarized radiation and the left circularly polarized radiation. We then obtain

$$\theta_F = \frac{\omega d}{2c} \frac{\partial n}{\partial \mathcal{E}} \Delta\mathcal{E}_{ij}. \tag{5.64}$$

This is a useful expression for analyzing Faraday rotation in terms of the involved transition energy $\Delta\mathcal{E}_{ij}$.

5.7.2 *Interband Faraday rotation*

It was shown in the previous section that the Faraday rotation is determined by the difference in the dielectric constant between the left circular and right circular polarizations. In semiconductors, the dielectric constant is largely affected by the virtual optical transition between the valence band and the conduction band. Boswarva, Howard, and Lidiard calculated the Faraday rotation angle in semiconductors with a band-gap $\mathcal{E}_g$ based on a simple model of the Landau level structure and the dielectric constant determined by the virtual interband transitions [489]. Although the valence band structure in magnetic fields is complicated, as seen in Section 2.4.3, Boswarva *et al.* assumed that the energies of the conduction band and the valence band are expressed as follows.

$$\begin{aligned}
\mathcal{E}_c(N, k_z, m_J) &= \mathcal{E}_{c0} + \left(N + \frac{1}{2}\right)\hbar\omega_{cc} + g_c\mu_B B m_J + \frac{\hbar^2 k_z^2}{2m_c}, \\
\mathcal{E}_v(N, k_z, m_J) &= \mathcal{E}_{v0} + \left(N + \frac{1}{2}\right)\hbar\omega_{cv} + g_v\mu_B B m_J + \frac{\hbar^2 k_z^2}{2m_v}.
\end{aligned} \tag{5.65}$$

If we separate the Landau level splitting, the Zeeman energy of the conduction band and the valence band is as shown in Fig. 5.53. The selection rules of the transition are $\Delta N = 0$, $\Delta k_z = 0$, and $\Delta m_J = +1$ for the σ^+ radiation and Δm_J

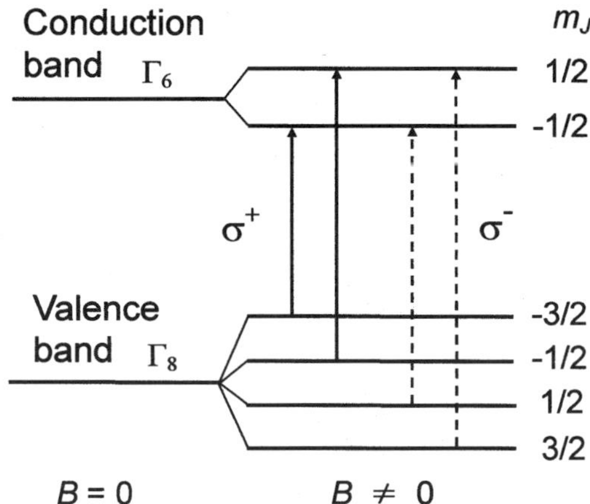

FIG. 5.53. Allowed transitions for the circularly polarized radiation σ^+ and σ^+ in the case of interband transition just focusing on the spin states neglecting the Landau level splitting.

$= -1$ for the σ^- radiation. The transition energy between the allowed transition is

$$\hbar\omega_N^{\pm} = \mathcal{E}_g + \left(N + \frac{1}{2}\right)\hbar(\omega_{cc} + \omega_{cv}) + \frac{\hbar^2 k_z^2}{2\mu^*} \pm \gamma B, \tag{5.66}$$

where

$$\frac{1}{\mu^*} = \frac{1}{m_c} + \frac{1}{m_v}, \qquad \gamma = \frac{g_c + g_v}{2\hbar}\mu_B. \tag{5.67}$$

The Faraday rotation angle is then

$$\theta_F(\omega) \propto \frac{1}{2\pi l^2}\sum_N \int dk_z \left[\frac{1}{\left(\hbar\omega_N^- + \frac{\hbar^2 k_z^2}{2\mu^*}\right)^2 - (\hbar\omega)^2} - \frac{1}{\left(\hbar\omega_N^+ + \frac{\hbar^2 k_z^2}{2\mu^*}\right)^2 - (\hbar\omega)^2}\right]. \tag{5.68}$$

The integration with respect to k_z can be readily executed. For the wavelength range longer than the absorption edge, $\hbar\omega_c \ll \mathcal{E}_g$, (5.68) can be expanded as a power series of γB, and taking the first term, we obtain,

$$\theta_F(\omega) \propto \frac{B}{\omega}\gamma B \sum_{N=0}^{\infty}\left[\frac{1}{(\omega_N - \omega)^{3/2}} - \frac{1}{(\omega_N + \omega)^{3/2}} - \frac{3\omega}{\omega_N^{5/2}}\right], \tag{5.69}$$

where we put

$$\hbar\omega_N = \mathcal{E}_g + (N + 1/2)\frac{\hbar e B}{\mu^*}, \tag{5.70}$$

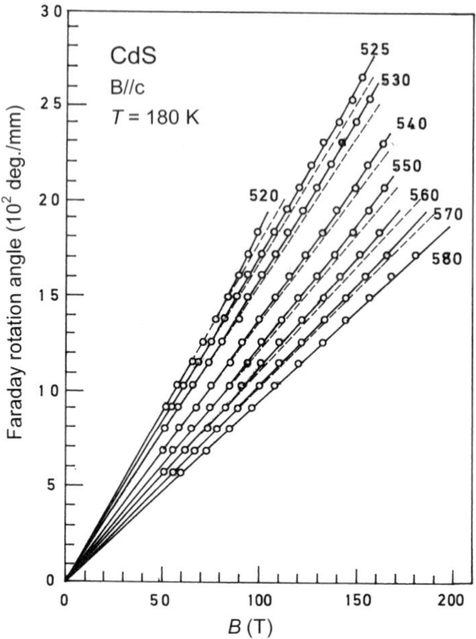

FIG. 5.54. Faraday rotation angle in CdS as a function of magnetic field at different wavelengths [491]. The broken lines show the linear extrapolations of the low field data. The number indicated for each curve shows the wavelength of the radiation in nm.

and added a constant term so that the θ should tend to 0 for $\omega \to 0$. For weak magnetic field, $eB/\mu^* \ll \omega_g - \omega$, the summation over N can be replaced by integration over dB and we obtain,

$$\theta_F(\omega) \propto \frac{\mu^*}{e\omega}\gamma B \left[\frac{1}{(\omega_g - \omega)^{1/2}} - \frac{1}{(\omega_g + \omega)^{1/2}} - \frac{\omega}{\omega_g^{3/2}} \right]. \tag{5.71}$$

Although the calculation by Boswarva *et al.* is based on an simple model of the Landau level structure, (5.71) is instructive, showing that the θ_F is proportional to magnetic field in the low field limit and that it increases as $\hbar\omega$ approaches the band gap $\mathcal{E}_g$. This feature can be conveniently used for the calibration of the magnetic field. However, for very high magnetic fields where $\hbar\omega_c$ becomes non-negligible in comparison to $\mathcal{E}_g$, the higher order terms give a non-linear contribution. From (5.71), we can expect that $\theta_F(B)$ deviates from the linear dependence showing an up-turn. Fowler *et al.* found such a non-linearity in the Faraday rotation in CdS [490]. Figure 5.54 shows the data of the Faraday rotation observed by Imanaka *et al.* for CdS in very high magnetic fields [491]. We can see that for small $\hbar\omega$, the rotation angle *vs.* field curve is almost linear, but as it is increased, there is a tendency to become superlinear. The deviation from the linear dependence becomes larger as the wavelength is decreased approaching the absorption edge.

When $\hbar\omega$ exceeds $\mathcal{E}_g$, interband transition between the Landau levels takes place, and we can expect that the rotation angle shows oscillatory behavior as a function of photon energy like the absorption spectra. The oscillatory Faraday rotation in such a regime was investigated by Nishina, Kolodziejczak and Lax in detail for Ge [492, 493]. Oscillatory Faraday rotation was also investigated in InSb by Pidgeon and Brown [13].

6

DILUTED MAGNETIC SEMICONDUCTORS

6.1 Diluted magnetic semiconductors in high magnetic fields

6.1.1 *Diluted magnetic semiconductors*

When semiconductor crystals are doped with magnetic impurity ions, the crystals possess both semiconducting and magnetic properties. Thus many possibilities for application to useful devices are opened up. A number of early attempts to clarify the properties of magnetic impurities in semiconductors were made in order to study the electron spin resonance in magnetic ions. In the early 1970s, a Warsaw-based group developed a new class of diluted magnetic semiconductors, II-VI semiconductors doped with transition metal ions such as Mn^{2+} or Fe^{2+} [494–496]. The magnetic ions are substituted for the cations of the mother crystals isoelectrically, and act as randomly distributed paramagnetic ions. This series of crystals was named "Semimagnetic semiconductors" or "Diluted magnetic semiconductors" (DMS) [497]. The magnetic ions greatly modify the electric and optical properties, particularly the spin Zeeman splitting of the crystals. As the Zeeman energy becomes enormous (giant Zeeman splitting), the crystals show very large Faraday rotation (giant Faraday rotation). This can be utilized for various optical device applications. The magnetism of the doped crystals themselves is of interest in the light of random magnetism. For instance, antiferromagnetic exchange interaction between the magnetic ions causes complicated magnetization. Magnetic ion pairs sitting in the nearest neighbor sites bring about successive step-wise magnetization increase. Triples and larger clusters cause more complicated magnetization. A spin glass phase was found in crystals with some concentration of magnetic ions at low temperatures [498]. In high magnetic fields, we can study these interactions by optical and magnetization measurements.

Many different crystals of DMS have been grown and investigated so far. At first, the research effort was devoted to II-VI compounds [497], but later good crystals of IV-VI compounds such as PbMnTe or PbMnSe have also been developed [499]. More recently, highly doped DMS based on III-V compounds such as GaAs or InAs have been attracting much attention because some of them undergo the ferromagnetic transition [500–502]. Now study of this class of semiconductors has become a very important research field in connection with many possible future applications such as "spintronics" [503].

6.1.2 *Energy levels in the conduction band and the valence band*

Magnetic interactions of conduction electrons and holes with magnetic ions are characterized by the $s-d$ or $p-d$ interaction represented by

$$\mathcal{H}'_{\rm sd} = \sum_{R_n} J(r - R_n)\boldsymbol{\sigma} \cdot \boldsymbol{S}_n, \tag{6.1}$$

where $\boldsymbol{\sigma}$ is the spin operator of a conduction electron or a hole and $\boldsymbol{\sigma}$ is the spin operator of the magnetic ions at a site $\boldsymbol{R}_n$. The energy band structure is influenced by such a magnetic interaction. For zinc blende crystals, we can take the basis functions as defined in Eq. (2.149). The interaction (6.1) has matrix elements between the lattice periodic functions,

$$\alpha = < S|J|S >, \tag{6.2}$$
$$\beta = < X|J|X >=< Y|J|Y >=< Z|J|Z > . \tag{6.3}$$

The parameters α and β represent the strength of the incorporation of the effect of magnetic interaction in the energy levels. The $s,\, p-d$ interaction introduces an additional term in the $\boldsymbol{k} \cdot \boldsymbol{p}$ Hamiltonian of the Pidgeon-Brown model,

$$D' = \begin{pmatrix} D'_a & 0 \\ 0 & D'_b \end{pmatrix}.$$

D'_a and D'_b are for the a-set and b-set in the Pidgeon-Brown model. The additional term is represented using α and β [504–506, 277],

$$D'_a = \begin{pmatrix} \frac{1}{2}N_0\beta < S_z > & 0 & 0 & 0 \\ 0 & \frac{1}{2}N_0\alpha < S_z > & 0 & 0 \\ 0 & 0 & -\frac{1}{6}N_0\beta < S_z > & -\frac{i\sqrt{2}}{3}N_0\beta < S_z > \\ 0 & 0 & \frac{i\sqrt{2}}{3}N_0\beta < S_z > & \frac{1}{6}N_0\beta < S_z > \end{pmatrix},$$
$$D'_b = \begin{pmatrix} -\frac{1}{2}N_0\beta < S_z > & 0 & 0 & 0 \\ 0 & \frac{1}{6}N_0\beta < S_z > & 0 & \frac{i\sqrt{2}}{3}N_0\beta < S_z > \\ 0 & 0 & -\frac{1}{2}N_0\beta < S_z > & 0 \\ 0 & -\frac{i\sqrt{2}}{3}N_0\beta < S_z > & 0 & -\frac{1}{6}N_0\beta < S_z > \end{pmatrix}. \tag{6.4}$$

Here $< S_z >$ stands for the expectation value of the z-component of the spin of the magnetic ions, and N_0 is the number of unit cells per unit volume in the crystal. The magnetic energy levels are obtained by adding (6.4) to the Pidgeon-Brown Hamiltonian. The Hamiltonian including the $s, p-d$ interaction (6.4) is called the modified Pidgeon-Brown Hamiltonian.

Solving the Schrödinger equation with the Hamiltonian including the $s, p-d$ interaction, we can obtain the Landau levels in diluted magnetic semiconductors. In the low field range at low temperatures, the spin Zeeman effect is enhanced

Table 6.1 Exchange interaction parameter α and β in Mn-doped II-VI compounds [507]

Material	$N_0\alpha$(eV)	$N_0\beta$(eV)
$Cd_{1-x}Mn_xTe$	0.22	−0.88
$Zn_{1-x}Mn_xSe$	0.18	−1.05
$Cd_{1-x}Mn_xSe$	0.26	−1.11
$Zn_{1-x}Mn_xSe$	0.26	−1.31
$Cd_{1-x}Mn_xS$	0.22	−1.80

due to the contribution of $< S_z >$ and the Zeeman splitting of the magnetic ions is larger than the Landau level splitting $\hbar\omega_c$ of the host crystal. The energies of the conduction band and the valence band $\mathcal{E}_N^{c,v}$ are roughly expressed as

$$\mathcal{E}_N^c \approx \alpha x N_0 \sigma < S_z > + \left(N + \frac{1}{2}\right)\hbar\omega_c, \tag{6.5}$$

$$\mathcal{E}_N^v \approx \beta x N_0 M_J < S_z > + \mathcal{E}_N(0), \tag{6.6}$$

where $\mathcal{E}_N(0)$ denotes the Landau level splitting of the valence bands. Therefore, the spin splitting as shown in Fig. 5.53 predominates the Landau level formation in each spin-split level. However, as the field is increased, the spin splitting and the Landau level splitting compete and the levels show complicated structure. In the high field limit where $< S_z >$ saturates, the levels tend to the Landau level energies of the host crystal.

The effect of the doping of magnetic ions is determined by the $s, p-d$ interaction characterized by $N_0\alpha$ and $N_0\beta$. The magnitudes of these parameters in some representative DMS are listed in Table 6.1 [507]. It should be noted that the sign is opposite between α and β, and the absolute magnitude of β is considerably larger than that of α.

6.2 Giant Faraday rotation

6.2.1 *Faraday rotation in diluted magnetic semiconductors*

As we have seen in Section 5.7, Faraday rotation is caused by virtual electronic transitions between energy levels. In magnetic substances, normally the term $\Delta\mathcal{E}_{ij}$ in (5.63) is proportional to the magnetization of the material. Therefore, the rotation angle θ_F has a term proportional to the magnetization M, and is expressed as follows:

$$\begin{aligned}\theta_F &= \theta_{\text{dia}} + \theta_M \\ &= V_{\text{dia}} B d + V_M M d.\end{aligned} \tag{6.7}$$

Here the first term is the ordinary diamagnetic Faraday rotation as discussed in Section 5.7, and the second term is the magnetic term. As it contains the term proportional to M, the Faraday rotation is a convenient method for measuring the magnetization optically. This is especially useful in pulsed high magnetic fields.

In DMSs, the Landau level splitting is usually smaller than the spin Zeeman splitting up to some field, so that the optical transitions between the magnetic energy levels can be roughly approximated by those as shown in Fig. 5.53 in the low field limit. Such a selection rule gives rise to a component of Faraday rotation proportional to $< S_z >$, and this component provides the dominant contribution to the Faraday rotation. Faraday rotation in DMSs containing $3d$ transition metals is then represented as

$$\theta_F = \chi(\hbar\omega)Bd + \theta_0(\hbar\omega)B_S\left(\frac{Sg\mu_B B}{kT}\right)d. \tag{6.8}$$

Here, the first and the second terms correspond to $\theta_{\rm dia}$ and θ_M in Eq. (6.7). B_S is the Brillouin function,

$$B_S(x) = \frac{2S+1}{2S}\coth\left(\frac{2S+1}{2S}x\right) - \frac{1}{2S}\coth\left(\frac{1}{2S}x\right). \tag{6.9}$$

and S is the spin of the magnetic ions. The coefficients $\chi(\hbar\omega)$ and $\theta_0(\hbar\omega)$ are functions of photon energy. Reflecting the large effect of the magnetization in the second term, DMSs generally show a very large Faraday rotation, called "giant Faraday rotation" [508]. Near the wavelength of exciton absorption line, the rotation is enhanced further. This may be useful for some devices as Faraday rotators. Figure 6.1 shows the Faraday rotation signal of $Cd_{1-x}Mn_xTe$ ($x = 0.1$) in high fields up to 150 T generated by the single turn coil technique [509]. By using a Wallaston prism as an analyzer, we can obtain two signals with a 90° difference in α in Eq. (5.50) simultaneously. In this case, $\alpha = \pm$ 45°.

It should be noted that the direction of the rotation by the two terms in (6.8), is just opposite to each other. Corresponding to the sign of the magnetization, we call the first term the diamagnetic rotation, and the second term the paramagnetic rotation. In a relatively low magnetic field, the absolute value of the second term is much larger than that of the first term, so the rotation angle increases sharply in proportion to $< S_z >$. As the field is increased, $< S_z >$ tends to saturate and then the first term whose absolute value is almost proportional to external magnetic field increases its importance. Then the rotation angle starts decreasing.

Such a competition between the diamagnetic and paramagnetic Faraday rotations was actually observed in $Zn_{1-x}Mn_xSe$ [510–512]. In order to investigate the wavelength dependence of Faraday rotation as well as the field dependence in pulsed magnetic fields, it is convenient to use streak spectroscopy techniques, as will be mentioned in Section 7.3.2. Figure 6.2 shows the streak spectra of Faraday rotation in $Zn_{0.985}Mn_{0.015}Se$ at low temperatures. The left-hand spectrum was obtained by using a CCD system in a long pulse field up to 40 T at

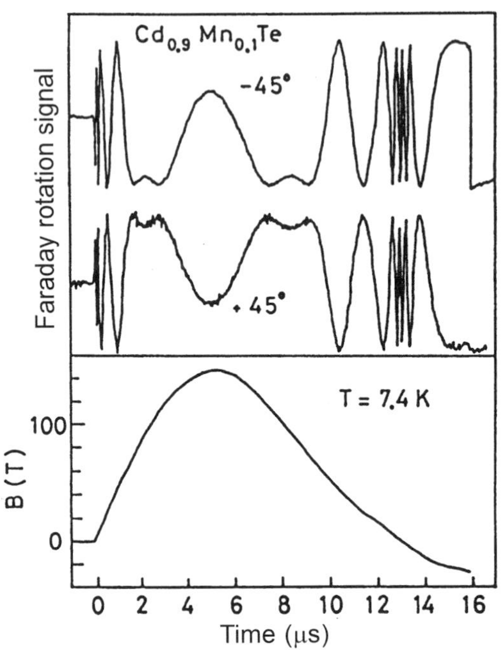

FIG. 6.1. Faraday rotation in $Cd_{0.90}Mn_{0.10}Te$ in high magnetic fields for $\alpha_0 = 45°$ and –45° [509].

1.6 K. The right-hand figure was obtained by using an image converter system in a short high pulse field up to 150 T at 7 K. The figures show only the signal intensity, with the abscissa indicating the time, and the ordinate indicating the photon energy. From these streak spectra, we can derive the signal intensity and the Faraday rotation angle as functions of magnetic field and photon energy. Figure 6.3 shows the Faraday rotation angle as a function of magnetic field for different wavelengths. We can see clearly a remarkable feature of the Faraday rotation in DMS. In the whole photon energy range, Faraday rotation first increases rapidly with field in the low field range, but at some field around 5 T, it takes a maximum and starts decreasing in higher fields. This feature can be explained by Eq. (6.8) in terms of the competition between the paramagnetic component (magnetic term) and the diamagnetic component, as mentioned above. Actually, the observed rotation angle is in good agreement with the curve calculated from Eq. (6.8) by taking suitable $\chi(\hbar\omega)$ and $\theta_0(\hbar\omega)$.

Regarding the photon energy dependence of the Faraday rotation, the coefficients $\chi(\hbar\omega)$ and $\theta_0(\hbar\omega)$ derived by curve fitting as shown in Fig. 6.3 are plotted in Fig. 6.4 [512]. The coefficient $\theta_0(\hbar\omega)$ increases as the photon energy $\hbar\omega$ approaches the band gap. This tendency is in reasonably good agreement with the theoretical expression for DMS [513],

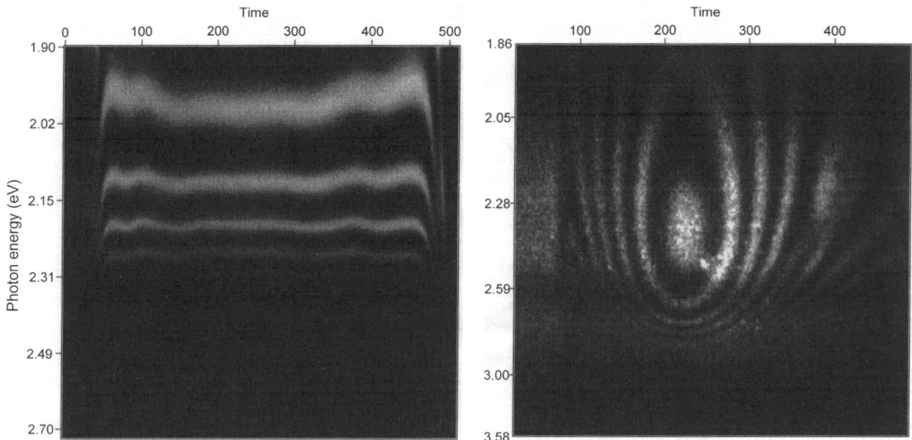

FIG. 6.2. Streak spectra of the Faraday rotation in $Zn_{0.985}Mn_{0.015}Se$ [512]. Left: up to 40 T at 1.6 K. Right: up to 150 T at 7 K. The abscissa indicates the time axis, and the magnetic field rises from left to right nearly sinusoidally and returns to zero in the right edge, taking a maximum in the middle. The ordinate indicates the photon energy. The light and shade demonstrate the light intensity detected by a CCD or an image converter camera as a function of wavelength and magnetic field. The interval between the stripes indicates the rotation of 180°.

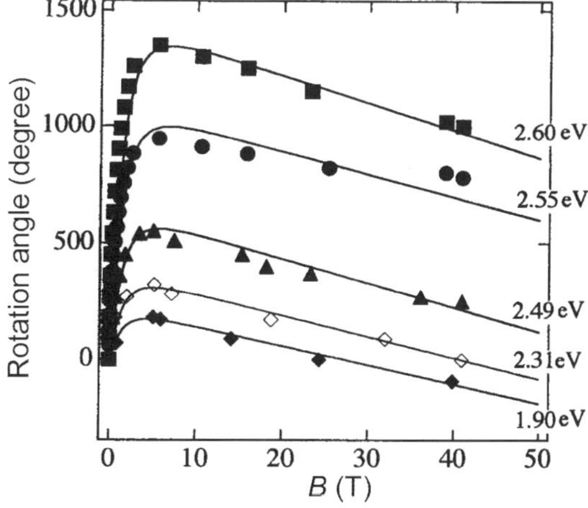

FIG. 6.3. Faraday rotation angle of $Zn_{0.985}Mn_{0.015}Se$ up to 40 T at 1.6 K. The points are experimental data and the solid lines were calculated from Eq. (6.10) [511].

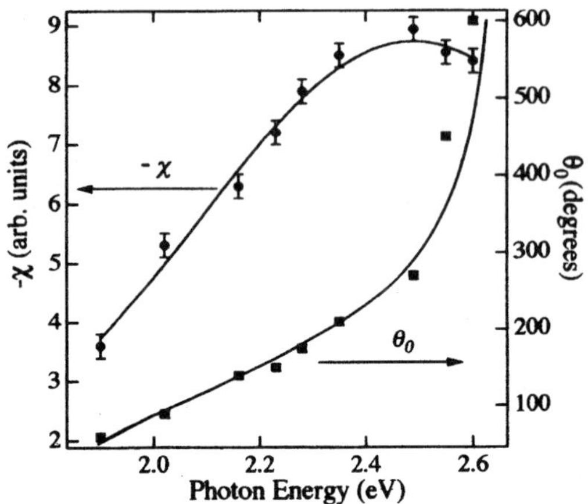

FIG. 6.4. Photon energy dependence of the coefficients of the Faraday rotation angle, $\chi(\hbar\omega)$ and $\theta_0(\hbar\omega)$ in $Zn_{0.985}Mn_{0.015}Se$ [512].

$$\begin{aligned} \theta_M &= \left(\frac{\sqrt{F_0}d}{2\hbar c} \frac{\beta - \alpha}{g_M \mu_B} M \right) \frac{1}{\mathcal{E}_0} \frac{y^2}{(1-y^2)^{3/2}}, \\ y &= \frac{\hbar\omega}{\mathcal{E}_0}, \end{aligned} \tag{6.10}$$

where F_0 is a constant involving the oscillator strength of the excitonic transition, g_M is the Lande g-factor of the Mn^{2+} spins and $\mathcal{E}$ is a single oscillator energy.

The coefficient $-\chi(\hbar\omega)$ also increases as the energy approaches the band gap, which is expected from the nature of the interband Faraday rotation. However, an interesting point is that it reaches a maximum at around 2.4 eV and starts decreasing with increasing energy. This anomaly is directly visible in the right-hand figure of Fig. 6.2 as an egg-shaped light region centered at around 2.4 eV. The unusual energy dependence may be explained in terms of a contribution of an electronic transition within Mn^{2+} ions which exists in the vicinity of 2.3 eV [514].

In the left-hand figure of Fig. 6.2, the stripes show small wavy changes. This is due to the magnetization steps as will be discussed in Section 6.3.2. From the magnetic field where the changes take place we can determine the antiferromagnetic exchange interaction constant between Mn^{2+} ions [510].

6.2.2 *Exciton spectra in diluted magnetic semiconductors*

Besides the giant Faraday rotation, measurement of the exciton spectra is also a powerful tool for studying the magnetic properties of DMS. In relatively wide gap DMS, such as CdMnTe, CdMnSe or ZnMnSe, the exciton peak is clearly seen in the photoluminescence spectra. The magnetic field dependence of the energy of the 1s exciton in DMS is given by

$$\Delta\mathcal{E}(B) = -\frac{1}{2}N_0(\alpha - \beta) < S_z > -\frac{1}{2}|g_e - g_h|B + \sigma B^2. \qquad (6.11)$$

The first term comes from the change of the band gap. The second term is from the Zeeman splitting and the last term is from the diamagnetic shift. The magnetic field dependence of the first term is very large at low temperatures, so that the exciton line shows a large shift to the low energy side in a magnetic field. From such a shift, we can obtain the information of the spin state $< S_z >$. Experimental data of excitons in DMS will be discussed for quasi-two-dimensional systems in Section 6.4.

6.3 Exchange interactions in DMS

6.3.1 *Magnetization curve in magnetic semiconductors*

In DMS, there are possibilities for two magnetic ions to occupy lattice sites closely located with each other. The nearby magnetic ions interact with each other by some magnetic interactions. The magnetic interactions of randomly distributed magnetic ions have been investigated for many years. It is known that the $3d$ ions in II-VI compounds interact with each other antiferromagnetically. This results in the zero cross of the $1/\chi - T$ curve at a negative temperature [515], and the appearance of the spin glass phase at low temperatures in highly doped crystals [497, 498].

The antiferromagnetic interaction between magnetic ions substitutionally doped in II-VI compounds is the direct superexchange via the anions. The superexchange interaction constant J is determined by the bond angle and thus by the ionic radii of the magnetic ions and ions in host crystals [507].

When the concentration of the magnetic ions is small, each magnetic ion is almost independent and the crystals show almost ideal paramagnetic behavior represented by

$$M = Ng\mu_B S B_S\left(\frac{gS\mu_B B}{kT}\right), \qquad (6.12)$$

where N is the number of magnetic ions and $B_S(x)$ is the Brillouin function. As the concentration is increased, the probability for magnetic ions to come to the nearest neighbor (NN) sites increases rapidly. This results in the decrease of the magnetization due to their antiferromagnetic interactions. The probability of an ion being separated from other magnetic ions in a zinc blende crystal with a concentration of x is

$$P_1 = (1 - x)^{12}, \qquad (6.13)$$

as there are 12 NN sites. Similarly, the probability of ions forming a NN pair is

$$P_2 = 12x(1 - x)^{18}. \qquad (6.14)$$

As regards the probability for three ions to come to the NN, we have to think of two cases depending on the sites for the third ion to sit, the open triangle and the closed triangle,

Table 6.2 Probability for single ions, pairs, and triangles.

x	0.1	0.2	0.3	0.4	0.5
P_1	0.282	0.0687	0.0138	2.18×10^{-2}	2.44×10^{-4}
P_2	0.180	0.0432	5.86×10^{-3}	4.87×10^{-4}	2.29×10^{-5}
P_{3a}	0.104	0.0255	2.44×10^{-3}	1.13×10^{-4}	2.41×10^{-6}
P_{3b}	0.0236	7.08×10^{-3}	8.45×10^{-4}	5.05×10^{-5}	1.43×10^{-6}

$$P_{3a} = 18x^2(1-x)^{23}(7-5x), \tag{6.15}$$

for an open triangle (three ions with two outer ions not as NN), and

$$P_{3b} = 24x^2(1-x)^{22}, \tag{6.16}$$

for closed triangles (three ions with each ion having the other ions as NN). Table 6.2 shows the probabilities P_1, P_2, P_{3a}, and P_{3b} as a function of x. P_1 is a rapidly decreasing function of x, and for large x, the probability of ions of forming pairs, triangles, and larger clusters becomes very large. Actually, P_1 is surprisingly small for substantial amount of x. As the concentration of the magnetic ions x increases, the simple magnetization of the single ions is decreased and the magnetization of groups of ions contributes to the total magnetization. Because of the antiferromagnetic interactions, the magnetization of the large clusters does not show saturation up to high magnetic fields. Heiman *et al.* found in $Cd_{1-x}Mn_xTe$ with relatively large x that the effect of the Mn clusters on the magnetization is approximately represented by a linear function of magnetic field, and the total magnetization can be written as,

$$M(B) = M_s B_{\frac{5}{2}}\left(\frac{5\mu_B B}{k_B T_{\text{eff}}}\right) + \chi_{\text{cluster}} B, \tag{6.17}$$

for $x > 0.2$ [516]. Nicholas *et al.* derived an empirical formula [517],

$$\begin{aligned} < S_z > &= \left[S_0 + \left(\frac{5}{2} - S_0\right) B_{\frac{5}{2}}(\xi_2)\right] B_{\frac{5}{2}}(\xi_1), \\ \xi_i &= \frac{5g\mu_B H}{2k_B(T+T_i)} \quad (i = 1, 2). \end{aligned} \tag{6.18}$$

Here, S_0 is an effective spin of an isolated single magnetic ion, T_1 is a parameter to represent the weak antiferromagnetic interaction in a low magnetic field, and T_2 is a parameter to represent stronger antiferromagnetic interaction of the clusters.

Equations (6.17) and (6.18) show that the magnetization of DMS with a substantial fraction of x does not saturate easily even in high magnetic fields. Isaacs *et al.* measured the magnetization M of $Cd_{1-x}Mn_xTe$ by means of the Faraday rotation up to 150 T for different x up to 0.5. The experimental results

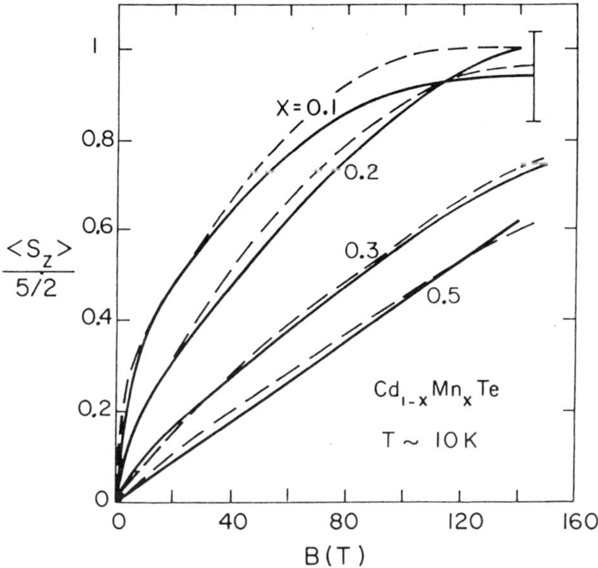

FIG. 6.5. Faraday rotation angle in $Cd_{1-x}Mn_xTe$ for different x [518].

of $< S_z >$ deduced from the rotation angle are shown in Fig. 6.5 [518]. Here the diamagnetic term θ_{dia} in (6.7) is subtracted from the rotation angle to extract θ_M from θ_F. As clearly seen in the figure, the magnetization does not show a saturation even in a field of 150 T for large x. The curves are in reasonably good agreement with the theoretical calculation of $< S_z >$,

$$< S_z > = < S_z >_1 + < S_z >_2 + < S_z >_{3a} + < S_z >_{3b} + < S_z >_{\text{cluster}}, \quad (6.19)$$

where $< S_z >_1$, $< S_z >_2$, $< S_z >_{3a}$, $< S_z >_{3b}$ and $< S_z >_{\text{cluster}}$ denote the spin alignment of the single ions, the pairs, the open triangles, the closed triangles, and other magnetic ions in clusters of four or more ions.

6.3.2 *Magnetization steps*

The spin alignment of the pairs, the triangles and the clusters shows complicated magnetic field dependences, because of the crossing of the spin levels with increasing field. The change of the total magnetization arising from the level crossing is generally small, but a change of the spin alignment of the pairs can be clearly observed at low temperatures as a step-wise change of the magnetization.

In the case of Mn ions in II-VI compounds, the spin S of an ion is 5/2. Two ions in the nearest neighbor sites $\boldsymbol{S}_1$ and $\boldsymbol{S}_2$ are coupled to form a pair by the antiferromagnetic exchange interaction. The energy of the pair is

$$\mathcal{E}_{\text{ex}} = -2J_{NN}\boldsymbol{S}_1 \cdot \boldsymbol{S}_2. \quad (6.20)$$

Here J_{NN} is the antiferromagnetic exchange interaction constant. The total spin of the pair is

$$\boldsymbol{S}_T = \boldsymbol{S}_1 + \boldsymbol{S}_2. \tag{6.21}$$

Since $S_T^2 = S_1^2 + S_2^2 + 2\boldsymbol{S}_1 \cdot \boldsymbol{S}_2$,

$$\begin{aligned} \boldsymbol{S}_1 \cdot \boldsymbol{S}_2 &= \frac{1}{2}\left[S_T^2 - (S_1^2 + S_2^2)\right] \\ &= \frac{1}{2}\left[S_T(S_T+1) - 2 \cdot \frac{5}{2}\left(\frac{5}{2}+1\right)\right], \quad (S_T = 0,1,2,3,4,5). \end{aligned} \tag{6.22}$$

The exchange energy (6.20) is

$$\mathcal{E}^0_{\mathrm{ex},S_T} = -J_{NN}\left[S_T(S_T+1) - \frac{35}{2}\right]. \tag{6.23}$$

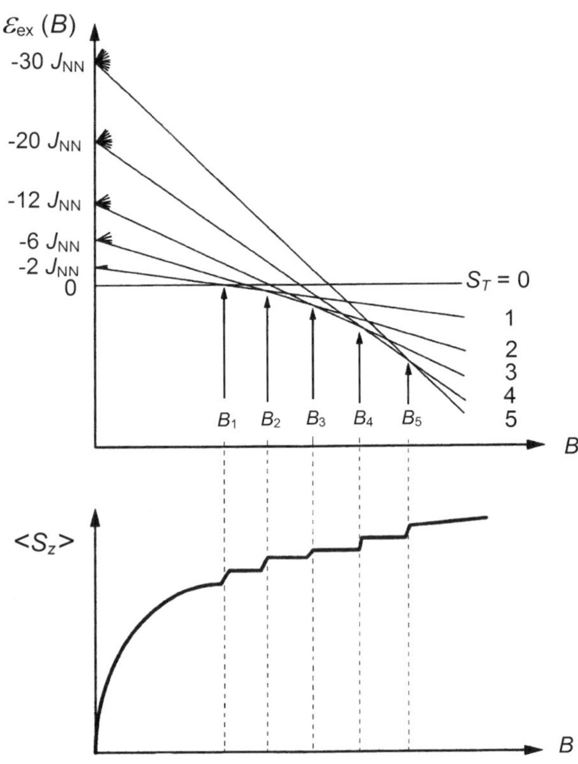

FIG. 6.6. (Upper panel) Energy levels of a magnetic ion pair with $S = 5/2$ in a magnetic field. (Lower panel) Variation of $< S_z >$ as a function of B. Steps appear at B_1, B_2, B_3, B_4, B_5.

The ground state is a non-magnetic singlet state with $S_T = 0$, and the excited states $S_T = 1$–5 are arranged as shown in Fig. 6.6. Under a magnetic field, the excited states show the Zeeman splitting and their energies are written as,

$$\mathcal{E}_{\text{ex}}(B) = \mathcal{E}^0_{\text{ex},S_T} + g\mu_B S_z B, \quad (S_z = -S_T, S_T + 1, \cdots, S_T). \tag{6.24}$$

As the magnetic field is increased, the higher excited states cross the lowest state successively. As the new ground state has a magnetic moment in the field direction, $< S_z >$ of the ground state shows a step-wise increase by 1 at each level crossing. The first level cross-over takes place between the $S_T = 0$ and ($S_T = 1$, $S_z = -1$) state at a field,

$$B_1 = \frac{2|J_{NN}|}{g\mu_B}. \tag{6.25}$$

Similarly, the N-th cross-over occurs between the ($S_T = N$–1, $S_z = N$–1) state and the ($S_T = N$, $S_z = -N$) at

$$B_N = \frac{2N|J_{NN}|}{g\mu_B}, \quad (N = 1, 2, 3, 4, 5). \tag{6.26}$$

The interval between the adjacent cross-over fields has essentially a uniform value $\Delta B_N = 2|J_{NN}|/\mu_B B$. Corresponding to the change of the ground state, the magnetization shows step-wise changes at B_N. The change is very sharp at $T = 0$, but as the temperature is raised, the steps show thermal broadening. The magnetization steps due to the ion pairs have been observed in many II-VI compounds doped with transition metal magnetic ions. From the field positions of the steps, the antiferromagnetic exchange constant J_{NN} between the magnetic ions can be obtained [519]. Due to the antiferromagnetic interaction of each member of the pair with other spins in the crystal, the first step occurs at a field slightly higher than the value given by (6.25), but the interval between the steps is almost $\Delta B_N = 2|J_{NN}|/\mu_B B$ [520]. The steps were observed not only in magnetization, but also in magneto-resistance [521], Faraday rotation [510], and in photoluminescence [522]. Table 6.3 lists the values of J_{NN} in various II- VI and IV-VI compounds obtained from the magnetization steps and from various other techniques, which have been reported in different literatures.

6.3.3 *Narrow gap DMS*

Narrow gap semiconductors with many free carriers such as HgTe or HgSe-based DMS are promising materials for practical applications, because transport properties can be controlled by magnetic ions. It is interesting to see whether the steps can be observed in transport properties as well. Yamada *et al.* actually observed the change of the magneto-resistance at the position of the steps in $Hg_{1-x-y}Cd_xMn_yTe$ [521]. In narrow gap DMS, there is a possibility of the contribution of the virtual intraband and interband transitions of the valence electrons to the antiferromagnetic exchange interaction besides the direct exchange. These

Table 6.3 Exchange interaction constant J_{NN} in II-VI and IV-VI compounds reported in different literatures.

Material	$-J_{NN}/k_B$ (K)
$Zn_{1-x}Mn_xS$	16.9 ±0.6
$Zn_{1-x}Mn_xSe$	9.9 ±0.9, 12.2±0.3
$Zn_{1-x}Mn_xTe$	10.8 ±0.8, 10.1±0.4
$Cd_{1-x}Mn_xS$	8.6 ±0.9, 10.5 ±0.3
$Cd_{1-x}Mn_xSe$	8.3 ±0.7, 7.9 ±0.5
$Cd_{1-x}Mn_xTe$	6.3 ±0.3, 6.1 ±0.3, 6.2 ±0.2
$Hg_{1-x}Mn_xSe$	6.0 ±0.5, 5.3 ±0.5
$Hg_{1-x}Mn_xTe$	5.1 ±0.5, 4.3 ±0.5
$Pb_{1-x}Mn_xSe$	≈ 1
$Pb_{1-x}Mn_xTe$	≈ 1
$Zn_{1-x}Fe_xSe$	22 ±2

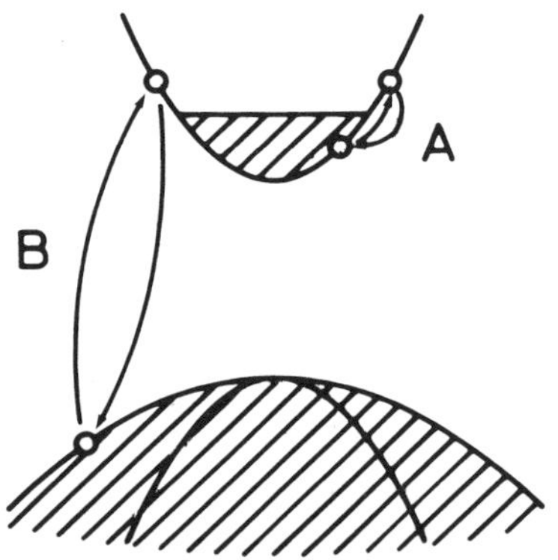

FIG. 6.7. Virtual electron transitions contributing the indirect exchange interaction. Process A (intraband transition) gives the RKKY interaction [524], and B (interband) the Bloembergen-Rowland interaction [525].

antiferromagnetic interactions are called indirect exchange interactions [523]. Figure 6.7 shows the virtual electron transitions which induce the exchange interactions via valence electrons. The transitions are caused by the s-d interaction given by (6.1). The second order perturbation energy of the s-d interaction is

$$\Delta\mathcal{E}^{(2)} = \sum_{ifkk'} \frac{f(\mathcal{E}_{ik})[1 - f(\mathcal{E}_{fk'})]}{\mathcal{E}_{ik} - \mathcal{E}_{fk'}} | < ik|\mathcal{H}_{s-d}|fk' > |^2. \tag{6.27}$$

When we write the exchange energy between $\boldsymbol{S}_i$ and $\boldsymbol{S}_j$ as

$$\mathcal{E}_{\text{ex}} = \sum_{ij} I_{ij}(\boldsymbol{R}_{ij})\boldsymbol{S}_i \cdot \boldsymbol{S}_j, \tag{6.28}$$

the exchange interaction constants I_{ij} for the transition A and B are

$$I_{ij}^{\text{intra}} = \frac{1}{\Omega^2}\sum_{kk'} \frac{f(\mathcal{E}_{ck} - f(\mathcal{E}_{ck'})}{\mathcal{E}_{ck} - \mathcal{E}_{ck'}} \exp[i(\boldsymbol{k} - \boldsymbol{k}') \cdot \boldsymbol{R}_{ij}] \times \sum_{\sigma\sigma'=\pm} | < u_{ck\sigma}|J(\boldsymbol{r})\sigma_z|u_{ck'\sigma'} > |^2, \tag{6.29}$$

$$I_{ij}^{\text{inter}} = \frac{2}{\Omega^2}\sum_{kk'} \frac{f(\mathcal{E}_{vk'}[1 - f(\mathcal{E}_{ck})]}{\mathcal{E}_{vk'} - \mathcal{E}_{ck}} \exp[i(\boldsymbol{k} - \boldsymbol{k}') \cdot \boldsymbol{R}_{ij}] \times \sum_{\sigma\sigma'=\pm} | < u_{ck\sigma}|J(\boldsymbol{r})\sigma_z|u_{ck'\sigma'} > |^2. \tag{6.30}$$

Here $u_{nk\sigma}$ is the Bloch part of the valence electron, $\boldsymbol{k}$ and $\boldsymbol{k}'$ are the wave vector, $\boldsymbol{R}_{ij}$ is the displacement vector between the ions i and j, and f is the distribution function.

The intraband term I_{ij}^{intra} gives the well-known RKKY (Ruderman-Kittel-Kasuya-Yosida) interaction and results in

$$I_{ij}^{\text{intra}} = \frac{4J^2 m^* k_F^4}{(2\pi)^3} F(2k_F R_{ij}), \quad F(x) = \frac{x\cos x - \sin x}{x^4}, \tag{6.31}$$

where k_F is the Fermi wave vector [524]. It is an oscillating function of $k_F R_{ij}$. The interband term I_{ij}^{inter} is called the Bloembergen-Rowland interaction [525]. In quaternary compound $Hg_{1-x-y}Cd_xMn_yTe$, the band gap $\mathcal{E}_g$ can be systematically changed by controlling x, fixing the Mn concentration y almost constant.

The exchange constant of $Hg_{1-x-y}Cd_xMn_yTe$ was measured for different band gaps by observing the magnetization steps [526]. The fields of the magnetization steps are shown in Fig. 6.8 up to $N = 3$. It was found the step positions showed no significant dependence on $\mathcal{E}_g$. Thus it was concluded that the Bloembergen-Rowland mechanism provides only a negligibly small contribution to the exchange constants in DMS. The carrier concentration dependence of the step position is also indiscernible in DMS. Therefore, the RKKY interaction is negligibly small in DMS in the relatively small x range where the steps are clearly observed ($x < 0.1$).

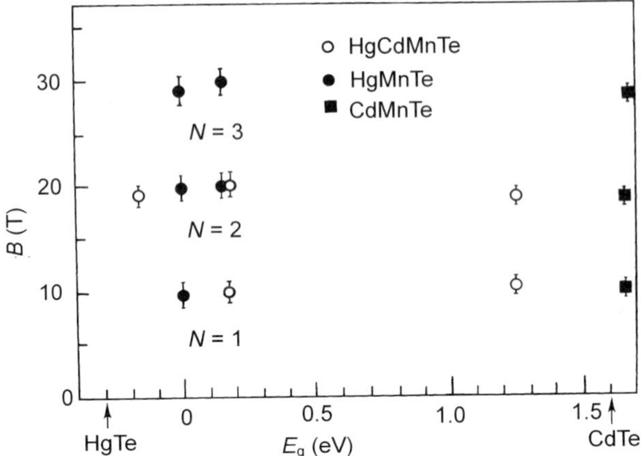

FIG. 6.8. Energy gap dependence of the magnetic field positions B(T) of the steps (N = 1–3) in $Hg_{1-x-y}Cd_xMn_yTe$ [526].

Among narrow gap DMS, Fe-doped Hg-compounds, HgTe:Fe and HgSe:Fe are known to have quite unique properties [527]. The localized Fe^{2+} states are degenerate in the conduction band. In the case of HgSe:Fe, the Fe level is located 210 meV above the bottom of the conduction band and provide carriers in the conduction band, by changing the valency of Fe ions from Fe^{2+} to Fe^{3+} up to a critical density of $n_{\mathrm{Fe}} = 5\times10^{18}$ cm^{-3}, the electron density corresponding to $\mathcal{E}_F = 210$ meV. Above this Fe density, the Fermi level is pinned to the Fe^{2+} state, and this situation brings about a number of interesting properties, such as the three-dimensional analogue of the quantum Hall effect [528]. The carrier mobility is enormously enhanced to the order of 10^6 cm^2/V·s above this critical density and the Shubnikov-de Haas effect can be clearly observed. The reason for the mobility enhancement is not totally clarified, but a few theoretical models have been proposed, such as the formation of a space charge superlattice by Fe^{+3} ions [529] or the localized zero-scattering potential model [530]. Figure 6.9 (a) shows the Shubnikov-de Haas effect observed in the transverse magneto-resistance of $Hg_{0.99}Mn_{0.1}Se$:Fe [530]. Due to the high mobility of the samples, the magneto-resistance oscillation can be clearly observed up to the highest measured temperature (92 K). Also, a tail is observed in the low field side of each peak reflecting the density of states of the Landau levels. The Fe doping does not induce any remarkable increase of the spin splitting, but because of the presence of the Mn ions, the effect of the enhanced spin splitting is clearly seen. In addition, we can see that the spin splitting is decreased as the temperature is increased.

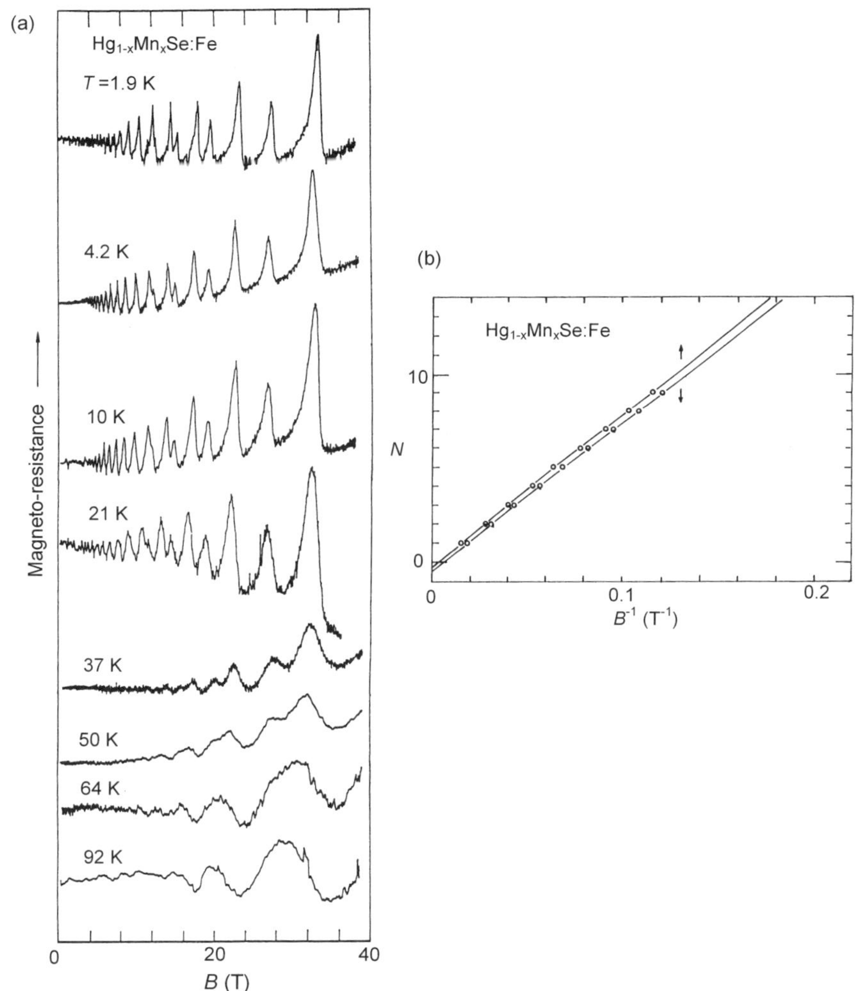

FIG. 6.9. Shubnikov-de Haas effect in $Hg_{1-x}Mn_xSe$:Fe ($x = 0.01$) [530]. $n_{Fe} = 5 \times 10^{19}$ cm^{-3}. (a) Experimental traces of the transverse magneto-resistance at different temperatures. (b) Plot of the relation between the quantum number N and the inverse of the peak positions of the Shubnikov-de Haas oscillation for the spin-up and spin down-states.

6.4 Properties of two-dimensional DMS systems

6.4.1 *Exchange interaction in DMS quantum wells*

In quantum wells comprising DMS layers, we can observe characteristic features different from bulk crystals. There are three different types of DMS quantum wells. The magnetic barrier type has a non-magnetic quantum well sandwiched by DMS layers as barriers. DMS such as $Cd_{1-x}Mn_xTe$ has a larger band gap than its mother crystal (such as CdTe), so that they can be used as barrier layers to form a quantum well structure such as $Cd_{1-x}Mn_xTe/CdTe/Cd_{1-x}Mn_xTe$. Wave functions of electrons and holes confined in the non-magnetic layers partly penetrate into the magnetic barriers, so that they can have magnetic properties. Uniform quantum wells (uniform type) have a structure in which magnetic ions are doped uniformly both in well layers and barrier layers. An example of this type is $Cd_{1-y}Mn_yTe/Cd_{1-x}Mn_xTe/Cd_{1-y}Mn_yTe$ (x and y are different). The third type is the structure with magnetic ions only in quantum wells (magnetic well type). An example is $Cd_{1-y}Mg_yTe/Cd_{1-x}Mn_xTe/Cd_{1-y}Mg_yTe$. We can investigate the confinement effect of DMS in such a structure.

As mentioned in Section 6.2, the giant Zeeman shift of excitons in the magneto-photoluminescence spectra provides information about the magnetization $< S_z >$. Ossau *et al.* found that in CdTe/CdMnTe quantum wells $< S_z >$ per volume increases and it becomes larger and larger as the well width is decreased [531]. They attributed this effect to the decrease of the number of pairs and larger clusters near the interface and the interface effect becomes larger as the well width is decreased. In this situation, $< S_z >$ per volume increases and the value of T_1 (effective temperature representing the antiferromagnetic exchange interaction as that appearing in Eq. (6.18)) decreases as the well width L_w is reduced. This model has now been generally accepted.

It is an intriguing question how the $s-d$ and $p-d$ exchange interaction constants $N_0\alpha$ and $N_0\beta$ depend on the dimensionality or well width. There have been several reports on this problem both from theory and from experiments [532–534]. In the uniform type, the portion of the wave function that penetrates into the barrier layers responds to the exchange interaction in a similar manner to that in the well layer so that the penetration effect should be very different from that in the magnetic barrier type. Mackh *et al.* found in the giant Zeeman shift of the uniform type quantum well $Cd_{1-y}Mn_yTe/Cd_{1-x}Mn_xTe/Cd_{1-y}Mn_yTe$ ($x = 0.05$ and $y = 0.25$) $N_0\alpha$ and $N_0\beta$ have a clear well width dependence as the well width is decreased [533]. Regarding the well width dependence of $N_0\alpha$ and $N_0\beta$, Bhattacharjee made theoretical calculations based on the effective mass approximation including the wave vector dependence of the exchange interaction constant [532]. To represent the modification of these parameters by the quantum well potential, Bhattacharjee introduced the reduction factor,

$$\rho^{QW} = \frac{\alpha\rho^e - \beta\rho^h}{\alpha - \beta}, \tag{6.32}$$

as a function of L_w. $\alpha\rho^e$ and $\beta\rho^h$ are the modified values of $s-d$ and $p-d$ exchange parameters (α and β) in a quantum well [532]. The ratio of the Zeeman splitting of the heavy hole exciton to the magnetization is reduced by this factor. ρ^{QW} is a quantity which is unity at the $L_w \to 0$ and $L_w \to \infty$ (3D) limits. ρ^{QW} first decreases as L_w is increased from 1, and after reaching a minimum increases again tending to 1. It was asserted that the theoretical calculation well explained

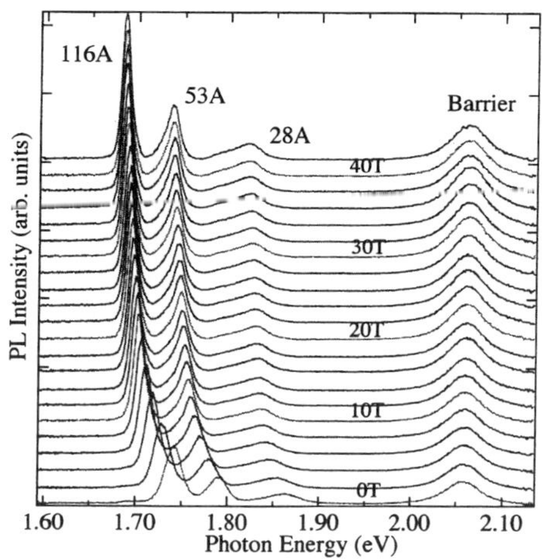

FIG. 6.10. Magneto-photoluminescence spectra for a magnetic well sample $Cd_{1-x}Mn_xTe/Cd_{1-y}Mg_yTe$ ($x = 0.08$, $y = 0.24$) with different well widths at $T = 1.5$ K [522].

the experimental results of the L_w dependence of ρ^{QW} obtained by Mackh *et al.* [533]. In this treatment, the effect of the penetration of the wave function is not taken into account as the experiment was done for the uniform samples.

Studies of the magnetic well samples $Cd_{1-x}Mn_xTe/Cd_{1-y}Mg_yTe$ have been made by Yasuhira *et al.* in pulsed high magnetic fields [522,535,536]. Figure 6.10 shows the magneto-photoluminescence spectra for a magnetic quantum well sample $Cd_{0.92}Mn_{0.08}Te/Cd_{0.76}Mg_{0.24}Te$. Three single quantum wells with different well widths were fabricated on the same substrate. Exciton peaks were observed from each quantum well. They show a shift to lower energies as the magnetic field is increased. Figure 6.11 shows plots of the peak energies of excitons in samples with slightly different compositions from Fig. 6.10. The peak energy shows a large red shift reflecting the giant Zeeman energy shift of the exciton line in quantum wells of the magnetic well sample $Cd_{1-x}Mn_xTe/Cd_{1-y}Mg_yTe$ for different well widths.

The contribution of the ordinary diamagnetic shift of the excitons in host crystals was subtracted using the data for $CdTe/Cd_{1-y}Mg_yTe$ samples with the same well widths. It is clear that the Zeeman shift of the excitons has a significant well width dependence. The resulting well width dependence of the reduction factor ρ^{QW} extracted from such data after correcting the penetration effect is shown in Fig. 6.12. It was found that the experimental data of ρ^{QW} have a general tendency similar to the theoretical curve by Bhattacharjee [532]. However, the absolute value of ρ^{QW} obtained from the experiment is about 20% smaller than the theoretical values. Apart from the quantitative agreement, qualitatively the experimental results are thus well explained by such a model.

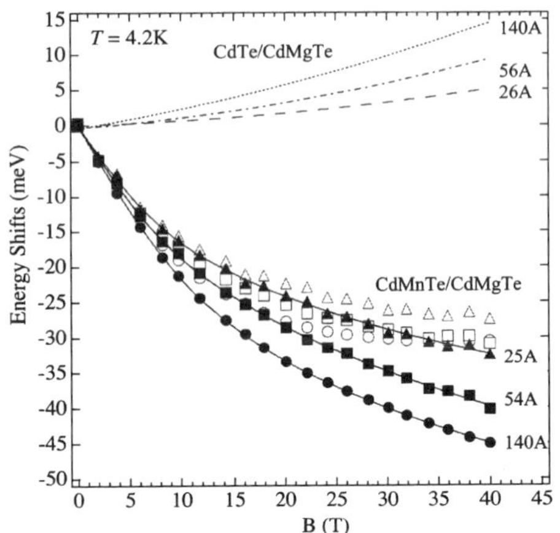

FIG. 6.11. Giant Zeeman energy shift of the exciton line in quantum wells of the magnetic well sample $Cd_{1-x}Mn_xTe/Cd_{1-y}Mg_yTe$ ($x = 0.049$, $y = 0.37$) for different well widths as a function of magnetic field [522]. The open points indicate the raw data. Thin lines showing a positive shift are the data in non-magnetic quantum well with nearly the same well width. Closed points denote the energy shift due to the Zeeman effect after subtraction of the positive shifts.

6.4.2 *Magnetization step in quantum wells*

In the previous subsection, it was shown that the $s-d$ or $p-d$ exchange interaction is subjected to a significant modification by the well width. Naturally, a question arises whether the $d-d$ nearest neighbor interaction will be affected by the quantum confinement. In fact, it is an intriguing question whether J_{NN} depends on the well width or not. It was found in $Cd_{1-x}Mn_xTe/Cd_{1-y}Mg_yTe$ quantum wells, the magneto-photoluminescence spectra exhibit structures at magnetic fields corresponding to the magnetization steps due to the Mn^{2+} pairs. The structure becomes clearer by taking the derivative of the energy of the peaks with respect to magnetic field. Figure 6.13 shows this derivative $d\mathcal{E}/dB$ as a function of B in $Cd_{1-x}Mn_xTe/Cd_{1-y}Mg_yTe$ ($x = 0.08$, $y = 0.24$) [522]. The structure due to the magnetization steps are clearly at B_1—B_4 corresponding to the magnetization steps. Table 6.4 shows the exchange interaction constant J_{NN} determined from the exciton photoluminescence peaks. From the fields of the structures, the antiferromagnetic exchange constants were obtained for each quantum well as listed in Table 6.4. Apparently, there is no significant well width dependence of the exchange constant including the bulk value. In other words, the direct exchange interaction is not much affected by the quantum confinement. This is in sharp contrast to the $s, p-d$ exchange interaction, which is largely reduced by the confinement.

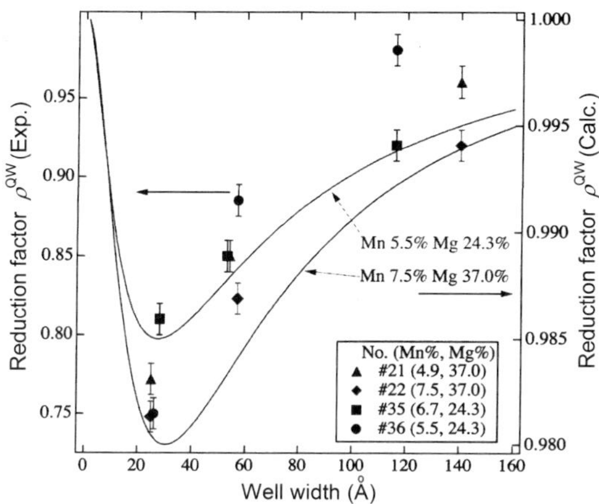

FIG. 6.12. Well width dependence of ρ^{QW} extracted from the giant Zeeman shift of the excitons in $Cd_{1-y}Mg_yTe/Cd_{1-x}Mn_xTe/Cd_{1-y}Mg_yTe$ after subtracting the contribution of the ordinary diamagnetic shift of the host crystal CdTe [535, 536]. The data points were obtained from the shift of the exciton peaks and the solid lines were calculated from (6.32). Note that there is a difference in the vertical axes between theory (right) and experiments (left).

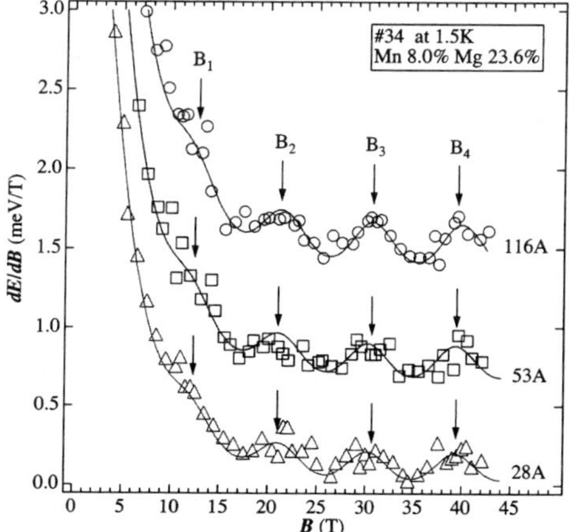

FIG. 6.13. Derivative of the energy of the photoluminescence exciton peaks with respect to magnetic field in $Cd_{0.92}Mn_{0.08}Te/Cd_{0.76}Mg_{0.24}Te$ [522].

6.4.3 *Magnetic field-induced type I - type II transition*

Quantum wells and superlattices with magnetic barrier layers have the unique feature that the barrier height is greatly modified by applying a magnetic field B and varying temperature T, since the g factor is proportional to $< S_z >$ which

Table 6.4 Dependence of the exchange interaction constant J_{NN} on quantum well width in $Cd_{1-x}Mn_xTe/Cd_{1-y}Mg_yTe$ ($x = 0.08$, $y = 0.24$) [522, 536].

L_w (nm)	2.8	5.3	11.3	bulk
$-J_{NN}/k_B$ (K)	6.4±0.3	6.4±0.3	6.5±0.3	6.1±0.3

is sensitively dependent on B and T [537]. For example, when we make quantum wells with CdTe as a well layer and $Cd_{1-x}Mn_xTe$ as a barrier layer, the barrier height rapidly decreases with increasing magnetic field due to the large effective g factor in $Cd_{1-x}Mn_xTe$, and the magnetic field-induced type I-type II transition would occur at some transition magnetic field when the barrier height crosses

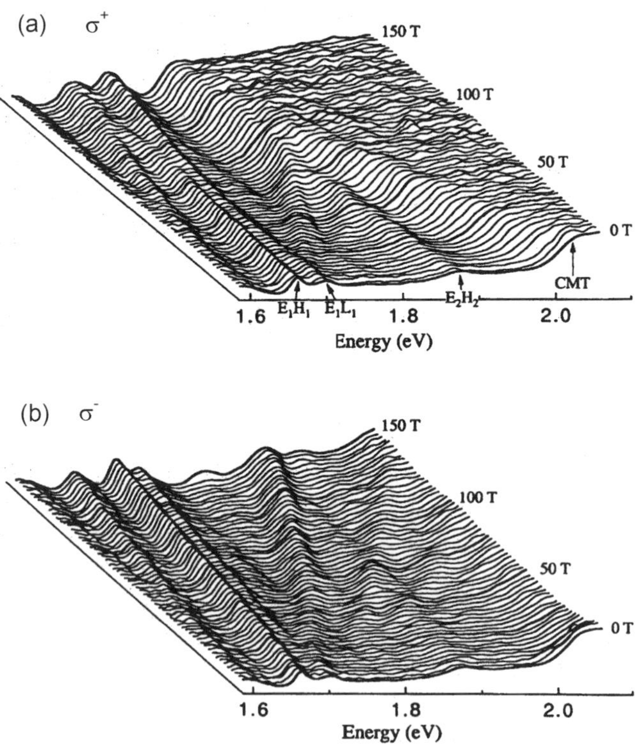

FIG. 6.14. Magneto-absorption spectra in $CdTe/Cd_{1-x}Mn_xTe$ MQW ($x = 0.27$) plotted on the plane of magnetic field and the photon energy [540]. The well width L_w and the barrier layer width L_b are 5.0 nm and 13.1 nm, respectively. (a) σ^+ (b) σ^-. $T \sim 16$K.

zero. Such phenomena have attracted much interest, and many investigations have been made in samples in small x range (0.055–0.1) [538, 539].

In quantum wells with relatively large x (~ 0.3), the transition field becomes very high but the type I-type II transition can be observed very distinctly. Figure 6.14 shows magneto-absorption spectra in CdTe/$Cd_{1-x}Mn_xTe$ MQW ($x = 0.27$) in high magnetic fields up to 150 T [540]. Circularly polarized radiation σ^+ and σ^- were employed in the measurements. At zero magnetic field, we can see the absorption peaks arising from the heavy-hole exciton (E_1H_1) at 1.66 eV, the light-hole exciton (E_1L_1) at 1.70 eV, the heavy-hole exciton of the second subband (E_2H_2) at 1.87 eV, and the absorption of the $Cd_{1-x}Mn_xTe$ cladding layer (CMT) at 2.1 eV were observed. As the magnetic field is increased, the CMT peak shows a large red shift for the σ^+ polarization due to the enhanced Zeeman splitting. The E_1H_1 peak position shows a blue shift and when the magnetic field is increased over 90 T, the absorption intensity for σ^+ is reduced and diminished at 120 T. This decrease corresponds to the type I-type II transition. In contrast, the (E_1H_1) peak for σ^- grows with increasing field. The transition was observed only for the σ^+ polarization, since it takes place only for the heavy holes with $m_J = -3/2$.

6.5 Cyclotron resonance in DMS

6.5.1 *Effect of magnetic ions on the effective mass*

As discussed in Chapter 5, cyclotron resonance (CR) is a powerful tool for obtaining the effective mass and other information concerning the electronic states in semiconductors. In practice, however, it is difficult to employ this technique for DMS crystals, because the mobility of DMSs is generally low (less than a few hundreds $cm^2/V{\cdot}s$ in most cases), due to the high doping of the magnetic impurity ions. Actually, the concentration of the magnetic impurities is much higher (usually 0.1% or more) than the normal doping of donors and acceptors (usually 0.01% at most). For this reason, there have been only a few reports on CR in DMS. Use of high magnetic fields, especially in the megagauss range, solves this problem. Matsuda *et al.* succeeded in a systematic measurement of CR in $Cd_{1-x}Mn_xTe$ in very high magnetic fields up to 150 T [277].

Figure 6.15 shows the experimental traces of the CR in $Cd_{1-x}Mn_xTe$ for different x [277]. A well-defined CR absorption peak is observed in each sample in spite of the low mobility ($\mu \approx 200$–$300\ cm^2/Vs$) as x is increased. Figure 6.16 shows the effective mass of $Cd_{1-x}Mn_xTe$ as a function of x. The cyclotron mass increases with x according to $m^*_{CR} = (0.104 + 0.131x)m$. This is partly due to the increase of the band gap energy $\mathcal{E}_g$, since the doping of Mn is known to increase the band gap. According to the two band model discussed in Section 2.4.1, the effective mass m^* increases with $\mathcal{E}_g$ following an approximate relation,

$$\frac{m}{m^*} \sim \left(1 + 2\frac{P^2}{m\mathcal{E}_g}\right), \tag{6.33}$$

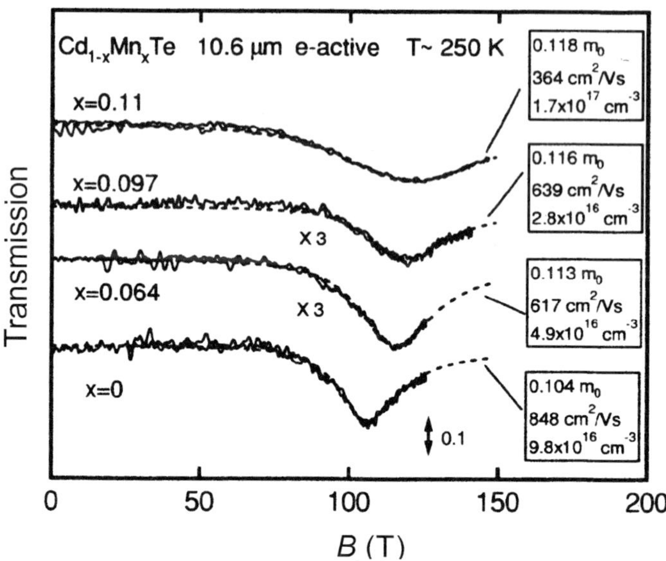

FIG. 6.15. Cyclotron resonance spectra in n-type $Cd_{1-x}Mn_xTe$ for different x in high magnetic fields [277]. Wavelength is 10.6 μm, $T = 250$ K. The obtained effective mass, the mobility and the electron concentration of the samples are shown in the right of the graphs.

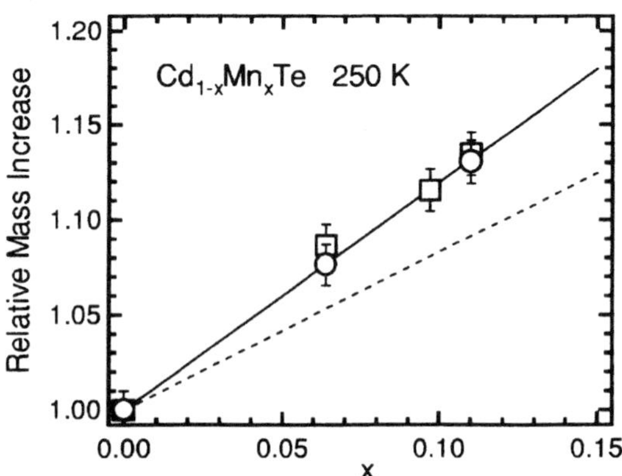

FIG. 6.16. Relative increase of the effective mass of n-type $Cd_{1-x}Mn_xTe$ as a function of x. Circles and squares denote the band edge mass and the cyclotron mass at 10.6 μm, respectively. Solid and dashed lines represent theoretical lines obtained from [541] with and without the $s, p - d$ hybridization effect, respectively.

where P is the $\boldsymbol{k} \cdot \boldsymbol{p}$ matrix element. The observed increase of the effective mass is significantly larger than expected from the increase of the band gap. In fact, in the cyclotron resonance in non-magnetic alloy crystals $Cd_{1-x}Mg_xTe$, the doping of Mg also increases the band gap, but the effective mass was found to increase just following the theoretically expected $\mathcal{E}_g$-dependence. As the origin of the large increase of the effective mass with x, we can deduce that the $s,p-d$ hybridization effect might effectively reduce P^2 in (6.33). Hui, Ehrenreich, and Hass calculated the band structure by the $\boldsymbol{k} \cdot \boldsymbol{p}$ method including the $s,p-d$ hybridization effect [541]. They showed that P is actually reduced with x by the hybridization effect. Thus in the conduction band of DMS, the $s,p-d$ hybridization effect plays a significant role in determining of the effective mass. A similar energy gap dependence was observed also in $Pb_{1-x}Mn_xSe$ [542].

6.5.2 *Cyclotron resonance in III-V magnetic semiconductors*

DMS have increased importance since ferromagnetism was discovered in III-V based compounds, $In_{1-x}Mn_xAs$ [500], and $Ga_{1-x}Mn_xAs$ [501]. Although this class of materials is very important for practical application, the mechanism of the ferromagnetism has not been totally understood in spite of many theoretical investigations [543–545]. It is expected that detailed studies of CR would provide a clue for clarifying the mechanism of the ferromagnetism. The attempt of CR measurement in a ferromagnetic $Ga_{1-x}Mn_xAs$ revealed that the mobility is too low even in the megagauss range due the high doping ($x \geq 0.2$) [546].

In narrow gap DMS crystals $In_{1-x}Mn_xAs$, we have a better chance to observe CR since the mobility is higher than in $Ga_{1-x}Mn_xAs$. Mn doping usually supplies acceptors in III-V compounds as Mn is a group III element, but $In_{1-x}Mn_xAs$ can be either p-type or n-type depending on the growth temperature [547].

In n-type $In_{1-x}Mn_xAs$, reasonably sharp cyclotron resonance peaks for electrons were observed [548–550]. It was found that the effective mass decreases with increasing x. This tendency is opposite to the case of CdMnTe, but it is reasonable if we recall that the band gap energy $\mathcal{E}_g$ decreases with x in the case of $In_{1-x}Mn_xAs$. However, the amount of decrease of the effective mass is smaller than expected from the relation (6.33). This is again due to the $s,p-d$ hybridization. In narrow gap semiconductors like InMnAs, it is essential to treat the conduction band and the valence band together in order to calculate the Landau levels. By solving the modified Pidgeon-Brown model, the observed CR spectra were well explained by assuming proper values of α and β [549,550].

In p-type DMS in which hole-mediated ferromagnetism is realized, cyclotron resonance should shed light on the problem of the mechanism of the ferromagnetism. Figure 6.17 shows cyclotron resonance spectra in p-type $In_{1-x}Mn_xAs$ for different x in the field up to 150 T. Well-defined peaks were obtained for heavy holes (HH) and light holes (LH). It should be noted that the HH peak is observed at lower fields than the LH peak (see discussion in Section 2.4.3). Figure 6.18 shows the cyclotron resonance spectra at even higher magnetic fields [551]. Besides the HH and LH cyclotron resonance peaks a broad peak was

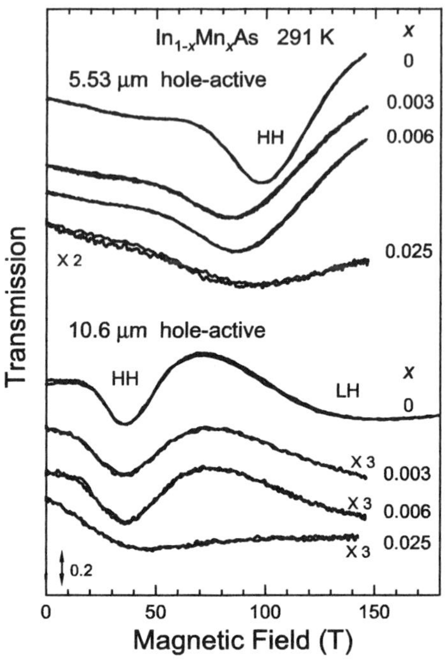

FIG. 6.17. Experimental traces of cyclotron resonance in p-type $In_{1-x}Mn_xAs$ with different x. The wavelength is 5.53 μm and 10.6 μm [551].

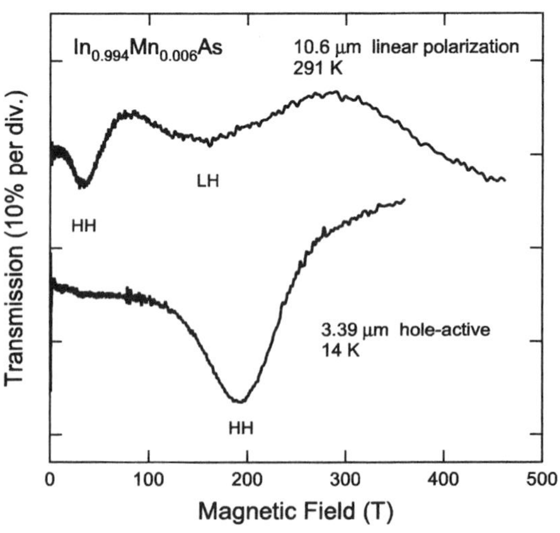

FIG. 6.18. Cyclotron resonance in p-type $In_{0.994}Mn_{0.006}As$ at very high magnetic fields [551]. The wavelength and the temperature are indicated for each figure.

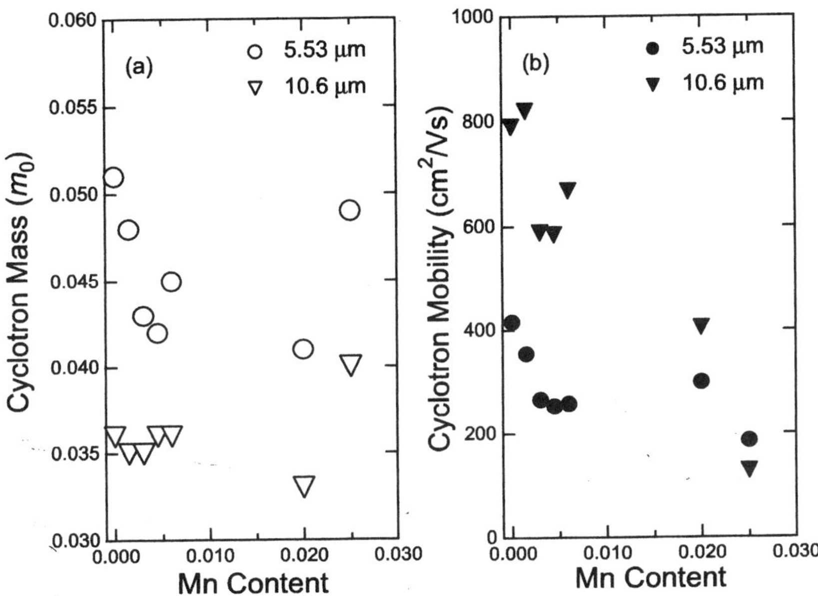

FIG. 6.19. Mn concentration dependence of (a) cyclotron effective mass and (b) mobility for $In_{1-x}Mn_xAs$ [551]. $T = 291$ K.

also observed near 450 T. This can be explained by the multiple transitions of holes between complicated Landau levels in the valence band. Figure 6.19 shows the cyclotron mass and the mobility which were obtained from the cyclotron resonance. Although there is a scattering of the data, it can be concluded that the effective mass does not depend so much on the Mn concentration. This is contrary to the case of the n-type samples, and gives insight into the mechanism of the ferromagnetism. If we are based on the double exchange model [545], the holes should partly possess the d-band nature, and then the effective mass should be very x dependent. On the other hand, according to the Zener type model [543, 544], the holes are almost p-like and the effective mass would be rather insensitive to x. The experimental results seem to be favorable for the latter model. As regards the mobility, it decreases remarkably as the Mn concentration is increased.

In the case of p-type InMnAs, it is an intriguing question whether the cyclotron mass should be affected by the appearance of the ferromagnetism. Figure 6.20 shows CR spectra for an $In_{0.91}Mn_{0.09}As$/GaSb heterostructure sample which exhibits ferromagnetism below $T_c = 55$ K at various temperatures [552]. In the high temperature range down to slightly above the T_c, a broad feature (labeled "A") is observed. with almost no change. At $T \sim 68$ K, which is sill above the T_c, abrupt and dramatic changes are observed in intensity, position and width with decreasing temperature. A significant reduction in line-width

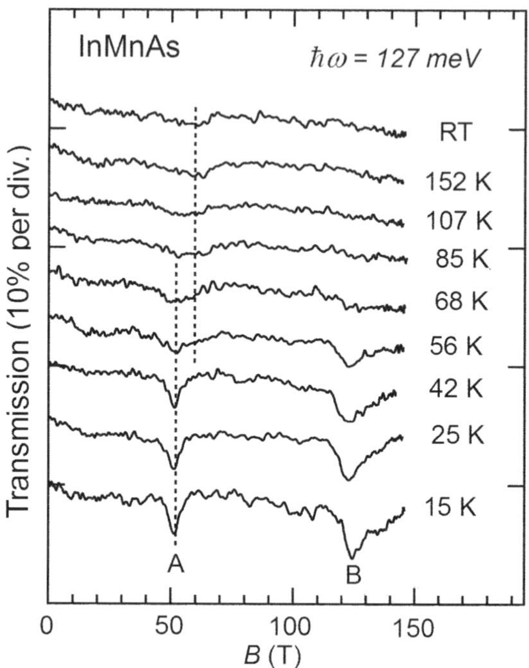

FIG. 6.20. CR spectra for ferromagnetic p-type $In_{0.91}Mn_{0.09}As/GaSb$ single heterostructure (T_c = 55 K) [552]. The transmission of hole-active circular polarized radiation at a wavelength of 10.6 μm is plotted as a function of magnetic field at different temperatures.

and a sudden shift to a lower B occur simultaneously. In addition, it increases in intensity rapidly with decreasing temperature. Furthermore, a second feature (labeled "B") suddenly appears at ~125 T, which also grows rapidly in intensity with decreasing temperature and saturates, similarly to feature of "A".

Peak A was identified as due to the transition between the heavy hole levels. The observed features of the cyclotron resonance have not been understood completely, but the temperature-dependent peak shift is attributed to the increase of carrier-Mn exchange interaction resulting from the increase of magnetic ordering at low temperatures [552, 553].

7

EXPERIMENTAL TECHNIQUES FOR HIGH MAGNETIC FIELDS

In this chapter, we describe key aspects of experimental techniques, particularly those for generating pulsed high magnetic fields and for measurements of solid state physics.

7.1 Generation of high magnetic fields

7.1.1 *Steady magnetic fields*

Today, high magnetic fields up to 10–16 T can easily be obtained in ordinary laboratories by using commercial superconducting magnets [554,555]. Superconducting wires such as Nb-Ti alloy, Nb_3Sn, Nb_3Ge, *etc.* are used for high field superconducting magnets. The maximum field of superconducting magnets is limited by the upper critical field B_{c2} and the critical current density J_c. To date, the highest field which is available by using a superconducting magnet is 21 T. Oxide high Tc superconductors have very high B_{c2} and in future may be employed for high field magnets, but it will be some time before these materials are used for actual magnets, because of still rather low critical current.

Steady magnetic fields higher than 20 T are now available at several magnet laboratories in the world, where they employ water-cooled resistive magnets and hybrid magnets, which are combinations of superconducting magnets and resistive magnets. A large amount of electric power, of the order of 5–40 MW, is consumed to supply large enough DC current to such magnets. The running of such magnets also needs a large amount of water flow to cool the magnets. Therefore, the facilities are inevitably of large scale and run by national facilities. The necessary power practically limits the maximum available field. Table 7.1 lists large magnet laboratory facilities around the world [556]. In this chapter, we will focus in pulsed high magnetic fields including very high short pulse fields (megagauss fields), so that if necessary, readers are requested to refer to the recent reviews on steady high magnetic fields in [GT-7].

7.1.2 *Pulsed high magnetic fields*

In comparison to DC fields, pulsed high magnetic fields can be generated more easily, by using facilities on a much smaller scale. Long pulsed high magnetic fields up to 50–60T can be generated by discharging a pulsed current to solenoid coils from condenser banks. A typical electrical circuit is shown in Fig. 7.1 (a). In

Table 7.1 Major facilities for static high magnetic fields around the world.

Facility	Founded year	Power MW	Resistive T	Resistive mm	Hybrid T	Hybrid mm
Braunschweig (Germany)	1972	5.6	18	32		
Grenoble (France)	1970	24	30	50	[40]	[34]
Nijmegen (The Netherlands)	1972	6	20	32	30	32
Nijmegen [under construction]	2002	[20]	[33]	[32]	[40]	[32]
Sendai (Japan)	1981	8	15	82	31	32
Tallahassee (Florida)	1990	10	33	32	45	32
Tsukuba (Japan)	1988	15	29	32	35	32

[ ] indicates the future plan.

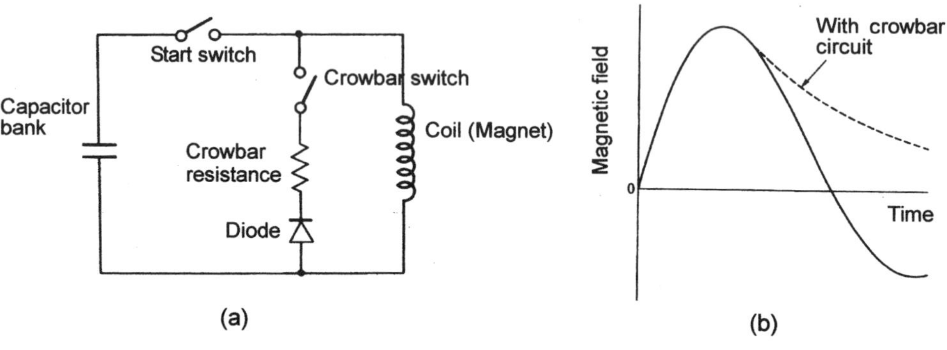

FIG. 7.1. (a) Electric circuit for generating pulsed high magnetic fields. (b) Typical wave form of a pulsed magnetic field.

a condenser bank, we usually use paper oil condensers which allow discharge of large pulsed current of several tens of kA. A condenser bank with a storing energy of 100–1000 kJ and voltage of 5–10 kV is usually used in ordinary laboratories. A typical wave-form for pulsed magnetic fields is shown in Fig. 7.1 (b). A so-called crowbar circuit consisting of a diode and a resistor is used to prolong the decay time of the pulsed current. The pulse form is nearly sinusoidal when the crowbar switch is off, as the current is through an LCR circuit. When the crowbar switch is on, the current after the peak flows through the crowbar circuit, instead of returning to the capacitor. Thus the current up to the peak is the same as that without the crowbar circuit, but after the peak it shows an exponential decay with a time constant L/R, where L and R are the inductance and the resistance of the circuit, respectively. We can also make a crowbar circuit by using a switch that is closed just at the top of the current, instead of a diode. A crowbar circuit is useful for measurements during the long decay time. On the other hand, the sinusoidal pulse shape provides nearly the same sweep rate in the rising slope and the falling slope, so that it is useful for checking the reproducibilily of the data between the two slopes or for investigating hysteresis phenomena if any.

A coil is made by winding a strong wire on a bobbin. The available field strength depends on the strength of the coil. The electromagnetic force exerted on a coil by a magnetic field is proportional to the square of the magnetic field. This force is called Maxwell stress and expressed as

$$T = \frac{B^2}{2\mu_0}. \tag{7.1}$$

The stress is towards the outer direction radially and contracting direction axially. It grows rapidly when the field is increased, and it reaches 10 kbar at 50 T and 40 kbar at 100 T [557,558,401]. Any magnet is destroyed when the magnet is subjected to such a strong force that it exceeds the strength of the magnet material. Therefore, we have to make strong coils in order to generate high magnetic fields.

The normal procedures for making an ordinary type of pulse magnet are as follows. After winding coils with a tensile stress of a few tens of kg, the spacing between the windings is impregnated with epoxy resin. Before the impregnation, the epoxy is pumped to remove any bubbles and the solidification is done at a high pressure in order to avoid any voids between the windings. The coil is then machined and mounted in a cylinder of stainless steel (or a strong material like Maraging steel) for reinforcement. Such a structure is a standard type of pulse magnet, which can produce magnetic fields of up to 50–60 T with a duration of about 10–20 ms. Magnets are usually immersed in liquid nitrogen in order to decrease the initial temperature.

Many different techniques for constructing strong magnets have been reported. The first breakthrough in the generation of modern long pulse field was achieved by Foner [559]. By using a strong wire with fine filaments embedded in a Cu matrix, he produced a field of up to 61 T with a duration of 10 ms. The Nb filaments are effective to prevent the motion of the dislocations in Cu wires.

Herlach *et al.* used a reinforcement by glass fibers [560]. Instead of using strong wires they used soft Cu wires, but they wound fine glass fibers in between the layers of the windings. The technique depends on the fact that the tensile strength of the glass fiber is much larger than that of soft Cu wire. The thicknesses of the glass fiber layers between the wire layers were carefully calculated to achieve a uniform stress distribution throughout the coil. The cross section of the Leuven type coil is shown in Fig. 7.2. The uniform stress distribution rather than the uniform current distribution as in a conventional pulse magnet is an important concept to obtain high magnetic fields. If we wind a coil uniformly, the stress is the largest in the innermost winding, and the highest possible field

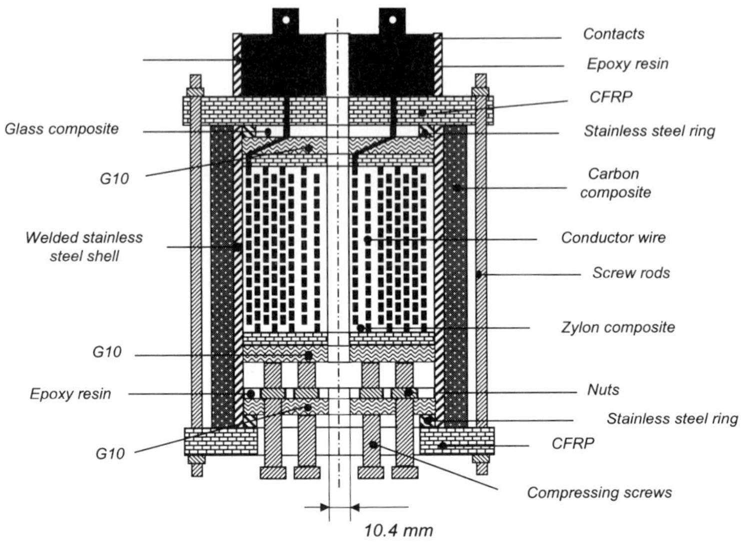

FIG. 7.2. Cross section of the Leuven type pulse magnet [560, 557].

is determined by this stress. By using this new idea, Herlach *et al.* generated long pulsed fields up to 71 T with a duration of 10 ms [560].

Boebinger, Passner and Bevk generated 70 T using a shell of "Vascomax" Maraging steel with zirconium beads as the transmitting medium with high strength and low compressibility [561]. This coil winding is made from different materials in order to optimize the distribution of stress and electrical properties. The inner layer is made from a ductile Cu-Be alloy, the next eight layers are wound with Cu-Nb microfilamentary wire, and the outer five layers are from Glidcop (a copper-alumina alloy).

Kindo recently succeeded in producing 80 T by using a coil wound very carefully by using a strong wire of Cu-Ag alloy [562].

At the Institute for Solid State Physics (ISSP), Tokyo, several different techniques have been attempted to construct strong pulse magnets. Some magnets were manufactured by impregnating not with epoxy but with water [563]. This idea was first invented by Motokawa *et al.* [564] The water is frozen to ice when the magnet is immersed in liquid nitrogen, and it becomes a good strong binder of the windings. The windings are wrapped firmly with glass tape and water penetrates into the glass fiber. When frozen, the ice increases the volume so that it stresses the coil. The advantage of using ice is that it easily penetrates into the gap in the coil due to a very small viscosity, and also when it is frozen, it is a good heat conductor. After each shot of the pulse, the temperature of the magnet is increased. We have to wait for about half an hour or so before the

Table 7.2 Pulsed field laboratories around the world. [556]

Laboratory location	Capacitor bank W [kJ]	V [kV]	I_m [kA]	User coil W [kJ]	B [T]	i.d. [mm]	Δt Duration [ms]
Berlin	42	2.5	20	42	51	10	3.5
	400	10	60	400	60	18	8.1
Dresden	1250	10	50	450	50	24	13
				1100	40	24	120
Frankfurt	800	7	100	800	36	22	600
				390	50	24	24
Kashiwa, Tokyo Univ.	900	5/10	50		50	22	50
	200	5	50	200	50	20	20
	200	4	50	200	50	20	18
Kobe	24	3	7.6	24	30	15.4	11
	100	3		34	36	16.5	9
Leuven	475	5	25	300	55	18	20
	600	10	50	260	70	10	8
Los Alamos	1600	10	50	565	60	15	35
				1400	50	15	350
Nijmegen	2000	16	40		55	23	
Osaka	1500	20	1000	240	60	18	7
					70	10	7
	1000	13	40	570	60	18	22
Oxford	800	7	20	150	60	12	10
				250	50	20	20
Sendai	100	5	35	70	40	17	10
				45	30	22	5
Sydney	800	7	60	400	60	22	25
Toulouse	14000	24	30	1250	61	11	150
				3300	58	26	285
Tsukuba	1600	5	100	500	48	16	20
	300	10	30	300	50	20	15
Vienna	75	2.5	50	20	40	25	10

W: stored energy, V: voltage, I_m: the maximum current, i.d.: the inner bore.

coil is cooled again. Magnets with ice impregnation allow a repetition of shots every 15 minutes or so while ordinary magnets with epoxy impregnation need at least half an hour after heavy shots. The principal large facilities for pulsed high magnetic fields around the world are listed in Table 7.2. While a combination of relatively small coils and modest size of capacitor banks usually produces pulsed fields with duration of about 10 ms, longer pulsed fields are obtained using a large generator or city power line as a power supply.

The prototype for these coils was constructed at the University of Amsterdam [565]. The city power line is used as the energy source and various pulse shapes of the current such as a flat top, stair-like, or triangular shape are produced by using thyristors and filters. A hard copper is used to wind a coil and a peak field of 40 T was obtained. The long pulse fields are particularly useful for measurements on conducting materials and phenomena with long relaxation. Unfortunately, the facilities were closed down to be amalgamated with those at Nijmegen. Recently, similar facilities which use the city power line were constructed in Vienna. Consuming a power of 10 MW, pulsed fields up to 40 T with a long duration (up to 1 s) are generated [566].

The Los Alamos branch of the National High Magnetic Fields (NHMFL) of USA is constructing a new large facility for generating pulsed fields up to 60–100T [567,568]. The magnet consists of nine separate coils which are energized in groups from three power supplies connected to a 1430 MVA flywheel-generator. At peak field, the inductive energy amounts to 90 MJ. Each sub-coil is tightly enclosed by a steel shell for reinforcement and for protection against magnet failure. In Rossendorf (Dresden), a large capacitor bank of 50 MJ is being constructed to generate long pulse fields up to 100 T [569].

When an excessively large current is supplied to a coil to generate too high fields, the coil is destroyed. Sometimes the destruction occurs very violently. The maximum field so far available for actual experiments by such magnets is limited to about 70 T, (or by the most advanced technology, ~90 T for a limited number of shots). At present, a considerable effort is being devoted in Europe, USA and Japan to start a project to generate a field of 100 T nondestructively.

7.1.3 *Ultra-high magnetic fields*

A magnetic field higher than 100 T can be produced only in a destructive way. Because 100 T is equal to 1 MG in cgs units, these high fields are often called "megagauss fields". Different techniques have been developed for generating the megagauss fields. Because of the short duration time (order of microseconds), application of the megagauss fields to experiments is not an easy task. Recently, however, several fine measuring techniques have been developed and various solid state experiments have become possible with a reasonable precision. Today, there are several facilities in the world where solid state experiments can be performed in megagauss fields. Table 7.3 lists such megagauss laboratories around the world.

A. Explosive flux compression

The first method which has been developed is the explosive-driven flux compression technique [557]. The technique was mainly developed in USA and the former Soviet Union. A schematic sketch of the explosive flux compression devices is shown in Fig. 7.3. In the cylinder type, a chemical explosive is set around a metal ring called a liner. By chemical explosion, the liner is squeezed rapidly inwards, and the magnetic flux which has been injected beforehand is compressed

Table 7.3 Megagauss field facilities around the world for solid state experiments

Laboratory	Country	Technique	Maximum field
Los Alamos	USA	Explosive (Cylidrical)	> 1000 T
		Explosive (Bellows)	200 T
		Single turn coil	∼300 T
Sarov	Russia	Explosive (Cylindrical, 3 cascades)	> 1000 T
ISSP	Japan	Electromagnetic flux compression	622 T
		Single turn coil	302 T
Berlin	Germany	Single turn coil	310 T

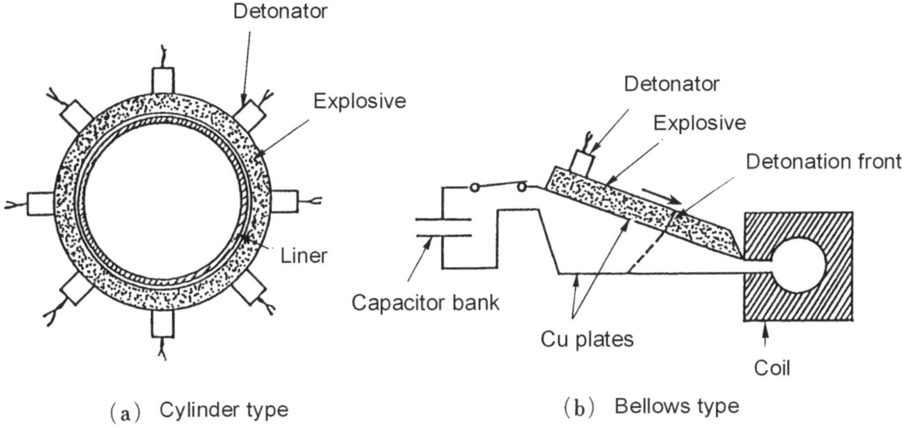

FIG. 7.3. Schematic sketch showing the principle of the explosive-driven flux compression technique. (a) Cylinder type. (b) Bellow type.

inside the liner, if the velocity of the liner is faster than the flux diffusion speed. The flux density B is increased in proportion to the inverse of the cross sectional area of the liner. In this way, magnetic fields over 1000 T have been produced [570, 571]. At VNIIEF in Sarov (the city was called Arzamas-16 in the former Soviet Union), a technique of three cascades has been developed to generate very high magnetic fields [571, 572]. At Los Alamos, this technique has been employed for various solid state experiments [490].

Besides the cylindrical flux compression, a bellows type technique was developed by the groups of Frascati [573] and Los Alamos [574]. In the bellows type, the magnetic field is first produced inside a large loop containing the metal plates and the coil by a capacitor discharge. When the explosion takes place, the detonation front shortens the circuit at the left edge of the plates in the figure. As the front proceeds from left to right, the magnetic flux inside the loop is pushed into the coil placed at the right edge. In this way the field in the coil is increased as the cross sectional area of the loop is decreased. The peak field is about 100 T with a one-stage system, if we use a two-stage system a field up to about 200 T was produced [490]. Although the available peak field is lower than in the cylinder type, the bellows type device is useful for solid state experiments, because the destruction around the sample space occurs in the final moment.

The explosive technique is a very powerful means of generating very high fields, as demonstrated in an international collaborative experimental series: the Dirac series at Los Alamos [575] and Kapitza series at Sarov [360]. However, it is very difficult to conduct indoor experiments for elaborate solid state measurements. More controllable methods available in laboratories are electromagnetic flux compression and the single turn coil technique, as described below.

B. Electromagnetic flux compression

Electromagnetic flux compression (EMFC) is a technique using an electromagnetic force instead of chemical explosives to squeeze a liner [557]. The technique was first invented by Cnare in 1968 [576] and later developed at ISSP [401]. The coil system of the EMFC is shown in Fig. 7.4. The clamping system of the primary coil actually used at ISSP is shown in Fig. 7.5. Inside a one turn primary coil, we place a liner (copper ring). When we supply a large pulse current (primary current) of the order of 4–6 MA to the primary coil, a secondary current is induced in the liner in the opposite direction. The magnitude of the secondary current is nearly the same as the primary, but the direction is just opposite to shield the field. Then the two large opposite currents repel each other and the liner is squeezed in the inward direction. In this process, a weak seed field (2–3 T) is injected in the liner beforehand. The seed field is a long pulse field produced by using a pair of coils on both sides of the primary coil. The primary current is started just at the top of the seed field. Then by the imploding motion of the liner, the seed flux is compressed as the inner cross section of the liner is decreased. When the cross section becomes sufficiently small, a very high magnetic field is generated in proportion to the inverse of the cross section.

The EMFC is based on a number of energy conversion processes: Electric energy stored in the bank → energy of primary current → kinetic energy of the liner → energy of magnetic field. The motion of the liner and the current are dependent sensitively on the experimental parameters. Therefore, a computer simulation is very important for designing the system [577]. In order to understand the process of the EMFC, it is helpful to see the basic equations of the

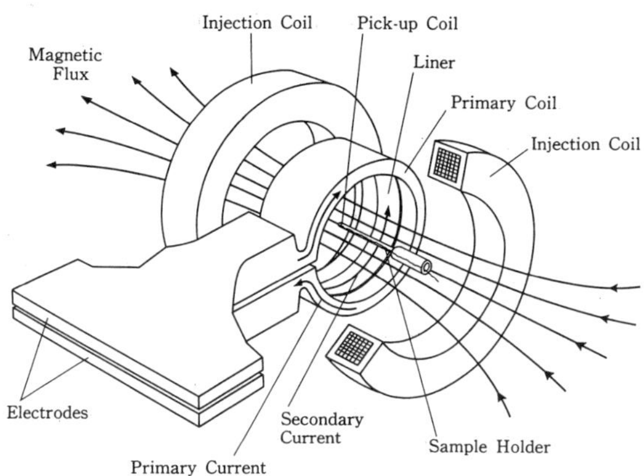

FIG. 7.4. Sketch of a primary coil and a liner for the EMFC [558]. The primary coil is made of steel. The actual dimension of the primary coil is 160 mm in inner diameter, 25 mm in wall thickness and 50–89 mm in length. The liner is made from copper. The actual size is 120–150 mm in outer diameter, 1.5–3 mm in thickness and 50–90 mm in length.

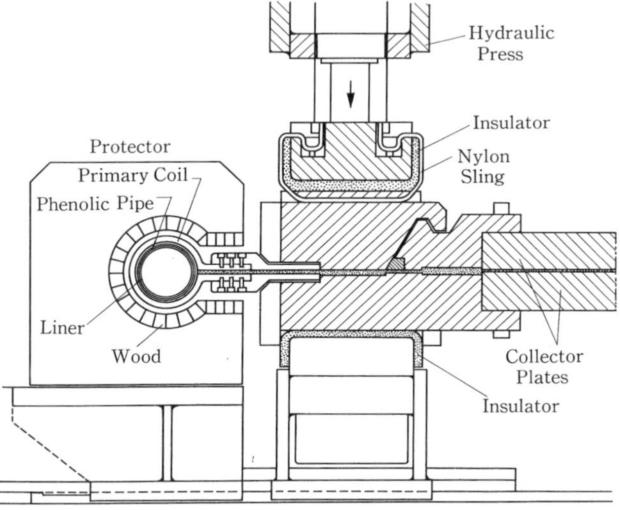

FIG. 7.5. Coil clamping system for the electromagnetic flux compression [558]. The coil is clamped by a hydraulic press. As the coil is destroyed, expanding outwards, a heavy protecting block made from steel is set around the primary coil. The inner surface of the protector is lined by pieces of wood to absorb the shock at the bump of the coil.

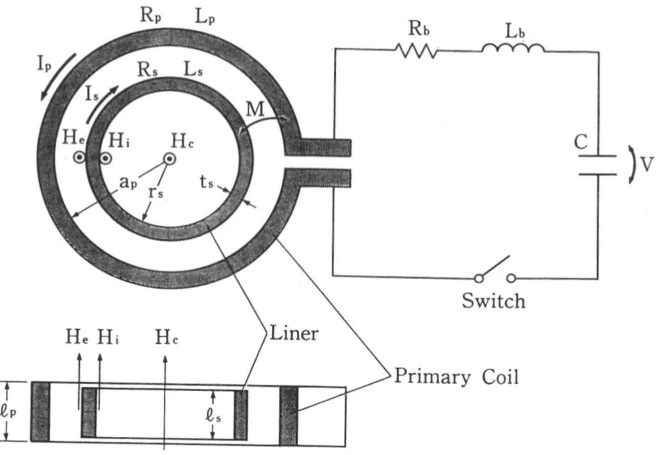

FIG. 7.6. Schematic electric circuit of the electromagnetic flux compression system.

computer simulation of EMFC. Figure 7.6 shows the equivalent circuit of EMFC. The capacitance C is connected to the primary coil that contains the liner. The inductance and resistance of the primary coil and the liner are L_p, R_p, L_s, and R_s respectively. I_p is the primary, I_s is the secondary current, V is the voltage across the capacitance, L_b is the residual inductance, R_b is the residual resistance of the capacitor bank and the cables, r_p is the radius of the primary coil, r_s is that of the liner, M is the mutual inductance between the primary coil and the liner. r_p and r_s as well as the liner temperature T are determined as a function of time t, and thus $M(t)$ and $L_s(t)$ are time dependent. The magnetic flux Φ in the liner and its time derivative are given by the following expressions:

$$\Phi = L_s I_s - M I_p - \mu_0 H_0 S \ , \tag{7.2}$$

$$\frac{d\Phi}{dt} = -R_s(r_s) I_s \ . \tag{7.3}$$

Here H_0 is the initial field and S the cross section inside the inner radius of the liner. From (7.2) and (7.3), the time derivatives of the currents are given by the following equation:

$$M\frac{dI_p}{dt} + \frac{dM}{dt} I_p - \left[L_s \frac{dI_s}{dt} + \frac{dL_s}{dt} I_s \right] + \mu_0 H_0 \frac{dS}{dt} - R_s I_s = 0 \ . \tag{7.4}$$

The primary current is obtained from

$$L_b \frac{dI_p}{dt} + R_b I_p - \left[M \frac{dI_s}{dt} + \frac{dM}{dt} I_s \right] = V(t), \tag{7.5}$$

where $V(t)$ is the voltage across the capacitor bank. The current induced in the liner gives rise to a difference in the Maxwell stress between the inner and outer surfaces of the liner and thus the liner is accelerated in the inward direction according to

$$-m\frac{d^2r_s}{dt^2} = 2\pi r_s l s \frac{\mu_0}{2}\left[(H_e + H_0)^2 - (H_i + H_0)^2\right] , \tag{7.6}$$

here m and l_s are the radius and the length of the liner. H_e and H_i are the magnetic fields at the outer and inner surfaces of the liner. Solving the combined equations (7.4), (7.5) and (7.6), we can obtain r_s, I_p, and I_s. The magnetic field at the center of the liner is obtained from

$$H_c = H_p + H_s + H_0, \tag{7.7}$$

where H_p and H_s are the magnetic fields due to I_p and I_s, respectively. In the crude approximation described above, the motion of the primary coil was neglected. Also, the wall thickness of the liner and of the primary coil were neglected for simplicity. In reality, however, the primary coil and the liner have thicknesses and the current distribution is not uniform throughout the cross section. Actually, the current has a tendency to be concentrated on the surface and the edge. In order to take into account the current distribution throughout the cross section of the liner and the primary coil, Miura and Nakao performed a two-dimensional calculation, dividing the cross section into many segments and calculating the currents through these thin rings [577]. The motion of the primary coil was also taken into account. Figure 7.7 shows an example of calculated results of the total primary current I_p, the total secondary current in the liner I_s, the radii r_p and r_s of the primary coil and the liner (the outer and the inner radii), the magnetic field B, and the liner velocity v_s as a function of time. The calculation was performed using parameters from an actual experiment. The experimental points are shown for comparison; the agreement between calculation and experiment is excellent, except for the final phase that is dominated by the compressibility of the liner material.

At ISSP, the project to generate megagauss fields by EMFC was first started in 1971. A main capacitor bank of 285 kJ (30 kV) and a sub capacitor bank of 32 kJ (3.3 kV, to generate the seed field) were installed. A peak field of 280 T was obtained [578]. Based on the success of this prototype research, a second phase was started in 1982. A new large main capacitor bank of 5 MJ (40 kV) and a sub capacitor bank of 1.5 MJ (10 kV) were installed. A higher magnetic field up to 550 T was generated and successfully employed for a variety of experiments [401,579]. On the occasion of the move of the ISSP from the Roppongi campus to the Kashiwa campus, new capacitor banks were constructed for the third phase of project. The sizes of the capacitor banks were the same as in the second phase but a faster rise time has become available by reducing the stray inductance of the circuit. Owing to the higher performance of the new capacitor banks, a peak field of 622 T has been achieved [580, 581].

A typical wave-form of the generated fields is shown in Fig. 7.8 together with the primary current I. The field rises from about 10 T level to the maximum

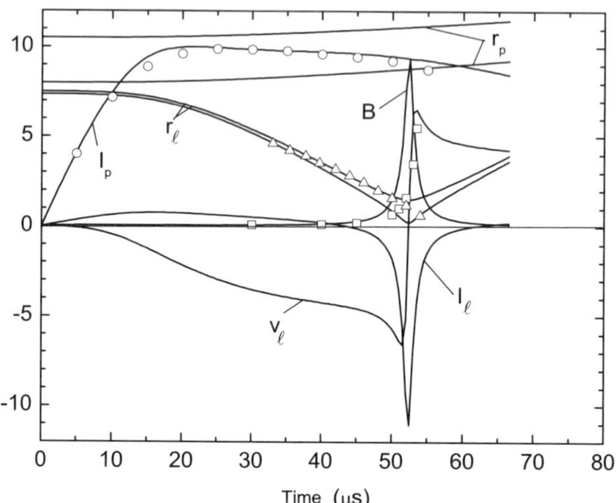

FIG. 7.7. Comparison of the result of the computer simulation (lines) with experimental data (points) [577]. The parameters are as follows: Energy 4 MJ, voltage 40 kV, inner diameter, thickness and length of the primary coil 160 mm, 25 mm, and 50 mm, outer diameter, thickness and length of the liner 150 mm, 1.5 mm, and 55 mm respectively, seed field 2.1 T.

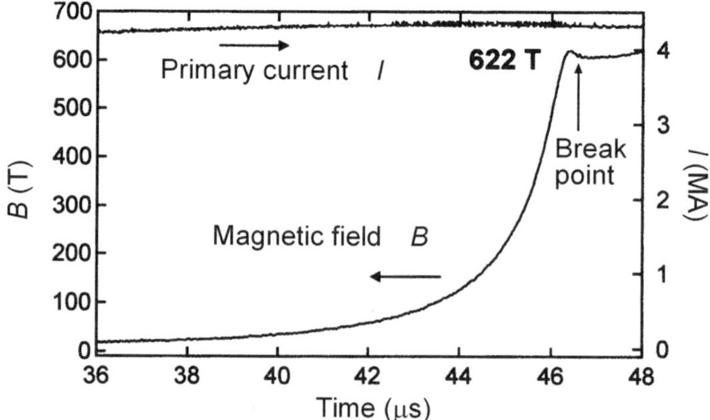

FIG. 7.8. Experimental trace of the magnetic field generated by electromagnetic flux compression [581]. The primary current I has a nearly sinusoidal form but in this expanded scale, it looks almost flat. In this shot, a peak field of 622 T was obtained and this is the highest field that has ever been generated by indoor experiments.

field in several microseconds. At the break point indicated by an arrow, the field probe is destroyed by the impact of the liner. At this break point, the sample is also destroyed, but we can apply the field to the sample during the rise time. The bore of the liner at the highest field is about 9 mm.

One of the difficulties in the EMFC is the problem of the feed gap. In order to feed the current, there must be a feed gap in the coil so that the coil is not an entire circle. It means that the field produced by the primary coil is weaker near the edge than in other parts. Therefore, the acceleration of the liner is also weaker near the gap, and this causes a non-uniform deformation of the liner. As a result, the liner shows a bulge towards the feed gap side in the final stage, and this causes violent plasma discharge which sometimes breaks the sample. In order to improve the feed gap effect, a feed gap compensator was invented and it was found to be very effective in solving the problem of the feed gap effect [580, 581]. More details of the technique of EMFC are described in [401, 558, 580, 581].

C. Single turn coil technique

The single turn coil (STC) technique provides a means of generating very high magnetic fields up to a few hundred tesla by using a disposal single turn coil which is destroyed every time but without destroying samples [557]. A compact system for the STC technique using a small light coil was invented by Herlach and McBroom at IIT in Chicago [582] and later developed by the ISSP [583] and von Ortenberg's group at the Humboldt University in Berlin [584]. Recently, new STC facilities have been built at Los Alamos [585]. Figure 7.9 shows the coil system of the STC technique used at ISSP. The single-turn coil is made from a copper plate with a thickness of 2–3 mm, which is bent to form a small single turn coil at the end. It is firmly clamped by a hydraulic press to collector plates. A large pulse current of about 2–3 MA is supplied by a discharge from a fast condenser bank. A very high magnetic field is generated by this current. The coil is destroyed by a large Maxwell stress. However, if the discharge is fast enough, high magnetic fields are generated before the coil destruction occurs, while the coil stays at its position due to its inertia. Therefore, we need a fast capacitor bank which allows a large current discharge in a very short time.

Table 7.4 lists the characteristics of the capacitor banks so far employed for the STC system. In order to obtain the highest possible field, it is crucially important to reduce the residual inductance of the discharge circuit as much as possible ($<$20 nH) for obtaining a short rise time. It is also necessary to reduce the capacitance of the bank, so that the voltage is inevitably very high to keep the stored energy large. If the coil axis is horizontal, the system is useful for optical experiments since the alignment of the optical path becomes easier. On the other hand, the vertical system is more useful for inserting a cryostat to refrigeratc the samples to liquid He temperature.

At ISSP, a capacitor bank was first built in 1983 and very high fields were successfully generated; 150 T in a bore of 10 mm and 200 T in a bore of 6 mm.

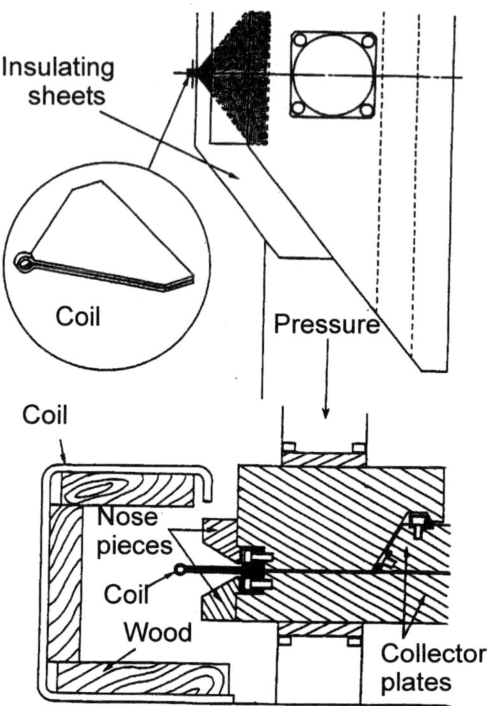

FIG. 7.9. Coil clamping system for the STC technique [583]. The inset shows a sketch of the STC which is made by bending a copper plate. The coil is clamped by a hydraulic press. A protecting device is placed around the coil to catch fragments of copper at the explosion of the coil.

On the occasion of the move of the Megagauss Laboratories to the Kashiwa campus, two larger capacitor banks of 200 kJ were installed (one with horizontal coil axis for optical experiments and one with vertical axis for transport and magnetization experiments), and consequently, even higher fields have been obtained. Examples of the experimentally recorded wave forms of the current I and the field B are shown in Fig. 7.10. In this case a current of about 2.5 MA was discharged to a coil with a bore of 6 mm and an axial length of 6 mm. The field reached 263 T in about 2.5 μs. As the coil radius increases with time, the maximum of the field occurs earlier than the current. The ratio B/I also decreases with time, as shown in the inset. The great advantage of this technique is that the samples and cryostat inserted in the coil are not damaged although the coil explodes violently. Therefore, the experiments can be repeated many times on the same sample. Another advantage is that the wave form of the field is nearly sinusoidal so that the field crosses the same magnitude twice, on the rising slope

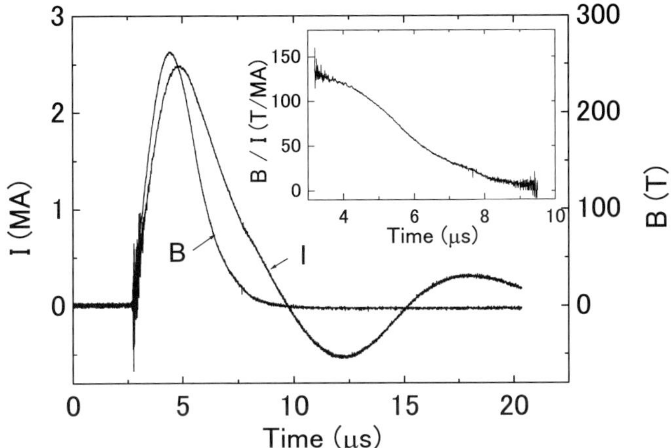

FIG. 7.10. A typical wave-form of the current I and magnetic field B in the STC technique. A capacitor bank of 200 kJ (50 kV) was employed. The coil diameter and the length are 6 mm. The inset shows B/I.

and the falling slope. Therefore we can check the reproducibility of the signal or any hysteresis effect.

The peak field depends on the coil size. Figure 7.11 shows the dependence of the peak field on the coil size. Here the bore diameter and the length of the coil are set to be the same. The coil size and the energy supplied to the coil are chosen in accordance with the requirement of the experiment.

Table 7.4 Capacitor banks for the single turn coil technique so far constructed around the world.

Laboratory	W (kJ)	V (V)	L_b (nH)	I_{max} (MA)	Operated period	Coil axis direction
Chicago	55	20	14	1.3	1970–1972	H
ISSP	100	40	18	2.5	1983–1999	H
ISSP	200	50	16.5	4.0	1999–	H
ISSP	200	40	13,1	4.0	1999–	V
Berlin*	200	60	11.2	2.6	1993–2006	V
Los Alamos	259	60	18.2		2005–	H

W: Stored energy, V: Voltage, L_b: Residual inductance, I_{max}: Maximum current
H and V denote the horizontal or vertical directions of the coil axis.
*After the shut down of the Berlin facilities, the bank has been moved to Toulouse.

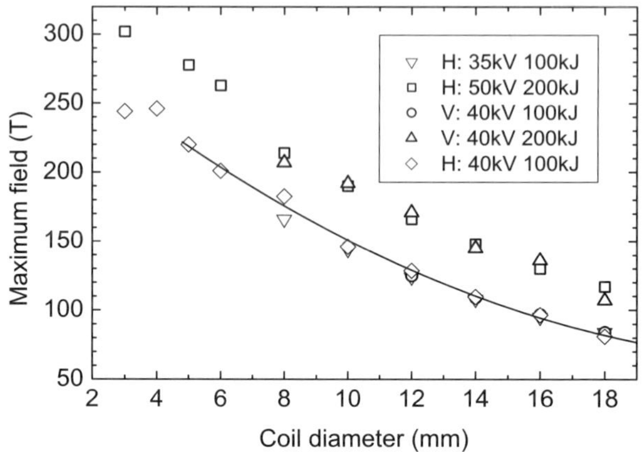

FIG. 7.11. Coil size dependence of the peak field obtained by the single turn coil technique. The horizontal axis denotes the bore and the length (set equal) of the coil. V and H indicate the two different systems at ISSP, vertical system (coil axis is vertical) and horizontal system (coil axis is horizontal). Results of the use of 100 kJ and 200 kJ of energy are plotted.

To refrigerate the sample, a helium-flow type cryostat is employed. As the time derivative of the magnetic flux is enormous, everything near the center of the field should be made from plastic materials in order to avoid the eddy current. Figure 7.12 shows the cross section of such a cryostat for optical measurement. The sample is mounted inside a double tube of epoxy placed in a vacuum near its edge. In the gap of the double tube there is another tube of Kapton separating the space for incoming and outgoing liquid helium. Thus we can circulate the liquid helium down to the edge of the double tube and cool the sample. The sample temperature is measured by an Au-Fe/Chromel thermocouple. The lowest temperature easily obtained by this cryostat is about 7 K. A similar cryostat can be employed for electromagnetic flux compression experiments as well. In that case, however, it is of course destroyed at every shot of the pulse. When the coil axis is vertical, we can insert a plastic cryostat containing liquid helium and immerse the sample in a liquid. Then we can cool the sample down to pumped liquid He temperature ($\sim$1.5 K), or even to 0.5 K by using a ^{3}He system.

7.2 Measurement of magnetic fields

For solid state measurements, accurate measurement and calibration of magnetic field is essential. In the steady fields produced by a superconducting magnet, water cooled magnet, or hybrid magnet, the magnetic field intensity is proportional to the current. Therefore the magnetic field is usually measured by the current

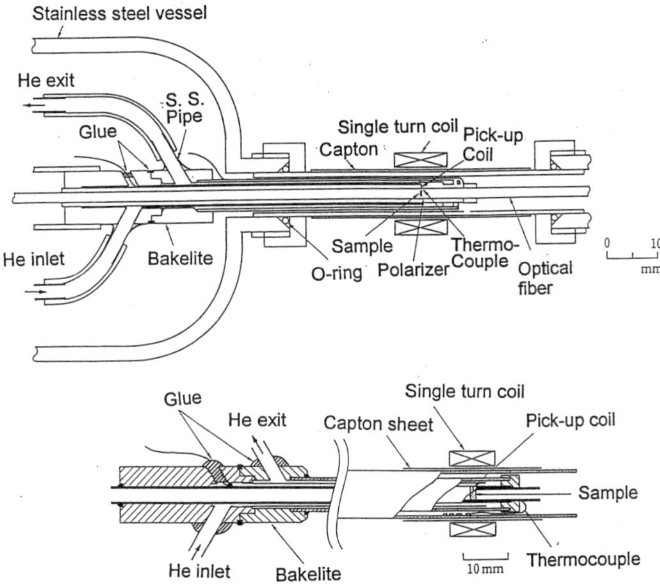

FIG. 7.12. The sample holder and the cooling system used for the single turn coil technique.

supplied to the magnet. When necessary, the magnetic field can be calibrated by a Hall probe or by magneto-resistance.

For pulsed fields, the field measurement is mostly done by using a pick-up coil. When a coil with a diameter D [m] and the number of turns of N is put in a magnetic flux density B [T], a voltage V induced across the coil is

$$V = \pi N \left(\frac{D}{2}\right)^2 \frac{dB}{dt}. \tag{7.8}$$

Hence by integrating the voltage with an RC integration circuit as shown in Fig. 7.13, we can obtain a signal V_o proportional to the flux density B according to the following equation,

$$V_o = \frac{r_s}{r + r_s} \pi \left(\frac{D}{2}\right)^2 \frac{A}{RC} B, \tag{7.9}$$

where R and C are the resistance and the capacitance of the integrator, A is the gain of the amplifier, r_s is the resistance of the cable termination (50 or 75 Ω), and r is the resistance of the pick-up coil and the lead wire. Needless to say, RC should be much larger than $1/\omega$ (ω is the angular frequency component of the field signal) for the complete integration. For the non-destructive long pulse

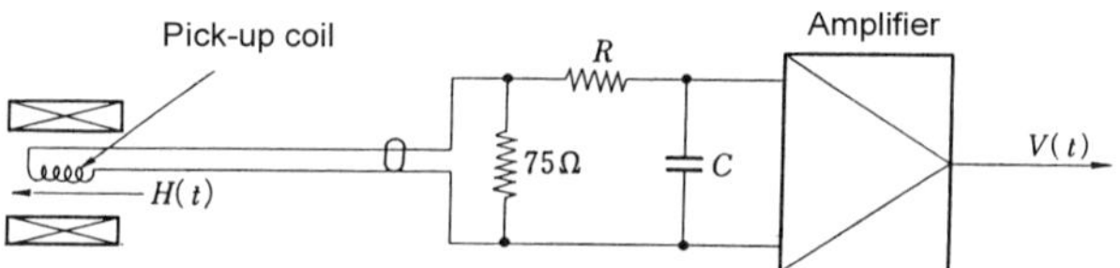

FIG. 7.13. A circuit to measure pulsed magnetic fields by means of a pick-up coil. A resistance of 75 Ω (sometimes 50 Ω) to terminate the cable is necessary for the impedance matching.

field, a typical coil diameter is 5–10 mm and the number of turns is 10–50. For short pulse fields in the megagauss range, the coil diameter is 1–3 mm and the number of turns is 1–2. Although the field is obtained from eq. (7.9), estimation of the coil diameter usually involves errors. Therefore, the sensitivity of the pick-up coil (coefficient of B in the equation) should be calibrated accurately once by some other means, such as ESR, cyclotron resonance, or Shubnikov-de Haas oscillation, whose resonance field is accurately known, or Faraday rotation for the accurate field measurement.

In the case of electromagnetic flux compression, a pick-up coil is destroyed every time, so that the sensitivity of each pick-up coil should be calibrated readily and quickly. For such a purpose, a convenient method is to compare the sensitivity of the pick-up coil with that of the standard coil whose sensitivity has been accurately calibrated beforehand. The sensitivity of the both coils can be measured by placing them in an AC magnetic field whose frequency is close to that of the actual magnetic field we want to measure [586].

7.3 Experiments in high magnetic fields

7.3.1 *Transport measurement*

The transport measurement is not so difficult in steady magnetic fields. By integrating the signal, with a reasonable time constant, we can reduce noises. In pulsed fields, however, the time constant should be very short. In addition, there is a problem of the induced voltage in the sample and the lead wire due to dB/dt. If there is a loop of effective area of S m^2, the induced voltage is

$$V = \frac{dB}{dt} S. \tag{7.10}$$

This spurious voltage becomes significant as compared with the bias voltage which we apply to the sample. Particularly, in the megagauss range, dB/dt is of the order of 10^8 T/s, and if the there is a loop with a diameter of 3 mm, V becomes about 1000 V. In order to eliminate the induced voltage, we usually use a “compensation circuit” as shown in Fig. 7.14. The idea is to pick up a voltage proportional to dB/dt by a small coil and add an adequate proportion of the voltage to the signal from the sample. The proportion is controlled by a

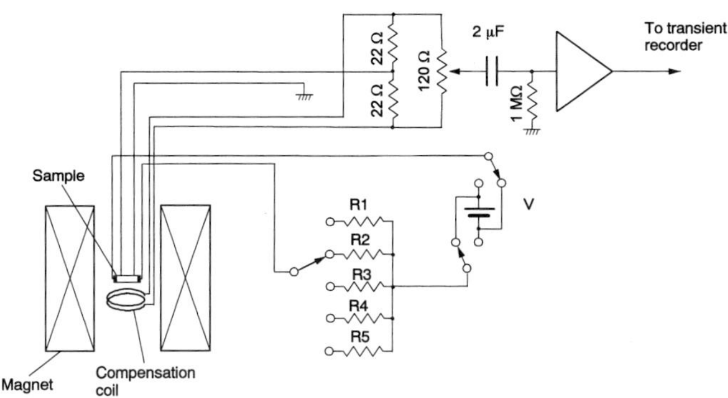

FIG. 7.14. Circuit for measuring the resistance of the sample in pulsed magnetic fields.

variable resistance R_v to minimize the spurious voltage at small preliminary shots before the high field shot is fired. The spurious voltage is not always completely proportional to dB/dt, and sometimes contains components of d^2B/dt^2, B and $\int Bdt$, *etc.* due to the stray inductance and capacitance of the measuring circuit, usually a small spurious voltage still remains even after the compensation by such a compensation circuit. This remaining noise can be further eliminated by making two shots in the same condition, either averaging signals obtained with opposite currents applied to the sample or by subtracting the signal without the current from that with the current. When the induced voltage is sufficiently small, the induced voltage can be compensated only by such an averaging without using the compensation circuit.

Another way to avoid the effect of the induced spurious voltage is to use a lock-in technique with an AC bias current. As the pulse duration of the non-destructive long pulse field is as long as 10 ms, we can use a frequency of the order of 100 kHz. The lock-in detection should be done with a short enough integration time.

Nowadays, very clean data of transport measurements can be obtained in long pulse fields with a duration longer than 1 ms by techniques as mentioned above. In short pulse fields (μs pulse) in the megagauss range, transport measurement is much harder due to a huge dB/dt. However, by reducing the sample size and any possible loop around the sample, and making the lead wire completely parallel with the field, DC and AC transport measurement becomes possible up to 150 T. To further reduce the induced spurious voltage, compensation circuits based on the principle shown in Fig. 7.14 were used for each of the four electrodes, not only for contacts for voltage but also for those for currents. Nakagawa *et al.* succeeded in measuring the magneto-resistance of $YBa_2Cu_3O_{7-\delta}$ and graphite up to 150 T in the single turn coil system [587, 588].

Another technique for the transport measurement is the contactless method. By measuring the transmission signal of the radio frequency electromagnetic

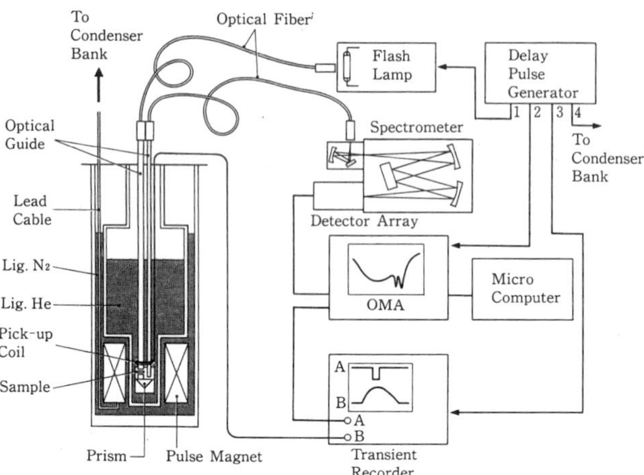

FIG. 7.15. Block diagram of the OMA system for the magneto-optical measurement in pulsed fields.

field through the sample, we can measure the conductivity (or resistivity) of the sample without using contacts, as the transmission is dependent on the AC conductivity [589]. Sekitani *et al.* succeeded in observing the super to normal transition of $YBa_2Cu_3O_{7-\delta}$ for $\boldsymbol{B} \perp$ c, in a field up to 500 T [590, 591].

7.3.2 *Optical measurement*

Magneto-optical measurements in pulsed fields can be carried out most conveniently by using an optical multichannel analyzer (OMA) or a streak spectrometer. The former allows the recording of the magneto-optical spectra in a short period of time by opening the gate of the optical sensitivity of the recording system at constant magnetic fields. In the latter case, we can obtain time-resolved spectra representing the entire magnetic field dependence of the optical spectra in one shot of the pulse fields. A CCD device or an image-converter camera is used for sweeping the optical image.

Figure 7.15 shows the schematic diagram of the magneto-optical spectrometer using an OMA. The sample is mounted in a cryostat. The optical signal from the sample is led to a monochromator by an optical fiber. At the position of the output slit of the monochromator, a microchannel plate is placed to intensify the image of the spectrum. In front of the microchannel plate, a Si diode array is mounted to detect the optical signal at different wavelengths. The gate pulse voltage is applied to the microchannel plate, at the top flat part of the magnetic field pulse. Then the optical signal is detected only during the gate pulse. The magnetic field can be regarded as almost constant if the gate pulse width is sufficiently short. For example, if the gate pulse width is 1/10 of the half period of the magnetic field pulse (time for magnetic field to start from zero and return

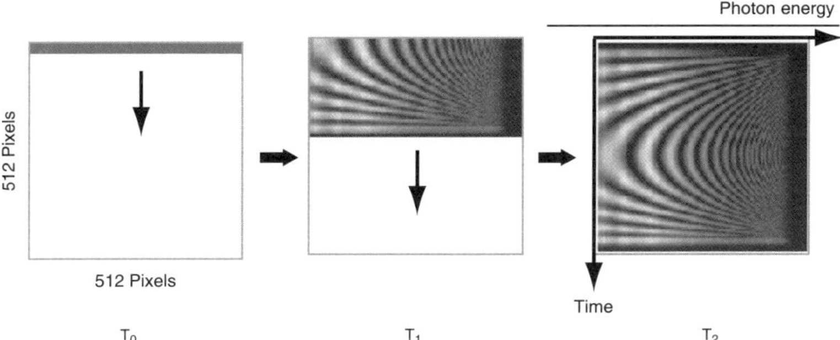

FIG. 7.16. Optical image recorded on a CCD frame in the measurement in pulsed magnetic fields. An example of the measurement of Faraday rotation is shown. The horizontal axis represents the wavelength and the vertical axis represents the time (the time proceeds from top to bottom). Three snapshots on the CCD frame are shown.

to zero), variation of the magnetic field during the gate pulse is only 0.3%. Thus we can obtain the optical spectra at almost constant magnetic fields. The diode array which we use at the ISSP has 1,024 channels, and each segment of the diode is several micrometers in size. The calibration of the wavelength is made by measuring the line spectrum of various lamps. In this way, we can obtain spectra of magneto-absorption, magneto-reflection, magneto-photoluminescence and Faraday rotation.

The above technique is a sort of point-by-point method. That is, we can obtain data corresponding to only one magnetic field point by one pulse shot. A different model of OMA (Model OMA-4) allows a time-resolved recording by using a CCD (charge coupled device) for a detector. The CCD consists of a two-dimensional matrix of optical detector segments. We focus the image of the optical spectra on the first line of the matrix, as shown in Fig. 7.16. This image is corresponding to spectra at one moment. The signal on the first line can be shifted to the second line and successively to 2, 3 $\cdots$ as the time proceeds, utilizing the shift register function of the CCD. During this process new data are always inputted on the first line. Hence we can obtain two-dimensional information containing the time dependence in the vertical direction. The CCD which is used at ISSP consists of a 512$\times$512 matrix of segments each of which has a dimension of 19 $\times$19 μm. The CCD is operated at liquid N_2 temperature.

The CCD is very useful for long pulse fields (ms pulse). However, the time needed for the shift register is a microsecond per line at shortest. Therefore, for short pulse fields in the megagauss range, we need a faster sweep of the optical image. For this purpose, we use an image converter camera. In Fig. 7.17, we depict a block-diagram of the streak spectromcter using an image converter camera. The image of the optical spectra is focused on a photo-cathode of the

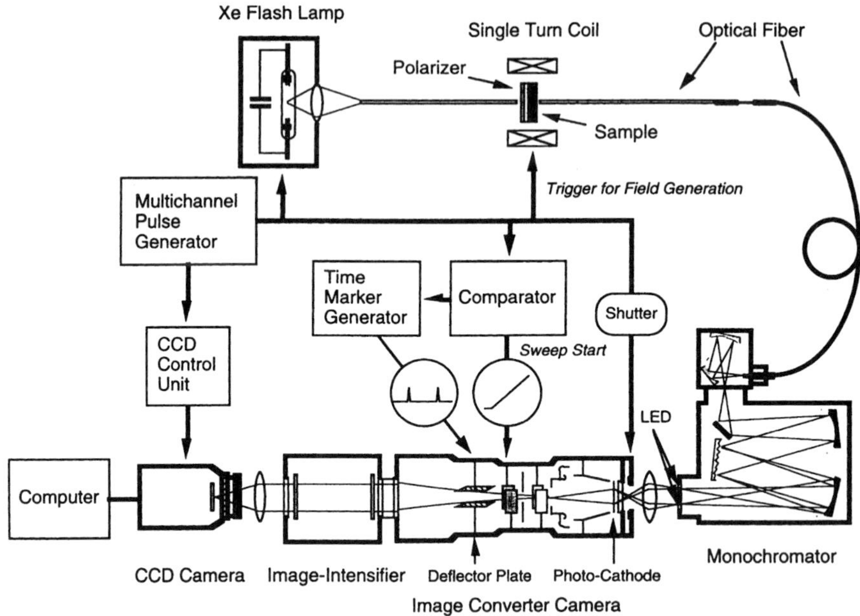

FIG. 7.17. System for measuring magneto-optical spectra using an image converter camera. Very fast sweep within a time of the order of microsecond range is possible.

image converter tube. Then photoelectrons are emitted from the photo-cathode in proportion to the optical intensity. The photoelectrons are accelerated and focused on the phosphor screen at the other end of the tube by a set of electrodes and an electric lens. The optical image is reproduced on the phosphor screen. During propagation, the electron beam is bent and swept by a voltage ramp applied to a pair of electrodes. Hence on the phosphor screen, the time-resolved spectrum is obtained. As the time sweep is made through an electron beam, very fast sweep is possible. The image on the phosphor screen is further intensified by an image intensifier and recorded by a CCD camera. In this case the CCD works simply as a medium to record a still image.

7.3.3 *Infrared and far-infrared laser spectroscopy*

A. Measuring system

Cyclotron resonance is a powerful means of investigating the effective mass and band structure of semiconductors. In high magnetic fields the wavelength of the resonant photon is in the infrared or far-infrared range. In this range, lasers are the most convenient radiation source in high magnetic fields. As the photon energy from a laser is constant, by sweeping magnetic field, we observe a reso-

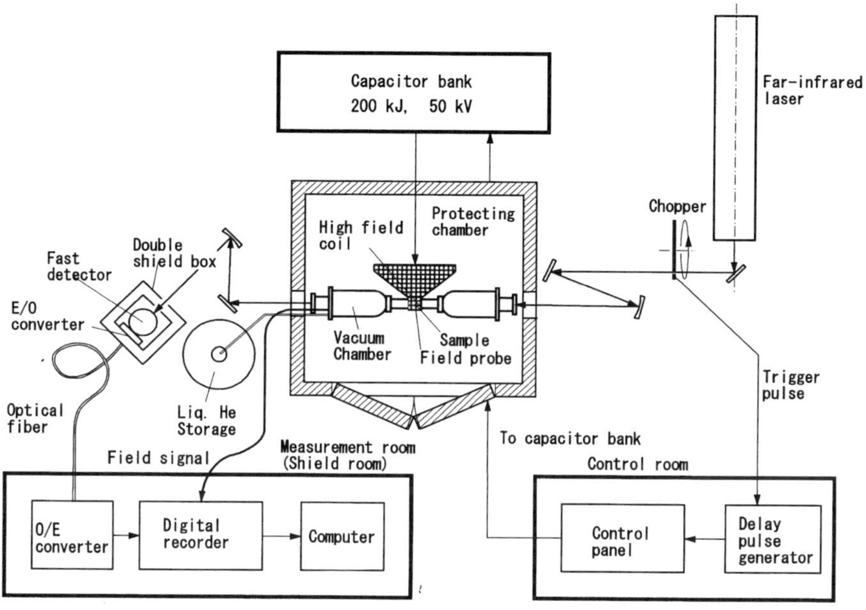

FIG. 7.18. Block diagram of the measurement system of cyclotron resonance or far-infrared spectroscopy in pulsed high magnetic fields. The case for the single turn coil technique is shown.

nance peak at some resonant magnetic field in the transmission of the radiation. The radiation is detected by an infrared or far-infrared detector. Cyclotron resonance can also be observed in the photoconductivity of the sample in the presence of the radiation. As the electron temperature is raised by strong absorption at the resonance, the conductivity is changed at the resonance field. This technique is called the cross modulation technique and is useful particularly in a steady field. Other than the cyclotron resonance, measurement of the AC conductivity in the millimeter or submillimeter wavelength range provides information similar to the DC conductivity. It is useful therefore when the DC transport measurement is difficult. For millimeter and submillimeter wavelengths, Carcinotron (back wave tube) or Gunn diodes are useful radiation sources.

Figure 7.18 shows an example of the block diagram of the system to measure cyclotron resonance in megagauss fields. The light beam is transmitted either by optics using mirrors or by light pipes. The former is convenient for the μs pulse field in the megagauss range, and the latter for the long pulse fields with ms duration and for DC fields.

B. Radiation sources

Table 7.5 Radiation sources conveniently used for far-infared and infrared spectroscopy in pulsed high magnetic fields

Radiation source	Wavelength (μm)	Photon energy (meV)
Gunn diode	2000–3000	0.4–0.6
Back wave tube (Carcinotron)	1000–3000	0.4–1.2
HCN laser	337	3.68
	311	3.99
H_2O laser (FIR)	119	10.4
CO_2 laser pumped molecular gas laser	33–570	2.18–37.6
H_2O laser (Near IR)	16.9	73.4
	23.0	53.9
	27.9	44.4
CO_2 laser	9–11	112–137
CO laser	5.3–5.7	217–233
He-Ne laser (Near IR)	3.39	365

Table 7.5 lists typical molecular gas lasers and other radiation sources which are used for cyclotron resonance experiments. For megagauss fields, intense infrared laser lines near 5.5μm produced by a CO laser, in the range of 9.2–11 μm produced by a CO_2 laser, and lines at 16.9, 23, 28, 119 μm produced by an H_20 laser are often employed. Also, using a CO_2 laser-pumped gas laser with CH_3OH *etc.*, many laser lines can be obtained in the range 30–500 μm. In the far-infrared range, lasers are usually operated in a pulsed mode to enhance the intensity. In this case, the pulse duration of the laser is made considerably longer than that of the magnetic field and the magnetic field pulse is started at the top of the magnetic field. The line at 16.9 μm of a H_2O laser is obtained only in a pulsed mode. In the pulsed oscillation of an H_2O laser, we can obtain intense and long laser oscillation lines, if He gas is mixed with H_2O gas [592]. When we use a CO_2 laser or a CO laser, they are usually operated in a continuous mode because the power is large and the stability is better in a continuous mode. However, the power is so intense that the radiation may damage the sample, if it is irradiated continuously. In such a case, the laser beam is chopped, reducing the duty ratio. For the μs pulse in the megagauss range, we produce an optical pulse of about 100 μs which makes the duty ratio of about 1/100.

C. Detectors

In the far-infrared range, the intensity from the light source is generally very

Table 7.6 Useful detectors for far-infared and infrared spectroscopy in pulsed high magnetic fields

Detector	Working temperature	Wavelength range
HgCdTe (Photovoltaic)	77 K	3–11 μm
HgMnTe (Photovoltaic)	77 K	3–11 μm
Ge(Cu)	4 K	10–30 μm
Ge(Ga)	4 K	30–300 μm
GaAs	4 K	200–700 μm
Ge diode	300 K	0.5–2 mm

weak, and moreover, in pulsed fields, the measurement is made in just one pulse without any integration. Therefore, very sensitive detectors are required. In the case of pulsed fields, the response time should be very fast as well. For detecting far-infrared radiation in DC fields, bolometers are usually used. However, the response time of bolometers is generally too slow for pulsed field measurements. In pulsed high magnetic fields the most commonly used detector utilizes the extrinsic or intrinsic photoconductivity of semiconductors. In order to use such a mechanism, electrons or holes should be kept in a low energy state, so that the device should be cooled to low temperatures. For example, doped Ge detectors should be refrigerated to liquid He temperature when we use their extrinsic photoconductivity. To represent the figure of merit of optical detectors, a parameter called detectivity D^* is employed. D^* is a quantity inversely proportional to minimum detectable power distinguishing from noise (noise equivalent power, NEP) and is measured in a unit of $\mathrm{cm{\cdot}Hz^{1/2}W^{-1}}$. Frequency and length are involved in the unit because the noise larger as the frequency band width is larger, and the light power is larger as the detector area is larger.

Table 7.6 lists the frequently used detectors in different wavelength ranges. For pulsed fields, it is important to choose a detector in terms not only of D^* but also of the response time. In the wavelength range 3–11 μm, the photovoltaic effect of HgCdTe is very useful for the fast detection of radiation. It can be used at liquid N_2 temperature and has a very fast response time ($\sim$ns). Infrared detectors usually have large resistances. Therefore, we have to use a wide band preamplifier to reduce the impedance and to avoid the integration of the signal through the transmission line.

7.3.4 *Magnetic measurement*

In semiconductor physics, magnetization measurement is particularly important for diluted magnetic semiconductors. There are several techniques for measuring magnetization, the Faraday method, the vibrating sample magnetometer (VSM) method, the induction method, the extraction method, and so on. In pulsed

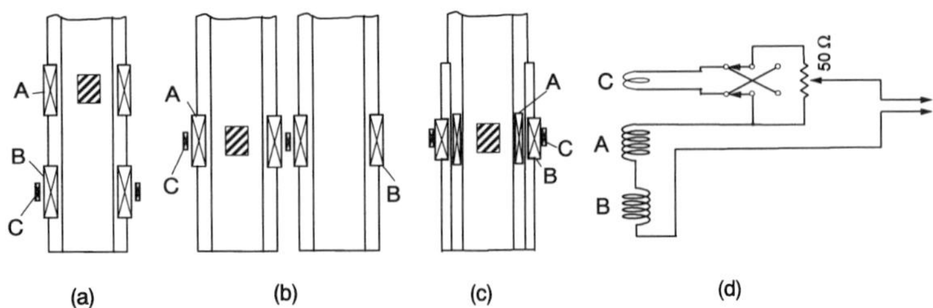

FIG. 7.19. Compensation system for the magnetization measurement. (a) longitudinally arranged type, (b) side by side type, (c) coaxially arranged type (d) circuit for the compensation.

fields, the induction method is the most convenient technique, because magnetic field is varied quickly and the induced voltage due to dB/dt is large. When we set a pick-up coil of N turns with a cross sectional area of S in a varying magnetic field, the voltage induced in the coil is

$$V = \frac{dB}{dt} SN. \tag{7.11}$$

and B is expressed as

$$B = \mu_0 H + M, \tag{7.12}$$

where H is the external magnetic field and M is the magnetization of the sample. As the term of external field is much larger than the magnetization term, it is necessary to compensate the first term to derive M. Figure 7.19 shows the pick-up coil system in which the first term can be compensated. In all the systems, two coils with different winding direction are connected in series and the sample is inserted in one of the coil. In system (a), coils are set side by side whereas in (b) they are axially aligned. System (c) is a coaxial type where the cross section S_1 and S_2 and the number of turns N_1 and N_2 of the coils are adjusted so as to fulfill a condition,

$$S_1 N_1 = S_2 N_2. \tag{7.13}$$

The choice of system should depend on the type of homogeneity of the field. In long pulse fields where the radial homogeneity is sufficiently good, system (c) generally gives the best results. In short pulse fields in the megagauss range, system (b) is most convenient, because it is symmetric both in the axial and radial direction and the total length can be smaller than (a).

The compensation of the voltage due to H is not always perfect since the two coils do not have exactly the same dimensions, and also there are some loops in the circuit. To eliminate the uncompensated voltage as much as possible, an appropriate portion of the voltage induced in another pick-up coil is added to the signal in a similar manner to the case of transport measurements. A compensation circuit as shown in Fig. 7.19 (d) is employed. The amount of the added voltage can be adjusted by many test shots in low magnetic fields. The single turn coil system is convenient for measuring magnetization in the megagauss range because we can repeat many shots for the same pick-up coil containing samples. The adjustment of the compensation circuit can be performed by repeating preliminary shots at low charge voltage in the capacitor bank within the limit of the non-deformation of the coil.

In a magnetic substance, there is a component proportional to magnetization in the Faraday rotation. Therefore, we can measure the magnetization indirectly through the Faraday rotation. In transparent substances, it is more convenient to measure the Faraday rotation instead of the direct magnetization measurement, because we can get rid of the problem of the induced spurious voltage.

8

GENERAL REFERENCES

I. Textbooks

Many excellent books have been published on the physics of semiconductors, and the science and technology of high magnetic fields. Here, we list only typical useful books related to the topics discussed in this book. Some of them are now rather old, but still very useful.

[GT-1] Houghton JT and Smith SD (1966) *Infra-red Physics* (Clarendon Press, Oxford, 1966).

[GT-2] Zeiger HH and Pratt GW (1973) *Magnetic Interaction in Solids* (Clarendon Press, Oxford, 1973).

[GT-3] Herlach F ed (1985) *Strong and Ultrastrong Magnetic Fields and Their Applications* (Springer-Verlag, 1985).

[GT-4] Chikazumi S (1997) *Physics of Ferromagnetism, Second Edition* (Clarendon Press, Oxford, 1997).

[GT-5] Seeger K (2004) *Semiconductor Physics*, 9th Edition (Springer-Verlag, 2004).

[GT-6] Yu PY and Cardona M (2001) *Fundamentals of Semiconductors* 3rd Edition (Springer-Verlag, 2001).

[GT-7] Herlach F and Miura N eds (2003-2006) *High Magnetic Fields, Science and Technology* Vol. 1-3 (World Scientific, 2003-2006).

II. Proceedings of international conferences on high magnetic fields

In recent years, many international conferences have been held on the topics of the physics of high magnetic fields. The proceedings of these conferences are very useful to overview the history, progress, and the present state of the art of the study of high magnetic fields.

A. International conferences on the generation and application of high magnetic fields (general)

The following proceedings are not necessarily focused in semiconductor physics, but are related to the topics in this book in some way.

[Pr-1] *High Magnetic Fields (Proceedings of the International Conference on High Magnetic Fields, MIT, Cambridge, MA, 1961)* eds Kolm H, Lax B,

Bitter F and Mills R (1962) (John Wiley & Sons, Inc., New York, London, 1962).

[Pr-2] *Physics of Solids in Intense Magnetic Fields (Lectures presented at the 1st Chania Conference held at Chania, Crete, 1967)*, ed. Haidemenakis ED (Plenum Press, New York, 1969).

[Pr-3] *Physique Sous Champs Magnetiques Intenses (Proceedings of the International Conference on the Physics in High Magnetic Fields, Grenoble, 1974)*, (C.N.R.S., Paris, 1975).

[Pr-4] *Solids and Plasmas in High Magnetic Fields (Proceedings of the International Conference on Solids and Plasmas in High Magnetic Fields, Cambridge, MA, 1978)*, eds Aggarwal RL, Freeman AJ and Schwartz BB [*J. Mag. Mag. Mater.* **11** (1979)].

[Pr-5] *Proceedings of the Honda International Symposium on the Application of High Magnetic Fields to Material Science (Sendai, 1989)* eds Nakagawa Y and Kido G, [*Physica B* **164** (1990)].

[Pr-6] *Frontiers in High Magnetic Fields (Proceedings of the Todai International Symposium on Frontiers in High Magnetic Fields, Tokyo, 1993)*, ed. Miura N, [*Physica B* **201** (1994)].

[Pr-7] *Advances in High Magnetic Fields (Proceedings of the International Workshop on Advances in High Magnetic Fields, Tsukuba, 1995)*, eds Maeda H, Kido G and Inoue K [*Physica B* **216** (1996)].

[Pr-8] *High Magnetic Fields: Applications, Generation, Materials (International Workshop on High Magnetic Fields: Industry, Materials and Technology (Tallahassee, 1996)*, ed. Schneider-Muntau HJ (World Scientific, 1997)

B. Application of high magnetic fields in semiconductor physics (SemiMag)

This conference series was started by Prof. G. Landwehr in 1972 at Würzburg as a satellite conference of the international conference on the physics of semiconductors (ICPS). Except for 1984, it has been held every two years associated with every ICPS. For the first four meetings, only work books compiling the lecture notes were printed, but since the fifth conference at Hakone, proceedings have been published.

[SM-1] *Lecture notes of the 1st International Conference on the Application of High Magnetic Fields in Semiconductor Physics*, ed. Landwehr G. (Würzburg, 1972).

[SM-2] *Lecture notes of the 2nd International Conference on the Application of High Magnetic Fields in Semiconductor Physics*, ed. Landwehr G. (Würzburg, 1974).

[SM-3] *Lecture notes of the 3rd International Conference on the Application of High Magnetic Fields in Semiconductor Physics*, ed. Landwehr G. (Würzburg, 1976).

[SM-4] *Lecture notes of the 4th International Conference on the Application of High Magnetic Fields in Semiconductor Physics*, ed. Ryan JF (Oxford, 1978).

[SM-5] *Physics in High Magnetic Fields (Proceedings of the 5th International Conference on the Application of High Magnetic Fields in Semiconductor Physics and Magnetic Materials)*, eds Chikazumi S and Miura N (1981) (Springer-Verlag, 1981) (Hakone, 1980).

[SM-6] *Application of High Magnetic Fields in Semiconductor Physics (Proceedings of the 6th International Conference in Application of High Magnetic Fields in Semiconductor Physics)*, ed. Landwehr G (Springer-Verlag, 1983) (Grenoble, 1982).

[SM-7] *High Magnetic Fields in Semiconductor Physics (Proceedings of the 7th International Conference in Application of High Magnetic Fields in Semiconductor Physics)*, ed. Landwehr G (Springer-Verlag, 1987) (Würzburg, 1986).

[SM-8] *High Magnetic Fields in Semiconductor Physics II (Proceedings of the 8th International Conference in Application of High Magnetic Fields in Semiconductor Physics)*, ed. Landwehr G (Springer-Verlag, 1989) (Würzburg, 1988).

[SM-9] *High Magnetic Fields in Semiconductor Physics III (Proceedings of the 9th International Conference in Application of High Magnetic Fields in Semiconductor Physics)*, ed. Landwehr G (Springer-Verlag, 1992) (Würzburg, 1990).

[SM-10] *The Application of High Magnetic Fields in Semiconductor Physics (Proceedings of the 10th International Conference in Application of High Magnetic Fields in Semiconductor Physics)*, ed. Miura N [*Physica B* **184** (1993)] (Tomiura, 1992).

[SM-11] *Proceedings of the 11th International Conference on High Magnetic Fields in Semiconductor Physics*, ed. Heiman D (World Scientific, 1995) (Cambridge (Massachusetts), 1994).

[SM-12] *Proceedings of the 12th International Conference on High Magnetic Fields in Semiconductor Physics*, eds Landwehr G and Ossau W (World Scientific, 1997) (Würzburg, 1996).

[SM-13] *Proceedings of the 13th International Conference on High Magnetic Fields in Semiconductor Physics*, ed. Maan YC (World Scientific, 1997) (Nijmegen, 1998).

[SM-14] *Proceedings of the 14th International Conference on High Magnetic Fields in Semiconductor Physics*, ed. Kido G [*Physica B* **298** (2001)] (Matsue, 2000).

[SM-15] *Proceedings of the 15th International Conference on High Magnetic Fields in Semiconductor Physics*, ed. Nicholas RJ (on CD-ROM, Institute of Physis, 2003) (Oxford, 2002).

[SM-16] *Proceedings of the 16th International Conference on High Magnetic Fields in Semiconductor Physics*, eds Wang YJ, Engel L and Bonensteel N (World Scientific, 2005) (Tallahassee, 2004).

[SM-17] *Proceedings of the 17th International Conference on High Magnetic Fields in Semiconductor Physics*, ed. Landwehr G [*Int. J. Modern Physics* (World Scientific, 2007)] (Wüzburg, 2006).

B. Research in High Magnetic Fields (RHMF)

This conference series was started by Prof. M. Date at Osaka in 1982 as a satellite conference of the International Conference of Magnetism (ICM). The second conference was revived by Prof. F. Herlach in 1988 at Leuven. Since then it has been held every three years at the occasion of every ICM.

[RH-1] *High Field Magnetism (Proceedings of the International Symposium on High Field Magnetism, Osaka, 1982)*, ed. Date M (North-Holland Publishing Co., 1983).

[RH-2] *High Field Magnetism (Proceeding of the 2nd International Conference on High Field Magnetism, Leuven, 1988)*, eds Herlach F and Franse JJM [*Physica B* **155** (1989)].

[RH-3] *Proceedings of the 3rd International Conference on Research in High Magnetic Fields, (Amsterdam, 1991)* eds de Boer FR, de Cahtel PF and Franse JJM [*Physica B* **177** (1992)].

[RH-4] *Proceedings of the 4th International Conference on Research in High Magnetic Fields*, (Nijmegen, 1994), ed. Perenboom JAAJ [*Physica B* **211** (1995)].

[RH-5] *Proceedings of the 5th International Conference on Research in High Magnetic Fields*, (Sydney, 1997), ed. Clark RG [*Physica B* **246-247** (1998)].

[RH-6] *Proceedings of the 6th International Conference on Research in High Magnetic Fields*, (Porto, 2000), eds Herlach F, de Boer F, Motokawa M and Sousa JB [*Physica B* **298** (2001)].

[RH-7] *Proceedings of the 7th International Conference on Research in High Magnetic Fields*, (Toulouse, 2003), eds Rikken GLJA, Portugall P and Vanacken J [*Physica B* **346-347** (2004)].

[RH-8] *Proceedings of the 8th International Conference on Research in High Magnetic Fields, Yamada Conference LX on Research in High Magnetic Fields (RHMF2006)*, (Sendai, 2006), eds Kobayashi N, Toyota N and Motokawa M [*J. Phys.: Condes. Matt. Conf. Ser.* **51** (2006)].

C. Physical Phenomena in High Magnetic Fields (PPHMF)

This is a conference series organized by the National High Magnetic Fields Laboratory at Tallahassee and started in 1991. The scope covers all the applications of high magnetic fields, mainly for solid state physics.

[PP-1] *Physical Phenomena at High Magnetic Fields* (PPHMF-I) (Tallahassee, USA, 1991), eds Manousakis E, Schlottmann P, Kumar P, Dedell K, and Mueller F (Addison-Wesley Pub. Co., 1992).

[PP-2] *Physical Phenomena at High Magnetic Fields-II* (PPHMF-II) (Tallahassee, USA, 1995), eds Fisk Z, Gor'kov LP, Meltzer D, and Schrieffer JR (World Scientific, 1996).

[PP-3] *Physical Phenomena at High Magnetic Fields-III* (PPHMF-III) (Tallahassee, USA, 1995), eds Fisk Z, Gor'kov LP and Schrieffer JR (World Scientific Pub. Co., 1999).

[PP-4] *Physical Phenomena in High Magnetic Fields-IV* (PPHMF-IV) (Santa Fe, USA, 2001), eds Boebinger G, Fisk Z, Gor'kov LP, Lacerda A and Schrieffer JR (World Scientific, 2002)

[PP-5] *Physical Phenomena in High Magnetic Fields-V* (PPHMF-V) (Tallahassee, USA, 2005) See the URL page
`http://magnet.fsu.edu/~bonestee/PPHMF_book.pdf`

In addition, there are international conference series on the
"Megagauss field generation and related topics (MG-series)",
and the
"Magnet technology (MT-series)".
The main scopes of these conferences are focused on technical problems of the high magnetic field science and technology, but important information can be obtained from the proceedings concerning the recent advances in very high magnetic fields in the megagauss range and superconducting magnets, respectively.

We can also see many valuable literatures related to the semiconductor physics in high magnetic fields in the proceedings of the following three large international conference series, which are held every 2 years:
"International Conference on the Physics of Semiconductors (ICPS)",
"International Conference on Electronic Properties of Two-Dimensional Systems (EP2DS)",
"International Conference on Modulated Semiconductor Structures (MSS)".

References

[1] Herlach F and Miura N (2003) in Vol. 1 of [GT-7].
[2] Kubo R, Miyake SJ and Hashizume N (1965) *Solid State Physics*, eds. Seitz F and Turnbull D, Vol. 17, (Academic Press, 1965) p. 270.
[3] Kittel C (1963) *Quantum Theory of Solids* (John Wiley & Sons, 1963)
[4] Callaway J (1964) *Energy Band Theory* (Academic Press, New York, London, 1964)
[5] Kane EO (1957) *J. Phys. Chem. Solids* **1** 249.
[6] Kane EO (1966) *Semiconductors and Semimetals*, eds. Willardson, RK and Beer AC (1966) (Academic Press, New York and London, 1966), Chap. 3.
[7] Luttinger JM and Kohn W (1955) *Phys. Rev.* **97** 869.
[8] Roth LM, Lax B and Zwerdling S (1959) *Phys. Rev.* **114** 90.
[9] Lax B, Mavroid JG, Zeiger HJ and Keys RJ (1961) *Phys. Rev.* **122** 31.
[10] Johnson EJ and Dickey DH (1970) *Phys. Rev.* B **1** 2676.
[11] Luttinger JM (1956) *Phys. Rev.* **102** 1030.
[12] Lawaetz P (1971) *Phys. Rev.* B **4** 3460.
[13] Pidgeon CR and Brown RN (1966) *Phys. Rev.* **146** 575.
[14] Ando T, Fowler AB and Stern F (1982) *Electronic properties of two-dimensional systems: Rev. Mod. Phys.* **54** 437-672.
[15] Fowler AB. Fang FF, Howerd WE and Stiles PJ (1966) *Phys. Rev. Lett.* **16** 901.
[16] Ando T and Uemura Y (1974) *J. Phys. Soc. Jpn.* **37** 1044.
[17] Luo J, Munekata H, Fang FF, and Stiles PJ (1990) *Phys. Rev.* B **41** 7685.
[18] Das B, Miller DC, Datta S, Reifenberger R, Hong WP, Bhattacharya PK, Singh J and Jaffe M (1989) *Phys. Rev.* B **39** 1411.
[19] Stein D, von Klitzing K and Weiman G (1983) *Phys. Rev. Lett.* **51** 130.
[20] Dresselhaus G (1955) *Phys. Rev.* **100** 580.
[21] Rashba EI (1960) *Sov. Phys. — Solid State* **2** 1109.
[22] Bychkov Yu A and Rashba EI (1984) *JETP Lett.* **39** 78.
[23] Ando T (1985) *J. Phys. Soc. Jpn.* **54** 1528.
[24] Ando T: *Private communication.*
[25] Iwasa Y (1985) Ph D Thesis, (1985, University of Tokyo).
[26] Esaki L and Tsu R (1970) *IBM J. Res. Develop.* **14** 61.
[27] Akiyama H (1998) *J. Phys. Condens. Matt.* **10** 3095.
[28] Bhat R, Kapon E, Hwang DM, Koza MA and Yun CP (1988) *J. Crys. Growth* **93** 850.
[29] Kapon E, Hwang DM and Bhat R (1989) *Phys. Rev. Lett.* **63** 430.
[30] Nagamune Y, Arakawa Y, Tsukamoto S, Nishioka M, Sakaki H and Miura N (1992) *Phys. Rev. Lett.* **69** 2963.
[31] Takeuchi M, Shiba K, Sato K, Huang HK, Inoue K and Nakashima H (1995) *Jpn. J. Appl. Phys.* **34** 4411.
[32] Demel T, Heitman D, Grambow P and Ploog K (1990) *Phys. Rev. Lett.* **64** 788.

[33] Tarucha S, Austing DG, HondaT, van der Hage RJ and Kouwenhoven LP (1996) *Phys. Rev.* **77** 3613.
[34] Stransky IN and Krastanow L (1938) *Sitz. Ber. Akad. Wiss., Math.-naturwiss. K1. Abt.* IIb **146** 797.
[35] Seifert W, Carisson N, Miller M, Pistol M-E, Samuelson L and Wallenberg LR (1996) *Prog. Crystal Growth* **33** 423.
[36] Sakuma Y, Sugiyama Y, Muto S and Yokoyama N (1995) *Appl. Phys. Lett.* **67** 256.
[37] Darwin CG (1930) *Prod. Cambridge Philos. Soc.* **27** 86.
[38] Fock V (1928) *Z. Phys.* **47** 446.
[39] Osada T and Miura N (1989) *Solid St. Commun.* **69** 1169.
[40] Harper PG (1955) *Proc. Phys. Soc.* (London) **A68** 874.
[41] Hofstadter DR (1976) *Phys. Rev.* B **14** 2239.
[42] Albrecht C, Smet JH, von Klitzing K, Weiss D, Umansky V and Schweizer H. (2001) *Phys. Rev. Lett.* **86** 147.
[43] Vogl P and Strahberger C (2002) *Proc. 26th Int. Conf. Physics of Semiconductors* (Edinburgh, 2002).
[44] Cohen MH and Falicov LM (1961) *Phys. Rev. Lett.* **7** 231.
[45] Falikov LM and Sievert PR (1965) *Phys. Rev.* **138** A88.
[46] Falikov LM and Zuckerman MJ (1967) *Phys. Rev.* **160** 372.
[47] Pippard AB (1962) *Proc. Roy. Soc.* **270** 1.
[48] Pippard AB (1962) *Phil. Tans. Roy. Soc.* **256** 317.
[49] Pippard AB (1969) in *Physics of Solids in Intense Magnetic Fields* ed. Haidemenakis ED (Plenum Press, New York, 1969) Chap. 18
[50] Priestley MG (1963) *Proc. Roy. Soc.* **270** 258.
[51] Blount EI (1962) *Phys. Rev.* **126** 1636.
[52] Lifshitz IM and Peschanskii VG (1959) *Sov. Phys. JETP* **35** 875.
[53] Lifshitz IM, Azbel MIa and Kaganov MI (1957) *Sov. Phys. JETP* **4** 41.
[54] Joseph AS and Thorsen AC (1963) *Phys. Rev. Lett.* **11** 67.
[55] Stark RW (1964) *Phys. Rev. Lett.* **135** A 1698.
[56] Mackintosh AR, Spanel LE and Young RC (1963) *Phys. Rev. Lett.* **10** 434.
[57] Priestley MG, Falikov LM and Weisz G (1963) *Phys. Rev.* **131** 617.
[58] Soret JC, Rosenman I, Simon CH and Battalan F (1985) *Phys. Rev.* B **32** 8361.
[59] Stradling RA, Eaves L, Hoult RA, Miura N, Simmonds PE and Bradley CC (1972) *Gallium Arsenide and Related Compounds (Proc. 4th Int. Symposium,* (Boulder, 1972) (Conference Series No. 17, The Institute of Physics, London and Bristol) p. 65.
[60] Simmons PE, Chamberlain JM, Hoult RA, Stradling RA and Bradley CC (1974) *J. Phys. C: Solid State Physics* **7** 4164.
[61] Cooke RA, Hoult RA, Kirkman RF and Stradling RA (1978) *J. Phys. D: Appl. Phys.* **11** 945.
[62] Yafet Y, Keyes RW and Adams EN (1956) *J. Phys. Chem. Solids* **1** 137.

[63] Beckman, O, Hanamura, E and Neuringer LJ (1967) *Phys. Rev. Lett.* **18** 774.
[64] Wallis RF and Bowlden HJ (1958) *J. Phys. Chem. Solids* **7** 78.
[65] Elliot RJ and Loudon R (1960) *J. Phys. Chem. Solids* **15** 196.
[66] Baldereschi A and Bassani F (1970) *Proc. 10th Int. Conf. Phys. Semiconductors* (Cambridge, 1970) p.191.
[67] Larsen DM (1968) *J. Phys. Chem. Solids* **29** 271.
[68] Aldrich C and Green RL (1979) *Phys. Stat. Sol.* (b) **93** 343.
[69] Makado PC and McGill NC (1986) **19** 873.
[70] Rösner W, Wunner G, Herold H and Ruder H (1984) *J. Phys.* B **17** 29.
[71] Praddande HC (1972) *Phys. Rev.* A **6** 1321.
[72] Cabib D, Fabri E and Fiorio G (1971) *Solid St. Commun.* **9** 1517.
[73] Cabib D, Fabri E and Fiorio G (1972) *Nuovo Cimento* **10B** 185.
[74] Shinada M, Akimoto O, Hasegawa H and Tanaka K (1970) *J. Phys. Soc. Jpn.* **28** 975.
[75] Lee N, Larsen DM and Lax B (1972) *J. Phys. Chem. Solids*, **34** 1959.
[76] Kirkman RF, Stradling RA and Lin-Chung PJ (1978) *J. Phys. C: Solid State* **11** 419.
[77] Knox, RS (1963) *Theory of Excitons, Solid State Physics,* Suppl. **5**, eds. Seitz F and Turnbull D (Academic Press, New York, 1963).
[78] Akimoto O and Hasegawa H (1967) *J. Phys. Soc. Jpn.* **22** 181.
[79] Van der Pauw LJ (1958) *Philips Res. Rep.* **13** 1.
[80] Van der Pauw LJ (1958/59) *Philips Res. Rep.* **20** 220.
[81] Hiruma K, Kido G and Miura N (1979) *Solid St. Commun.* **31** 1019.
[82] Hiruma K, Kido G. Kawauchi K and Miura N (1980) *Solid St. Commun.* **33** 257.
[83] Miura N, Hiruma K, Kido G and Chikazumi S (1982) *Phys. Rev. Lett.* **49** 1339.
[84] Roth LM and Argyres PN (1966) *Semiconductors and Semimetals*, eds. Willardson RK and Beer AC (Academic Press, New York and London, 1966), Chap. 6.
[85] Fowler AB, Fang FF, Howard WE and Stiles PJ (1966) *Phys. Rev. Lett.* **16** 901.
[86] Fowler AB, Fang FF, Howard WE and Stiles PJ (1966) *J. Phys. Soc. Jpn.* **21** Suppl. 331.
[87] Ando T (1974) *J. Phys. Soc. Jpn.* **37** 1233.
[88] Leadley DR, Nicholas RJ, Foxon C T and Harris JJ (1994) *Phys. Rev. Lett.* **72** 1906.
[89] Kawaji S and Wakabayashi J (1976) *Surf. Sci.* **58** 238.
[90] Ando T, Matsumoto Y, Uemura Y, Kobayashi M and Komatsubara KF (1972) *J. Phys. Soc. Jpn.* **32** 859.
[91] Kobayashi M and Komatsubara KF (1974) *Jpn. J. Appl. Phys.* Suppl. **2** Part. 2 343.
[92] Ando T (1974) *J. Phys. Soc. Jpn.* **37** 622.

[93] Anderson PW (1958) *Phys. Rev.* **102** 1008.
[94] Abrahams E, Anderson PW, Licciardello DC and Ramakrishnan TV (1979) *Phys. Rev. Lett.* **42** 673.
[95] von Klitzing K, Dorda G and Pepper M (1980) *Phys. Rev. Lett.* **45** 494.
[96] Von Klitzing reported the discovery of the quantum Hall effect at a specially organized rump session of the 5th International Conference on the Application of High Magnetic Fields in Semiconductor Physics and Magnetic Materials which was held at Hakone in 1980. This was the first report which he presented to the international community about the discovery.
[97] Kawaji S and Wakabayashi J (1981) in [SM-5] p.284.
[98] Paalanen MA, Tsui DC and Gossard AC (1982) *Phys. Rev.* B **25** 5566.
[99] Laughlin RB (1981) *Phys. Rev.* B **23** 5632.
[100] Aoki H and Ando T (1981) *Solid St. Commun.* **38** 1079.
[101] Büttiker M (1988) *Phys. Rev.* B **38** 9375.
[102] Nicholas RJ, Haug K, von Klitzing K and Weimann G (1988) *Phys.Rev.* B **37** 1294.
[103] Schmeller A, Eisenstein JP, Pfeiffer LN and West KW (1995) *Phys. Rev. Lett.* **75** 4290.
[104] Barret SE, Dabbagh G, Pfeiffer LN, West KW and Tycko Z (1995) *Phys. Rev. Lett.* **74** 5112.
[105] Aifer EH, Goldberg BB and Broid DA (1996) *Phys. Rev. Lett.* **76** 680.
[106] Tsui DC, Störmer HL and Gossard AC (1982) *Phys. Rev. Lett.* **48** 1559.
[107] Willet RL, Eisenstein JP, Störmer HL, Tsui DC, Gossard AC and English AC (1987) *Phys. Rev. Lett.* **59** 1776.
[108] Du RR, Störmer HL, Tsui DC, Yeh AS, Pfeiffer LN and West KW (1994) *Phys. Rev. Lett.* **73** 3274.
[109] Pan W, Störmer HL, Tsui DC, Pfeiffer LN, Baldwin KW and West KW (2002) *Phys. Rev. Lett.* **88** 176802.
[110] Laughlin RB (1983) *Phys. Rev. Lett.* **50** 1395.
[111] Haldane FDM (1983) *Phys. Rev. Lett.* **51** 605.
[112] Laughlin RB (1984) *Surf. Sci.* **142** 163.
[113] Halperin BI (1984) *Phys. Rev. Lett.* **52** 1583.
[114] Jain JK (1989) *Phys. Rev. Lett.* **63** 199, (1990) *Phys. Rev.* B **41** 7653.
[115] Simon SH and Halperin BI (1993) *Phys. Rev.* B **48** 17368.
[116] Willett RL, Paalanen MA, Ruel RR, West KW, Pfeiffer LN and Bishop DJ (1990) *Phys. Rev. Lett.* **65** 112.
[117] Willett RL, West KW and Pfeiffer LN (1995) *Phys. Rev. Lett.* **75** 2988.
[118] Kang W, Störmer HL, Pfeiffer LN, Baldwin KW and West KW (1993) *Phys. Rev. Lett.* **71** 3850.
[119] Wigner E (1934) *Phys. Rev.* **46** 1002.
[120] Platzman PM and Fukuyama H (1974) *Phys. Rev.* B **10** 3150.
[121] Grimes CC and Adams G (1979) *Phys. Rev. Lett.* **42** 795.
Fischer DS, Halperin BL and Platzman PM (1979) *Phys. Rev. Lett.* **42** 798.

[122] Monarkha Y and Kono K *Two-Dimensional Coulomb Liquids and Solids* (Springer, 2003).
[123] Fukuyama H and Lee PA (1978) *Phys. Rev.* B **18** 6245.
[124] Jiang HW, Willet RL, Störmer HL, Tsui DC, Pfeiffer LN and West KW (1990) *Phys. Rev. Lett.* **65** 633.
[125] Takamasu T, Dodo H and Miura N (1995) *Solid St. Commun.* **96** 121.
[126] Streda P and von Klitizing K (1984) *J. Phys. C* **17** L483.
[127] Cage ME, Dzuiba RF, Field BF, Williams ER, Girvin SM, Gossard AC, Tsui DC and Wagner RJ (1983) *Phys. Rev. Lett.* **51** 1374.
[128] Ebert G, von Klitzing K, Ploog K and Weimann G (1983) *J. Phys. C* **16** 5441.
[129] Kawaji S, Hirakawa K, Nagata M, Okamoto T, Fukase T and Goto T (1994) *J. Phys. Soc. Jpn* **63** 2303.
[130] Komiyama S, Takamasu T, Hiyamizu S and Sasa S (1986) *Solid St. Commun.* **54** 47.
[131] Takamasu T, Kido G, Ohno M, Miura N, Endo A, Kato M, Katsumoto S and Iye Y (1998) *Physica B* **246-247** 12.
[132] Esaki L (1958) *Phys. Rev.* **109** 603.
[133] Mendez EE, Esaki L and Wang WI (1986) *Phys. Rev.* B**33** 2893.
[134] Kamata N, Yamada K, Miura N and Eaves L (1993) *J. Phys. Soc. Jpn.* **62** 2120.
[135] Goldman VJ, Tsui DC and Cunningham JE (1987) *Phys. Rev.* B **36** 7635.
[136] Bando H (1987) *Jpn. J. Appl. Phys.* **26** Suppl. 26-3 765.
[137] Leadbeater ML, Alves ES, Eaves L, Henini M and Hughes OH (1989) *Phys. Rev.* **39** 3438.
[138] Wirner C, Awano Y, Yokoyama N, Ohno M, Miura N, Nakagawa T and Bando H (1997) *Semicond. Sci. Technol.* **12** 998.
[139] Osada T, Miura N and Eaves L (1992) *Solid St. Commun.* **81** 1019.
[140] Hayden RK, Takamasu T, Maude DK, Valadares EC, Eaves L, Ekenberg U, Miura N, Henini M, Portal JC, Hill G and Pate MA (1992) *Semicond. Sci. Technol.* **7** B413.
[141] Hayden RK, Eaves L, Henini M, Valadares EC, Kuhn O, Maude DK, Portal JC, Takamasu T, Miura N and Ekenberg U (1994) *Semicond. Sci. Technol.* **9** 298.
[142] Hayden RK, Takamasu T, Miura N, Henini M, Eaves L and Hill G (1996) *Semicond. Sci. Technol.* **11** 1424.
[143] Fromhold TM, Eaves L, Sheard FW, Leadbeater ML, Foster TJ and Main PC (1994) *Phys. Rev. Lett.* **72** 2608.
[144] Marcus CM, Rimberg AJ, Westervelt RM, Hopkins PF and Gossard AC (1992) *Phys. Rev. Lett.* **69** 506.
[145] Jensen RV (1992) *Nature* **355** 311.
[146] Fromhold TM, Wilkinson PB, Sheard FW, Eaves L, Miao J and Edwards G (1995) *Phys. Rev. Lett.* **75** 1142.

[147] Wilkinson PB, Fromhold TM, Eaves L, Sheard FW, Miura N and Takamasu T (1996) *Nature* **380** 608.

[148] Boebinger G, *Phys. Rev. Lett.* Müller, G, Boebinger, GS, Mathur, H, Pfeiffer LN and West KW (1995) *Phys. Rev. Lett.* **75** 2875.

[149] See for example [3] Chapt. 7.

[150] Gurevich VL and Firsov Yu A (1961) *Soviet Phys. JETP* **13** 137.

[151] Firsov Yu A, Gurevich VL, Parfe'nev RV and Shalyt SS (1964) *Phys. Rev. Lett.* **12** 660.

[152] Gurevich VL, Firsov Yu A and Efros AL (1963) *Soviet Phys. Solid St.* **7** 1331.

[153] Stradling RA and Wood RA (1968) J. Phys. C **1** (1968) 1711.

[154] Harper PG, Stradling RA and Hodby JW (1973) *Rep. Prog. Phys.* **36** No. 1.

[155] Nicholas RJ (1985) *Prog. Quant. Electr.* **10** 1.

[156] Kido G and Miura N (1983) *J. Phys. Soc. Jpn.* **52** 1734.

[157] Barker JR (1972) *J. Phys. C* **5** 1657.

[158] Parlmer RJ (1970) *D. Phil. Thesis* (University of Oxford, 1970) (unpublished).

[159] Stradling RA and Wood RA (1970) *J. Phys. C Solid St.* **3** L94.

[160] Mears AL, Stradling RA and Inall EK (1968) *J. Phys. C* **1** 821.

[161] Nakayama M (1969) *J. Phys. Soc. Jpn.* **27** 636.

[162] Eaves L, Stradling RA, Askenazy S, Leotin J, Portal JC and Ulmet JP (1971) *J. Phys. C* **4** L42.

[163] Maeda Y, Taki H, Sakata M, Ohta E, Yamada S, Fukui T and Miura N (1984) *J. Phys. Soc. Jpn.* **53** 3553.

[164] Ehrenreich H (1957) *J. Phys. Chem. Solids* **2** 131.

[165] Lang IG (1962) *Sov. Phys. Solid State* **3** 1871.

[166] Ravich Yu I (1965) *Sov. Phys. Solid State* **7** 1466.

[167] Miura N, Stradling RA, Askenazy S, Carrere G, Leotin J, Portal JC and Ulmet JP (1972) *J. Phys. C* **5** 3332.

[168] Nakao K, Doi T, and Kamimura H (1971) *J. Phys. Soc. Jpn.* **39** 1400.

[169] Couder Y (1969) *Phys. Rev. Lett.* **22** 890.

[170] Yoshizaki R and Tanaka S (1971) *J. Phys. Soc. Jpn.* **30** 1389.

[171] Miura N and Tanaka S (1970) *Phys. Stat. Solidi* **42** 257.

[172] Peterson RL (1972) *Phys. Rev. Lett.* **28** 431.

[173] Stradling RA (1972) *Proc. Int. Conf. Phys. Semiconductors*, (Warsaw, 1972) p. 261.

[174] Eaves L, Stradling RA, Askenazy S, Leotin, J, Portal, JC and Ulmet JP (1971) *J. Phys. C: Solid St. Phys.* **4** L42.

[175] Aksel'rod MM and Tsidil'kovskii IM (1969) *Sov. Phys. JETP Lett.* **9** 381.

[176] Tsui DC, Englert Th, Cho AY and Gossard AC *Phys. Rev. Lett.* **44** 341.

[177] Kido G, Miura N, Ohono H and Sakaki H (1982) *J. Phys. Soc. Jpn.* **51** 2168.

[178] Brummell MA, Nicholas RJ and Hopkins MA (1987) *Phys. Rev. Lett.* **58** 77.
[179] Noguchi H, Sakaki H, Takamasu T and Miura N (1992) *Phys. Rev.* B **45** 12148.
[180] Stradling RA, Eaves L, Hoult RA, Mears AL and Wood RA (1970) *Proc. Int. Conf. Physics of Semiconductors, (Boston, 1970)* p. 369.
[181] Eaves L, Guimaraes PSS and Portal JC (1984) *J. Phys. C.: Solid St. Phys.* **17** 6177.
[182] Mori N, Nakamura N, Tacniguchi K and Hamaguchi C (1987) *Semicond. Sci. Technol.* **2** 542.
[183] Yamada K, Miura N, Eaves L, Portal JC, Forte-Poisson MA and Brylinski C (1988) *Semicond. Sci. Technol.* **3** 895.
[184] Eaves L, Stradling RA and Wood RA (1970) *Proc. Int. Conf. Physics of Semiconductors*, (Boston, 1970) p.816.
[185] Portal JC, Eaves L, Askenazy S and Stradling RA (1974) *Solid St. Commun.* **14** 1241.
[186] Hamaguchi C, Hirose Y and Shimomae K (1983) *Jpn. J. Appl. Phys.* **22** Suppl. 22-3 190.
[187] Yamada K, Miura N and Hamaguchi C (1990) *Semicond. Sci. Technol.* **5** 159.
[188] Kajita K, Nishio Y, Takahashi T, Sasaki W, Kato R, Kobayashi H, Kobayashi A and Iye Y (1989) *Solid St. Commun.* **70** 1189.
[189] Kartsovnik MV, Kononovich PA, Laukhin VN and Schegolev IF (1988) *Sov. Phys. JETP Lett.* **48** 541.
[190] Yamaji K (1989) *J. Phys. Soc. Jpn.* **58** 1520.
[191] Yagi R, Iye Y, Osada T and Kagoshima S (1990) *J. Phys. Soc. Jpn.* **59** 3069.
[192] Yagi R, Iye Y, Hashimoto Y, Odagiri T, Noguchi H, Sakaki H and Ikoma T (1991) *J. Phys. Soc. Jpn.* **60** 3784.
[193] Osada T, Kawasumi A, Kagoshima S, Miura N and Saito G (1991) *Phys. Rev. Lett.* **66** 1525.
[194] Danner GM, Kang W and Chaikin PM (1992) *Phys. Rev. Lett.* **69** 2827.
[195] Osada, T, Kagoshima S and Miura N (1996) *Phys. Rev. Lett.* **77** 5261.
[196] Dresselhaus G, Kip AF and Kittel C (1953) *Phys. Rev.* **92** 827, (1955) *Phys. Rev.* **98** 368.
[197] Lax B, Zeiger HJ, Dexter RN and Rosenblum ES (1954) *Phys. Rev.* **93** 1418.
[198] Miura N, Kido G and Chikazumi S (1983) *J. Phys. Soc. Jpn.* **52** 2838.
[199] Narita S, Terada S, Mori S, Muro K, Akahama Y and Endo S, (1983) *J. Phys. Soc. Jpn.* **52** 3544.
[200] Takeyama S, Miura N, Akahama Y and Endo S (1990) *J. Phys. Soc. Jpn.* **59** 2400.
[201] Allen Jr. SJ, Tsui DC and Dalton JV (1974) *Phys. Rev. Lett.* **32** 107.

[202] Abstreiter G, Kneschanrek P, Kotthaus JP and Koch JF (1974) *Phys. Rev. Lett.* **32** 104.

[203] Abstreiter G, Kotthaus JP, Koch JF and Dorda G (1976) *Phys. Rev.* B **14** 2480.

[204] Pan W, Tsui DC and Draper BL (1999) *Phys. Rev.* B **59** 10208.

[205] Wilson BA, Allen Jr. SJ and Tsui DC (1980) *Phys. Rev. Lett.* **44** 479.

[206] Platzman PM and Fukuyama H (1974) *Phys. Rev.* B **10** 3150.

[207] Kohn W (1961) *Phys. Rev.* **123** 1242.

[208] Chitra R, Giamarchi T and Le Doussal P (1998) *Phys. Rev.* B **80** 3827.

[209] Imanaka Y, Miura and Kukimoto H (1994) *Phys. Rev.* B **49** 16965.

[210] Larsen DM (1979) *Phys. Rev. Lett.* **42** 742.

[211] Najda SP, Armistead CJ, Trager C and Stradling RA (1989) *Semicond. Sci. Technol.* **4** 439.

[212] Herlach F, Davis J and Schmidt R (1974) *Phys. Rev.* B **10** 682.

[213] Miura N, Kido G and Chikazumi S (1976) *Solid St. Commun.* **18** 885.

[214] Hopkins MA, Nicholas RJ, Brummell MA, Harris JJ and Foxon CT (1986) *Superlattices and Microstructures* **2** 319.

[215] Palik ED, Picus CS, Teitler S, and Wallis RF (1961) *Phys. Rev.* **122** 475.

[216] Zawadzki W, Pfeffer P and Sigg H (1985) *Solid St. Commun.* **53** 777.

[217] Hopkins MA, Nicholas RJ, Brummell MA, Harris JJ and Foxon CT (1987) *Phys. Rev.* B **36** 4789.

[218] Sig H, Perenboom JAAJ, Pfeffer P and Zawadzki W (1987) *Solid St. Commun.* **61** 685.

[219] Najda SP, Takeyama S, Miura N, Pfeffer P and Zawadzki W (1989) *Phys. Rev. B* **40** 6189.

[220] Miura N, Nojiri H, Pfeffer P and Zawadzki W (1997) *Phys. Rev.* B **55** 13598.

[221] Fukai M, Kawamura H, Sekido K and Imai I (1964) *J. Phys. Soc. Jpn.* **19** 30.

[222] Kawamura H, Saji H, Fukai M, Sekido K and Imai I (1964) *J. Phys. Soc. Jpn* **19** 288.

[223] Bagguley DMS, Stradling RA and Whiting JSS (1961) *Proc. Roy. Soc. A* **262** 40.

[224] Bagguley DMS, Stradling RA and Whiting JSS (1961) *Proc. Roy. Soc. A* **262** 365.

[225] Fink D and Braunstein R (1974) *Solid St. Commun.* **15** 1627.

[226] Meyer HJG (1962) *Phys. Lett.* **2** 259.

[227] Ito R, Fukai M and Imai I (1966) *J. Phys. Soc. Jpn.* **21**, Supplement 357.

[228] Suzuki A and Dunn D (1982) *Phys. Rev.* B **25** 7754.

[229] Dunn D and Suzuki A (1984) *Phys. Rev.* B **29** 942.

[230] Kobori H, Ohyama T and Otsuka E (1990) *J. Phys. Soc. Jpn.* **59** 2141.

[231] Miura N, Kido G, Suzuki K and Chikazumi S (1976) *Proc. of the International Conf. on the Application of High Magnetic Fields in Semiconductor Physics*, ed. G. Landwehr (Wurzburg, 1976) p. 44l.

[232] Apel JR, Poehler TO, Westgate CR and Joseph RI (1971) *Phys. Rev. B* **4** 436.
[233] Shin EEH, Argyres PN and Lax B (1973) *Phys. Rev. B* **7** 3572.
[234] Kawabata A (1967) *J. Phys. Soc. Jpn.* **23** 999.
[235] McCombe BD, Kaplan R, Wagner RJ, Gornik E and Müller W (1976) *Phys. Rev.* B **13** 2536.
[236] Kobori H, Ohyama T and Otsuka E (1990) *J. Phys. Soc. Jpn.* **59** 2164.
[237] Otsuka E (1986) *Jpn. J. Appl. Phys.* **25** 303.
[238] Arimoto H, Miura N and Stradling RA (2003) *Phys. Rev.* B **67** 155319.
[239] Sugihara K, Arimoto H and Miura N (2001) *Physica B* **298** 195.
[240] Pascher H, Appold G, Ebert R and Hafele HG (1976) *Optics Commun.* **19** 104.
[241] Potemski M, Maan JC, Fasolino A, Ploog K and Weiman G (1989) *Phys. Rev. Lett.* **63** 2409.
[242] Stolpe I, Portugall O, Puhlmann N, Mueller H-U, von Ortenberg M, von Truchsess M, Becker CR, Pfeuffer-Jeschke A and Landwehr G (2001) *Physica B* **294-205** 459.
[243] Singleton J, Pratt FL, Doporto, Janssen TJBM, Kurmoo M, Perenboom JAAJ, Hayes W and Day P (1992) *Phys. Rev. Lett.* **68** 2500.
[244] Summers GM, Warburton RJ, Michels JG, Nicholas RJ, Harris JJ and Foxon CT (1993) *Phys. Rev. Lett.* **70** 2150.
[245] Nicholas RJ, Barnes DJ, Miura N, Foxon CT and Harris JJ (1993) *J. Phys. Soc. Jpn.* **62** 1267.
[246] Engelhardt CM, Gornik E, Besson M, Bohm G and Weimann G (1994) *Sur. Sci.* **305** 23.
[247] Cooper NR and Chalker JT (1994) *Phys. Rev. Lett.* **72** 2057.
[248] Asano K and Ando T (1996) *J. Phys. Soc. Jpn.* **65** 1191.
[249] Asano K and Ando T (1998) *Phys. Rev.* B **58** 1485.
[250] Kublbeck H and Kotthaus JP (1975) *Phys. Rev. Lett.* **35** 1019.
[251] Miura N, Nojiri H and Imanaka Y (1995) *Proc. 22nd Int. Conf. Phys. Semiconductors*, ed. Lockwood DJ (World Sci., 1995) p. 1111.
[252] Michels JG, Hill S, Warburton RJ, Summers GM, Gee P, Singleton J, Nicholas RJ, Foxon CT and Harris JJ (1994) *Surf. Sci.* **305** 33.
[253] Stradling RA, Rowe AC, Ikaida T, Matsuda YH and Miura N (2001) *Proc. 10th Int. Conf. on Narrow Gap Semiconductors and Related Small Energy Phenomena, Physics and Application* (IPAP Conference Series 2), eds. Miura N, Yamada S and Takeyama S (IPAP, 2001) p. 13.
[254] Cole BE, Batty W, Singleton J, Chamberlain JM, Li L, van Bockstal L, Imanaka Y, Shimamoto Y, Miura N, Peeters FM, Henini M and Cheng T (1997) *J. Phys: Condes. Matter.* **9** 4887.
[255] Ando T (1975) *J. Phys. Soc. Jpn.* **38** 989.
[256] Englert Th, Maan JC, Uihlein Ch, Tsui DC and Gossard AC (1983) *Solid St. Commun.* **36** 545.
[257] Staguhn W, Takeyama S and Miura N (1989) *Solid St. Commun.* **71** 1113.

[258] Kaesen KF, Huber A. Lorenz H, Kotthaus JP, Bakker S and Klapwijk TM (1996) *Phys. Rev.* B **54** 1514.

[259] Thiele F, Hansen W, Horst M, Kotthaus JP, Maan JC, Merkt M, Ploog K, Weimann G Wieck AD (1987) in [SM-7] p. 252.

[260] Richter J, Sigg H, von Klitzing K and Ploog K (1989) *Phys. Rev.* B **39** 6268.

[261] Nicholas RJ, Hopkins MA, Barnes DJ, Brummell MA, Sigg H, Heitmann D, Ensslin K, Harris JJ, Foxon CT and Weimann G (1989) *Phys. Rev.* B **39** 10955.

[262] Besson M, Gornik E, Böhm G and Weimann G (1992) *Surf. Sci.* **263** 650.

[263] Kono J, McCombe BD, Cheng JP, Lo I, Mitchel WC and Stutz CE (1994) *Phys. Rev.* B **50** 12242

[264] Imanaka Y, Takamasu T, Kido G, Karczewski G, Wojtowicz T and Kossut J (1998) *Physics B* **256-258** 457.

[265] Miura N, Ikaida T, Ikeda S, Ostojic GN, Zaric S, Kono J, Wang YJ, Shiraki Y, Takano F, Kuroda S and Takita K (2003) *Bulletin of the March Meeting of APS (Austin, 2003).*
Ikaida T (2002) Ph D Thesis (University of Tokyo, 2002).

[266] Dickey DH, Johnson EJ and Larsen DM (1967) *Phys. Rev. Lett.* **18** 599.

[267] Larsen DM (1964) *Phys. Rev.* B **135** A419.

[268] Kaplan R and Wallis RF (1968) *Phys. Rev. Lett.* **20** 1499.

[269] Najda SP, Yokoi H, Takeyama S, Miura N and Pfeffer P (1991) *Phys. Rev.* B **44** 1087.

[270] Miura N, Kido G and Chikazumi S (1979) *Physics of Semiconductors (Proc. 14th Int. Conf. on Phys. Semiconductors, Edingburgh, 1978)* Inst. Phys. Conf. Ser. A No. 43 (1979), ed. B. L. H. Wilson, (The Institute of Physics, Bristol and London) p. l109

[271] Miura N, Kido G and Chikazumi S (1978) *Proc. Int. Conf. Application of High Magnetic Fields in Semiconductor Physics*, ed. Ryan, J. F., (Oxford, 1978) p. 233.

[272] Miura N, Imanaka Y and Nojiri H (1995) *Materials Science Forum* **182-184** 287.

[273] Miura N and Imanaka Y (2003) *Physica Status Solidi* (b) **237** 244-251.

[274] Imanaka Y, Takamasu T, Kido G, Matsuda YH, Ikaida T, Ikeda S and Miura N (2002) *Proc. 15th Int. Conf. High Magnetic Fields in Semiconductor Physics* (Institute of Physics, CD-ROM, Conference Series Number 171, 2002, ed. R. J. Nicholas) C25.

[275] Imanaka Y and Miura N (1994) *Phys. Rev. B* **50** 14065.

[276] Waldman J (1963) *Phys. Rev. Lett.* **23** 1033.

[277] Matsuda YH, Ikaida T, Miura N, Kuroda S, Takano F and Takita K (2002) *Phys. Rev. B* **65** 115202.

[278] Hodby JW, Borders JA and Brown FC (1967) *Phys. Rev. Lett.* **19** 952.

[279] Hodby JW (1971) *J. Phys. C: Solid State Physics* **4** L8.

[280] Jenkins GT, Hodby JW and Gross U (1971) *J. Phys. C: Solid State Physics* **4** L89.
[281] Enck RC, Saleh AS and Fan HY (1969) *Phys. Rev.* **182** 790.
[282] Das Sarma S (1984) *Phys. Rev. Lett.* **52** 859.
[283] Hopkins MA, Nicholas RJ, Brummell MA, Harris JJ and Foxon CT (1987) *Phys. Rev.* B **36** 4789.
[284] Grisar R and Preier H (1979) *Appl. Phys.* **20** 289.
[285] Yokoi H, Takeyama S, Miura N and Bauer G (1991) *Phys. Rev.* B **44** 6519.
[286] Pascher H (1984) *Appl. Phys. B* **34** 107.
Pascher H, Röthlein P, Bauer G and von Ortenberg M (1989) *Phys. Rev.* B **40** 10469.
[287] Hewes CR, Adler MS and Senturia SD (1973) *Phys. Rev.* B **7** 5195.
[288] Yokoi H, Takeyama S, Portugall O and Miura N (1993) *Physica B* **184** 173.
[289] McCombe BD, Bishop SG and Kaplan R (1967) *Phys. Rev. Lett.* **18** 748.
[290] McCombe BD (1969) *Phys. Rev.* **181** 1206.
[291] McCombe BD and Wagner RJ (1971) *Phys. Rev.* B **4** 1285.
[292] Yokoi H, Takeyama S, Miura, N and Bauer G (1991) *Proc. 20th Int. Conf. Phys. Semiconductors*, eds. Anastassakis EM, and Joannopoulos JD (World Scientific, Singapore, 1991) p.1779.
[293] Jantsch W (1983) *Dynamical Properties of IV-VI Compounds*, Vol. 99 of Springer Tracts in Modern Physics (Springer-Verlag, Berlin, 1983) p. 1.
[294] Murase K, Sugai S. Takaoka S and Katayama S (1976) *Proc. Int. Conf. Phys. Semiconductors*, (Rome, 1976) ed. Fumi F G p.305.
[295] Takaoka S and Murase K (1979) *Phys. Rev.* B **20** 2823.
[296] Yokoi H, Takeyama S, Miura N and Bauer G (1993) *J. Phys. Soc. Jpn.* **62** 1245.
[297] Fletcher RC, Yaeger WA and Merritt FR (1955) *Phys. Rev.* **100** 747.
[298] Hensel JC and Suzuki K (1974) *Phys. Rev. B* **9** 4219.
[299] Suzuki K and Miura N (1975) *J. Phys. Soc. Jpn.* **39** 148.
[300] Hensel JC (1962) *Proc. Int. Conf. Physics of Semiconductors* (The Institute of Physics and Physical Society, London) p. 281.
[301] Bradley CC, Button KJ and Lax B and Rubin LG (1968) *IEEE J. Quantum Electronics* **QE-4** 733.
[302] Stradling RA (1966) *Phys. Letters* **20** 217.
[303] Kotthaus JP and Ranvaud R (1977) *Phys. Rev.* B **15** 5758.
[304] von Klitzing K, Landwehr G and Dorda G (1974) *Solid St. Commun.* **14** 387, **15** 489.
[305] Bangert E, von Klitzing K and Landwehr G (1974) *Proc. 12th Int. Conf. Phys. Semiconductors, eds. M. H. Pilkuhn (Teubner, Stuttgart, 1974)* p. 714.
[306] Ohkawa FJ and Uemura Y (1975) *Prog. Theor. Phys.* **57** 164.
[307] Iwasa Y, Miura N, Tarucha S, Okamoto H and Ando T (1986) *Surf. Sci.* **170** 587.

[308] Iwasa Y, Miura N, Takeyama S and Ando T (1987) *High magnetic fields in smiconductor physics* ed. Landwehr G (Springer, 1987). p. 274.

[309] Leotin J, Ousset JC, Barbaste R, Askenazy S, Skolnick MS, Stradling RA and Poibland G (1975) *Solid St. Commun.* **16** 363.

[310] Miura N, Kido G, Suekane M and Chikazumi S (1983) *J. Phys. Soc. Jpn.* **52** 2838.

[311] Lay TS, Heremans JJ, Suen YW, Santos MB, Hirakawa K, Shayegan M and Zrenner A (1993) *Appl. Phys. Lette.* **52** 2431.

[312] Miura N, Yokoi H, Kono J and Sasaki S (1991) *Solid St. Commun.* **79** 1039.

[313] Rheinländer B, Neumann H, Fischer P and Kühn G (1972) *Phys. Stat. Sol.* (b) **49** K167.

[314] Maezawa K, Mizutani T and Yamada S (1990) *Institute of Physics Conference Series*, No. 112, p. 515.

[315] Hess E, Topol I, Schulze KR, Neumann H and Unger K (1973) *Phys. Stat. Sol.* (b) **55** 187.

[316] Kono J, Takeyama S, Yokoi H, Miura N, Yamanaka M, Shinohara M and Ikoma K (1993) *Phys. Rev.* B **48** 10909.

[317] Kaplan R, Wagner RJ, Kim HJ and Davis RF (1985) *Solid St. Commun.* **55** 67.

[318] Dean PJ, Lightowlers EC and Wright DR (1965) *Phys. Rev.* **140** A352.

[319] Herman F, Kuglin CD, Cuff KF and Kortum RL (1963) *Phys. Rev. Lett.* **11** 541.

[320] Serrano J, Cardona M and Ruf T (2000) *Solid St. Commun.* **113** 411.

[321] Rauch CJ (1961) *Phys. Rev. Lett.* **7** 83.
Rauch CJ (1962) *Proc. Int. Conf. Phys. Semiconductors*, (Exeter, 1962) ed. Stickland AC (The Institute of Physics and the Physical Society, London, 1962) p. 276.

[322] Kono J, Takeyama S, Miura N, Fujimori N, Nishibayashi Y, Nakajima T and Tsuji K (1993) *Phys. Rev. B* **48** 10917.

[323] Tanuma S, Inada R, Furukawa A, Takahashi O, Iye Y and Onuki Y (1981) *Physics in High Magnetic Fields*, Eds. Chikazumi S and Miura N (Springer, Berlin 1981) p. 316.

[324] Yoshioka D and Fukuyama H (1981) *Phys. Soc. Jpn.* **50** 725.

[325] Iye Y Tedrow PM, Timp G, Shayegan M, Dresselhaus MS, Dresselhaus G, Furukawa A and Tanuma S (1982) *Phys. Rev.* B **25** 5478.

[326] Ochimizu H, Takamasu T, Takeyama S, Sasaki S and Miura N (1992) *Phys. Rev.* B **46** 1986.

[327] Yaguchi H and Singleton J (1998) *Phys. Rev. Lett.* **81** 5193.

[328] Suematsu H (1976) *J. Phys. Soc. Jpn.* **40** 172.

[329] Doezema RE, Datars WR, Schaber H and Van Schyndel A (1979) *Phys. Rev* B **19** 4224.

[330] Shimamoto Y, Miura N and Nojiri H (1998) *J. Phys.: Condens. Matter* **10** 11289.

[331] Takada Y and Goto H (1998) *J. Phys.: Condens. Matter* **10** 11315.
[332] Funagai K, Miyahara Y, Ozaki H and Kulbachinskii VA (1996) *Proc. Int. Conf. Thermoelectrics, 1996, Pasadena, USA* p. 408.
[333] Köler H (1976) *Phys. Stat. Sol.* (b) **74** 591.
[334] Kulbachinskii VA, Inoue M, Sasaki M, Negishi H, Gao WX, Takase K, Giman Y, Lostak P and Horak J (1994) *Phys. Rev.* B **50** 16921.
[335] Köler H (1973) *Phys. Stat. Sol.* (b) **58** 91.
[336] Kulbachinskii VA, Miura N, Nakagawa H, Arimoto H, Ikaida T, Lostak P and Drasar C (1998) *Proc. Int. Conf. Physics Semiconductors, (Jerusalem, 1998)* p. 142.
[337] Kulbachinskii VA, Miura N, Arimoto H, Ikaida T, Lostak P, Horak H and Drasar C (1999) *J. Phys. Soc. Jpn.* **68** 3328.
[338] Kulbachinskii VA, Miura N, Nakagawa H, Arimoto H, kaida T, Lostak P and Drasar C (1999) *Phys. Rev.* B **59** 15733.
[339] Sasaki S, Miura N and Horikoshi Y (1990) *J. Phys. Soc. Jpn.* **59** 3374.
[340] Allen Jr. SJ. Duffield T, Bhat R, Koza M, Tamargo MC, Harbison MC, DeRosa F, Hwang DM, Grabbe P and Rush KM (1987) *Proc. Int. Conf. High Magnetic Fields in Semiconductor Physics*, (Würzburg, 1986), Ed. Landwher G (Springer, 1987) p. 184.
Duffield T, Bhat R, Koza M, DeRosa F, Hwang DM, Grabbe P and Allen Jr. SJ (1986) *Phys. Rev. Lett.* **56** 2724.
[341] Beinvogl W and Koch F (1978) *Phys. Rev. Lett.* **40** 1736.
[342] Schlesinger Z, Hwang JCM and Allen Jr. SJ (1983) *Phys. Rev. Lett.* **40** 2098.
[343] Rikken GLJA, Sigg H, Langerak CJGM, Myron HW, Perenboom JAAJ and Weimann G (1986) *Phys. Rev.* B **34** 5590.
[344] Wiek AD, Maan JC, Merkt U, Kotthaus JP, Ploog K and Weimann G (1987) *Phys. Rev.* B **35** 4145.
[345] Lorke A, Merkt U, Weimann G and Schlapp W (1990) Phys. Rev. B **42** 1321.
[346] Vasilyev Yu B, Ivanov SV, Meltser BYa, Suchalkin SD and Grambow P (1996) *Surf. Sci.* **361/362** 415.
[347] Arimoto H, Saku T, Hirayama Y and Miura N (2000) *Physica E* **6** 191.
Arimoto H (2000) Ph D Thesis (University of Tokyo, 2000).
[348] Demel T, Heitman D, Grambow P and Ploog K (1990) *Phys. Rev. Lett.* **64** 788.
[349] Heitman D, Demel T, Grambow P, Kohl M and Ploog K (1990) *Proc. 20th Int. Conf. Physics of Semiconductors*, eds. Anastassakkis, E. M. and Joannopoulos, J. D. (World Scientific, 1990) p. 13.
[350] Allen SJ, Störmer, HL and Hwang JCM (1983) *Phys. Rev. B* **28** 4875.
[351] Sikorski Ch and Merkt U (1989) *Phys. Rev. Lett.* **62** 2164.
[352] Ikaida T, Miura N, Tsujino S, Xomalin P, Allen Jr. SJ, Springholz G, Pinczolitz M and Bauer G (2001) *Proc. 10th Int. Conf. Narrow Gap Semiconductors and Related Phenomena* eds. Miura N, Yamada S and Takeyama

S (IPAP Conference Series 2, The Institute of Pure and Applied Physics (IPAP Conference Series 2) p. 48.
Ikaida T (2002) Ph D Thesis (University of Tokyo, 2002).
[353] Springholz G, Holy V, Pinczolits M and Bauer G (1998) *Science* **282** 734.
[354] Springholz G, Pinczolits M, Mayer P, Holy V, Bauer G, Kang HH and Salamanca-Riba L (2000) *Phys. Rev. Lett.* **84** 4669.
[355] Basinski J, Kwan CCY and Woolley JC (1972) *Can. J. Phys.* **50** 1068.
[356] Averous M, Bougnot G, Calas J and Chevrier J (1970) *Phys. Status Solidi* **37** 307.
[357] Piller H (1964) *Proc. 7th Int. Conf. Phys. Semiconductors, (Paris, 1964)* (Dunod, Paris, 1964) p. 297.
[358] Perrier P, Portal JC, Calibert J and Askenazy S (1980) *J. Phys. Soc. Jpn.* **49** Suppl. A113.
[359] Arimoto H, Miura N, Nicholas RJ, Mason NJ and Walker PJ (1998) *Phys. Rev. B* **58** 4560.
[360] Von Ortenberg M, Puhlman, N, Stolpe I, Kirste A, Müller H-U, Hansel S, Tatsenko OM, Markevtsev IM, Moiseenko NA, Platonov VV, Bykov AI and Selemir VD (2002) *Proc. 9th Int. Conf. Megagauss Magnetic Field Generation and Related Topics (MG-IX, Moscow and St. Petersburg, 2002)* eds. Selemir VD and Plyashkevich LN (VNIIEF, 2004) p. 454.
[361] Von Ortenberg M (2006) in [SM-17].
[362] Cingolani R, Holtz R, Muralidharan R, Ploog K, Reimann K and Syassen K (1990) *Surf. Sci.* **228** 217.
[363] Li G, Jiang D, Han H, Wang Z and Ploog K (1990) *J. Lumin.* **46** 261.
[364] Matsuoka T, Nakazawa T, Ohya T, Taniguchi K, Hamaguchi C, Kato H and Watanabe Y (1991) *Phys. Rev.* B **43** 11798.
[365] Ihm, J (1987) *Appl. Phys. Lett* **50** 1068.
[366] Nakayama M, Tanaka I, Kimura I and Nishimura H (1990) *J. Phys. Soc. Jpn.* **29** 41.
[367] Xia JB and Chang YC (1990) *Phys. Rev.* B **42** 1781.
[368] Fukuda T, Yamanaka K, Momose H, Hamaguchi C, Imanaka Y, Shimamoto Y and Miura N (1996) *Surf. Sci.* **361-362** 406.
[369] Yamanaka K, Momose H, Mori N, Hamaguchi C, Arimoto H, Imanaka Y, Shimamoto Y and Miura N (1996) *Physica B* **227** 356.
[370] Sasaki S, Miura N, Yagi T and Horikoshi Y (1993) *J. Phys. Soc. Jpn.* **62** 2490.
[371] Michels JG, Warburton RJ, Nicholas RJ, Harris JJ and Foxon CT (1993) *Physica B* **184** 159.
[372] Miura N, Shimamoto Y, Imanaka Y, Arimoto H, Nojiri H, Kunimatsu H, Uchida K, Fukuda T, Yamanaka K, Momose H, Mori N and Hamaguchi C (1996) *Semicond. Sci. Technol.* **11** 1586.
[373] Miura N and Nojiri H (1996) in [PP-2] p. 684.
[374] Hiruma K and Miura N (1983) *J. Phys. Soc. Jpn.* **52** 2118.

[375] Fukuyama H and Nagai T (1971) *Phys. Rev.* B **3** 4413.
Fukuyama H and Nagai T (1971) *J. Phys. Soc. Jpn.* **31** 812.
[376] Fenton EW (1968) *Phys. Rev.* **170** 816.
[377] Nakajima S and Yoshioka D (1976) *J. Phys. Soc. Jpn.* **40** 328.
[378] Yoshioka D (1978) *J. Phys. Soc. Jpn.* (1978) **45** 1165.
[379] Brandt NB and Chudinov SM (1972) *J. Low Temp. Phys.* **8** 339.
[380] Chang LL, Kawai N, Sai-Halasz J, Chang CA and Esaki L (1979) *Appl. Phys. Lett.* 939.
[381] Barnes DJ, Nicholas RJ, Warburton R J, Mason NJ, Walker PJ and Miura N (1994) *Phys. Rev.* B **49** 10474.
[382] Palik ED, Teitler S, Henvis BW and Wallis RF (1962) *Proc. Int. Conf. Phys. Semiconductors, (Exeter, 1962)*, (Institute Physics and Physical Society, London, 1962) P. 288.
[383] Hiruma K, Kido G and Miura N (1983) *J. Phys. Soc. Jpn.* **52** 2550.
[384] Nakamura K, Kido G and Miura N (1983) *Solid St. Commun.* **47** 349.
[385] Nakamura K, Kido G, Nakao K and Miura N (1984) *J. Phys. Soc. Jpn.* **53** 1164.
[386] Dingle R (1975) in *Festkörperprobleme (Advances in Solid State Physics)*, ed Queisser HJ (Pergamon-Verlag, Braunschweig), **Vol. XV** p. 21.
[387] Zwerdling S, Lax B, Roth LM and Button KJ (1959) *Phys. Rev.* **114** 80.
[388] Briggs HB and Fletcher RC (1952) *Phys. Rev.* **87** 1130.
[389] Newman R and Tyler WW (1957) *Phys. Rev.* **105** 885.
[390] Kahn AH (1955) *Phys. Rev.* **97** 1647.
[391] Caldwell RS and Fan HY (1959) *Phys. Rev.* **114** 664.
[392] Kamimura H, Nakao K and Doi T (1970) *Proc. Int. Conf. Phys. Semiconductors* (Boston, 1970) p. 342.
[393] Tarucha S, Okamoto H, Iwasa Y and Miura N (1984) *Solid St. Commun.* **52** 815.
[394] Aoyagi K, Misu A, Kuwabara G, Nishina Y, Kurita S, Fukuroi T, Akimoto O, Hasegawa H, Shinada M and Sugano S (1966) *J. Phys. Soc. Jpn.* Suppl. **21** 174.
[395] Miura N, Kido G, Kayayama H and Chikazumi S (1980) *J. Phys. Soc. Jpn.* **49** Suppl. A 409.
[396] Sasaki Y, Kuroda N, Nishina Y, Hori H, Shinada S and Date M (1981) in [SM-5] p. 195.
[397] Halpern J (1966) *J. Phys. Soc. Jpn.* **21** Supple. 180.
[398] Aldrich CH, Fowler CM, Caird RS, Garn WB and Witteman WG (1981) *Phys. Rev.* B **23** 3970.
[399] Watanabe K, Uchida K and Miura N (2003) *Phys. Rev.* B **68** 155312.
[400] Haken H and Schottky W (1958) *Z. Physik Chem.* (Frankfurt) **16** 218.
[401] Miura N (1994) *Physica B* **201** 40.
[402] Green RL and Bajaj KK (1983) *Solid St. Commun.* **45** 831.
[403] Shinozuka Y and Matsuura M (1983) *Phys. Rev.* B **28** 4878.
[404] Miura N, Iwasa Y, Tarucha S and Okamoto H (1985) *Proc. 17th Int. Conf.*

Physics of Semiconductors, eds. Chadi, J. D. and Harrison, W. A. (Springer-Verlag, 1985) 359.

[405] Miller RC, Kleinman DA, Tsang WT and Gossard AC (1981) *Phys. Rev.* B **24** 1134.

[406] Miura N, Kunimatsu H, Uchida K, Matsuda YH, Yasuhira T, Nakashima H, Sakuma Y, Awano Y, Futatsugi T and Yokoyama N (1998) *Physica B* **256-258** 308.

[407] Ruderman M (1971) *Phys. Rev. Lett.* **27** 1306.

[408] Chui ST (1971) *Phys. Rev.* **9** 3438.

[409] Salpeter EE (1998) *J. Phys.: Condens. Matt.* **10** 11285.

[410] Burstein E (1954) *Phys. Rev.* **93** 632.

[411] Bauer GEW and Ando T (1986) *J. Phys. C* **19** 1537.

[412] Kleinman DA (1985) *Phys. Rev.* B **32** 3766.

[413] Mahan GD (1967) *Phys. Rev. Lett.* **18** 448, (1967) *Phys. Rev.* **163** 612.

[414] Nozieres P and de Dominicis CT (1969) *Phys. Rev.* **178** 1097.

[415] Ohtaka Kand Tanabe Y (1983) *Phys. Rev.* B **28** 6833., (1984) **30** 4235.

[416] Tanabe Y and Ohtaka K (1984) *Phys. Rev.* B **29** 1653., (1986) **34** 3763.

[417] Miller RC and Kleinman DA (1985) *J. Lumin.* **30** 520.

[418] Iwasa Y, Lee JS and Miura N (1987) *Solid St. Commun.* **64** 597.

[419] Pinczuk A, Shah J, Miller JC, Gossard AC and Wiegmann W (1984) *Solid St. Commun.* **50** 735.

[420] Skolnick MS, Rorison JM, Nash KJ, Mowbray DJ, Tapster PR, Bass SJ and Pitt AD (1987) *Phys. Rev. Lett.* **58** 2130.

[421] Lee JS, Iwasa Y and Miura N (1987) *Semicond. Sci. Technol.* **2** 675.

[422] Lee JS, Miura N and Ando T (1990) *J. Phys. Soc. Jpn.* **59** 2254.

[423] Perry CH, Worlock JM, Smith MC and Petrou A (1986), in [SM-7] p. 202.

[424] Turberfield AJ, Haynes SR, Wright PA, Ford A, Clark RG, Ryan JF, Harris JJ and Foxon CT (1990) *Phys. Rev. Lett.* **65** 637.

[425] Goldberg BB, Heiman D, Pinczuk A, Pfeiffer L and West KW (1992) **263** 9.

[426] Buhmann H, Joss W, von Klitzing K, Kukushkin IV, Martinez G, Plaut AS, Ploog K and Timofeev VB (1990) *Phys. Rev. Lett.* **65** 1056.

[427] Buhmann H, Joss W, von Klitzing K, Kukushkin IV, Plaut AS, Martinez G, Ploog K and Timofeev VB (1991) *Phys. Rev. Lett.* **66** 926.

[428] Kunimatsu H, Takeyama S, Uchida K, Miura N, Karczewski G, Wojtowiez T and Kossut J (1998) *Physica B* **249-251** 951.

[429] Takeyama S, Karczewski G, Wojtowicz T, Kossut J, Kunimatsu H, Uchida K and Miura N (1999) *Phys. Rev.* B **59** 7327.

[430] Heiman D, Goldberg BB, Pinczuk A, Tu CW, Gossard AC and English JH (1988) *Phys. Rev. Lett.* **61** 605.

[431] Sasaki S, Miura N and Horikoshi Y (1991) *Appl. Phys. Lett.* **59** 96.

[432] Meynadier MH, Nahory RE, Worlock JM, Tamargo MC and de Miguel JL (1988) *Phys. Rev. Lett.* **60** 1338.

[433] Li G, Jiang D, Han H and Wang Z (1989) *Phys. Rev.* B **40** 10430.

[434] Nunnenkamp J, Reimann K, Kuhl J and Ploog K (1991) *Phys. Rev.* B **44** 8159.

[435] Miura N, Shimamoto Y, Imanaka Y, Arimoto H, Nojiri H, Kunimatsu H, Uchida K, Fukuda T, Yamanaka K, Momose H, Mori N and Hamaguchi C (1996) *Semicond. Sci. Technol.* **11** 1586.

[436] Minami F, Hirata K, Era K, Yao T and Masumoto Y (1987) *Phys. Rev.* B **36** 2875.

[437] Miura N, Kunimatsu H, Uchida K, Sasaki S and Yagi T (1995) *Proc. 11th Int. Conf. High Magnetic Fields in Semiconductor Physics*, (Cambridge, 1994) ed. Heiman, D. (World Scientific. Pub. Co. Pte. Ltd., 1995) p. 324.

[438] Usami N, Issiki F, Nayak DK, Shiraki Y and Fukatsu S (1995) *Appl. Phys. Lett.* **67** 524.

[439] Shiraki Y and Fukatsu S (1994) *Semicond. Sci. Technol* **9** 2017.

[440] Morii A, Okagawa H, Hara K, Yoshino J and Kukimoto H (1992) *J. Crystal Growth* **124** 772.

[441] Gnutzmann U and Clausecker K (1974) *Appl. Phys.* **3** 9.

[442] Asami K, Asahi H, Watanabe T, Gonda S, Okumura H and Yoshida S (1992) *Surf. Sci.* **267** 450.

[443] Asami K, Asahi H, Kim SG, Kim J H, Ishida A, Tkamuku S and Gonda S (1994) *Appl. Phys. Lett.* **64** 2430.

[444] Wang XL, Wakahara A and Sasaki A (1994) *J. Appl. Phys.* **76** 524.

[445] Issiki F, Fukatsu S and Shiraki Y (1995) *Appl. Phys. Lett.* **67** 1048.

[446] Shibata G, Nakayama T and Kamimura H (1994) *Jpn. J. Appl. Phys.* **33** 6121.

[447] Kobayashi Y and Kamimura H (1996) *Solid St. Commun.* **98** 957.

[448] Uchida K, Miura N, Kitamura J and Kukimoto H (1996) *Phys. Rev.* B **53** 4809.

[449] Uchida K, Miura N, Sugita T, Issiki F, Usami F and Shiraki Y (1998) *Physica* B **249-251** 909.

[450] Uchida K, Miura N, Issiki F and Shiraki Y (2001) *Physica* B **298** 310.

[451] Schubert EF and Tsang WT *Phys. Rev.* B **34** 2991.

[452] Belle G, Maan JC and Weimann G (1985) *Solid St. Commun.* **56** 63.
Belle G, Maan JC and Weimann G (1986) *Surf. Sci.* **170** 611.

[453] Fasolino A and Altarelli M (1988) *Proc. 19th Int. Conf. Phys. Semiconductors*, ed. Zawadzki, W. (Institute of Physics, Polish Academy of Science, Warsaw, 1988), p. 361, and the reference therein by the same authors.

[454] Wang PD, Merz JL, Fafard S, Leon RD, Leonard R, Medeiros-Riberiro G, Oestreich M, Petroff PM, Uchida K, Miura N, Akiyama H and Sakaki H (1996) *Phys. Rev.* B **53** 16458.

[455] Wang PD, Ledentsov NN, Sotommayer-Torres CM, Yassievich IN, Pakhomov A, Egovov AYu, Kopev PS and Ustinov VM (1994) *Phys. Rev.* B **50** 1604.

[456] Someya T, Akiyama H and Sakaki H (1995) *Phys. Rev. Lett.* **74** 3664.

[457] Hayden RK, Uchida K, Miura N, Polimeni A, Stoddart ST, Henini M, Eaves L and Main PC (1998) *Physica B* **246-247** 93.
[458] Hayden RK, Uchida K, Miura N, Polimeni A, Stoddart ST, Henini M, Eaves L and Main PC (1998) *Physica B* **249-251** 262.
[459] MacDonald AH and Ritchie DS (1986) *Phys. Rev.* B **33** 8336.
[460] Cingolani R, Rinaldi R, Lipsanen H, Sopanen M, Virkkala R, Maijala K, Tulkki J, Ahopelto J, Uchida K, Miura N and Arakawa Y (1999) *Phys. Rev. Lett.* **83** 4832.
[461] Yasuhira T, Uchida K, Miura N, Kurtz E and Klingshirn C (2001) *Proc. Int. Conf. Physics of Semiconductors* (Osaka, 2000) eds. Miura N and Ando T (Springer, 2001) p.1215.
[462] Miura N, Uchida K, Yasuhira T, Kurtz E, Klingshirn C, Nakashima H, Issiki F and Shiraki Y (2002) *Physica E* **13** 263.
[463] Nagamune Y, Arakawa Y, Tsukamoto S, Nishioka M, Sasaki S and Miura N (1992) *Phys. Rev. Lett.* **69** 2963.
[464] Kobayashi M, Kanisawa K, Misu A, Nagamune Y, Takeyama S and Miura N (1989) *J. Phys. Soc. Jpn.* **58** 1823.
[465] Hopfield JJ and Thomas DG (1961) *Phys.Rev.* **122** 35.
[466] Nagamune Y, Takeyama S and Miura N (1989) *Phys. Rev.* B **40** 8099.
[467] Taguchi S, Goto T, Takeda M. Kido G (1988) *J. Phys. Soc. Jpn.* **57** 3256.
[468] Goto T, Taguchi S, Nagamune Y, Takeyama S and Miura N (1989) *J. Phys. Soc. Jpn.* **58** 3822
[469] Goto T, Taguchi S, Cho K, Nagamune Y, Takeyama S and N. Miura N (1990) *J. Phys. Soc. Jpn.* **59** 773.
[470] Komatsu T, Kaifu Y, Takeyama S and Miura N (1987) *Phys. Rev. Lett.* **58** 2259.
[471] Komatsu T, Koike K, Kaifu Y, Takeyama S, Watanabe K and Miura N (1993) *Phys. Rev.* B **48** 5095.
[472] Kataoka T, Kondo T, Ito R, Sasaki S, Uchida K and Miura N (1993) *Phys. Rev.* B **47** 2010.
[473] Schlüter M (1973) *Nuovo Cimento* **13B** 313.
[474] Mooser E and Schlüter M (1973) *Nuovo Cimento* B **B18** 164.
[475] Nikitine S and Perry G (1955) *Compt. Rend.* **240** 64.
[476] Harbeke G and Tosatti (1976) *J. Phys. Solids* **37** 126.
[477] Baldini G and Franchi S (1971) *Phys. Rev. Lett.* **26** 503.
[478] Thanh Le Chi, Depeursinge C, Levy F and Mooser F (1975) *J. Phys. Chem. Solids* **36** 699.
[479] Gerlach B and Pollmann J (1975) *Phys. Stat. Sol.* (b) **67** 93.
[480] Lukovsky G and White RM (1977) *Nuovo Cimento* **38** 290.
[481] Nagamune Y, Takeyama S and Miura N (1991) *Phys. Rev.* B **43** 12401.
[482] Schlüter I Ch, and Schlüter M (1974) *Phys. Rev.* B **9** 1652.
[483] Harbeke G and Tosatti E (1972) *Phys. Rev. Lett.* **28** 1567.
[484] Gross EF, Perel VI and Shekhmametev RI (1971) *JETP Lett.* **13** 229, 357. [*Pis'ma Zh. Eksp. Teor. Fiz.* **13** (1971) 320. 503.]

[485] Tatsumi S, Karasawa T, Komatsu T and Kaifu Y (1985) *Solid St. Commun.* **54** 587.

[486] Watanabe K, Karasawa T, Komatsu T and Kaifu Y (1986) *J. Phys. Soc. Jpn.* **55** 897.

[487] Ishihara T, Takahashi J and Goto T (1990) Phys. Rev. B **42** 11099.

[488] Hanamura E, Nagaosa N, Kumagai M and Takagawara T (1988) *J. Material Sci. Eng.* **1** 255.

[489] Boswarva IM, Howard RE and Lidiard AB (1962) *Proc. Roy. Soc.* A **269** 125.
Boswarva IM and Lidiard AB (1964) *Proc. Roy. Soc.* A **278** 588.

[490] Fowler CM (1973) *Science* **180** 261.

[491] Imanaka Y and Miura N unpublished.

[492] Nishina Y, Kolodziejczak J and Lax B (1962) *Phys. Rev. Lett.* **9** 55.

[493] Kolodziejczak J, Lax B and Nishina Y (1962) *Phys. Rev.* **128** 2655.

[494] Galazka RP (1978) *Proc. 14th Int. Conf. Physics of Semiconductors* (Edinburgh, 1978) ed. Wilson BLH (IOP Conf. Ser. **43**, IOP, London, 1978) p.133.

[495] Gaj JA (1980) *Proc. 15th Int. Conf. Physics of Semiconductors* (Kyoto, 1980) *J. Phys. Soc. Jpn.* **49** Suppl. A, p.747.

[496] Dietl T (1981) in [SM-5] p.344.

[497] Furdyna JK (1982) *J. Appl. Phys.* **53** 7637.

[498] Nagata S, Galazka RR , Mullin DP and Akbarzaden H, Khattak GD, Furdyna JK and Keesom PH (1980) *Phys. Rev.* B **22** 3331.

[499] Bauer G, Pascher H and Zawadzki W (1992) *Semicond. Sci. Technol.* **7** 703.

[500] Munekata H. Ohno H, von Molnár S, Segmüller A, Chang LL and Esaki L (1989) *Phys. Rev. Lett.* **63** 1849.

[501] Ohno H, Munekata H, Penney T, von Molnár S and Chang LL. (1992) *Phys. Rev. Lett* **68** 2664.

[502] Ohno H (1998) *Science* **281** 951.

[503] Wolf SA, Awschalom DD, Buhrman RA, Daughton JM, von Molnar S, Roukes ML, Chtchelkanova AY and Treger DM (2001) *Science* **294** 1488.

[504] Jaczynski M, Kossut J and Galazka RR (1978) *Phys. Stat. Solidi (b)* **88** 73.

[505] Bastard G, Rigaux, Guldner Y, Mycielski A, Furdyna JK and Mullin DP (1981) *Phys. Rev.* B **24** 1961.

[506] Kossut J (1988) *Semiconductors and Semimetals* eds Furdyna JK and Kossut J (Academic Press, New York, 1988) Vol. 25 p.183.

[507] Larson BE, Hass KC, Ehrenreich H and Carson AE (1988) *Phys. Rev.* B **37** 4137.

[508] Gaj JA, Galazka RR and Nawrozki M (1978) *Solid St. Commun.* **25** 193.

[509] Miura N (1989) in [SM-8] p. 618.

[510] Yasuhira T, Uchida K, Matsuda, YH Miura N and Towardowski A (1999) *J. Phys. Soc. Jpn.* **68** 3436.

[511] Yasuhira T, Uchida K, Matsuda YH, Miura N and Twardowski A (1999) *Semicond. Sci. Technol.* **14** 1161.

[512] Yasuhira T, Uchida K, Matsuda Y, Miura N and Twardowski A (2000) *Phys. Rev.* B **61** 4685.

[513] Bartholomew DU, Furdyna JK and Ramdas (1986) *Phys. Rev.* B **34** 6943.

[514] Lee YR, Ramdas AK and Aggarwal RL (1988) *Phys. Rev.* B **38** 10600.

[515] Spalek J, Lewicki A, Tarnawski Z, Furdyna JK, Galazka RR and Obuszko Z (1986) *Phys. Rev.* B **33** 3407.

[516] Heiman D, Isaacs, ED, Becla P and Foner S (1987) *Phys. Rev.* B **35** 3307.

[517] Nicholas RJ, Lawless MJ, Chang HH, Ashenford DE and Lunn B (1995) *Semicond. Sci. Technol* **10** 791.

[518] Issacs ED, Heiman D, Wang X, Becla P, Nakao K, Takeyama S and Miura N (1991) *Phys. Rev.* B **43** 3351.

[519] Foner S, Shapira Y, Heiman D, Becla P, Kershaw R, Dwight K and Wold A (1989) *Phys. Rev.* B **39** 11793.

[520] Larson BE, Hass KC and Aggarwal RL (1986) *Phys. Rev.* B **33** 1789.

[521] Yamada N, Takeyama S, Sakakibara T, Goto T and Miura N (1986) *Phys. Rev.* B **34** 4121.

[522] Yasuhira T, Uchida K, Matsuda YH, Miura N, Kuroda S and Takita K (2002) *Physica* E **13** 568.

[523] Yu SS and Lee VC (1992) *J. Phys. Condens. Matter.* **4** 2961.

[524] Ruderman MA and Kittel C (1954) *Phys. Rev.* **96** 99.
Kasuya T (1956) *Prog. Theor. Phys.* **16** 45.
Yosida (1957) *Phys. Rev.* **106** 893.

[525] Bloembergen N and Rowland TJ (1955) *Phys. Rev.* **97** 1679.

[526] Miura N, Takeyama S, Sakakibara T, McNiff EJ Jr., and Karczewski G (1991) *Proc. 20th Int. Conf. Phys. Semiconductors*, eds. Anastassakis EM, and Joannopoulos JD (World Scientific, Singapore, 1991) p.767.

[527] von Ortenberg M (1989) in [SM-8] p.486.

[528] von Ortenberg M, Portugall O, Dobrowolski W, Mycielski A, Galazka RR and Herlach F (1988) *Solid St. Commun.* **21** 5393.

[529] Mycielski J (1986) *Solid St. Commun.* **60** 168.

[530] von Ortenberg M, Miura N and Dobrowolski W (1990) *Semicond. Sci. Technol.* **5** S275.

[531] Ossau W, Kuhn-Heinrich B, Waag A and Landwehr G (1996) *J. Crystal Growth* **159** 1046.

[532] Bhattacharjee AK (1998) *Phys. Rev.* B **58** 15660.

[533] Mackh, G, Ossau W, Waag A and Landwehr G (1989) *Phys. Rev.* B **54** R5227.

[534] Merkulov IA, Yakovlev DR, Keller A, Ossau W, Geurts J, Waag, A, Landwehr G, Karczewski, Wojtowicz T and Kossut J (1999) *Phys. Rev.* B **83** 1431.

[535] Yasuhira T, Matusda YH, Uchida K, Miura N, Kuroda S, Hara I and Takita K (2001) *Physica* B **298** 411.

[536] Yasuhira T (2001) Ph D Thesis (University of Tokyo).
[537] von Ortenberg M (1982) *Phys. Rev. Lett.* **49** 1041.
[538] Deleporte E, Berroir JM, Bastard G, Delalande C, Hong JM and Chang LL (1990) *Phys. Rev.* B **42** 5891.
[539] Ivchenko EL, Kavokin AV, Kochereshko VP, Posina GR, Uraltsev IN, Yakovlev DR, Bicknell-Tassius RN, Waag A and Landwehr G (1992) *Phys. Rev.* B **46** 7713.
[540] Kuroda S, Kojima K, Takita K, Uchida K and Miura N (1996) *J. Crystal Growth* **159** 967.
[541] Hui PM, Ehrenreich and Hass KC (1989) *Phys. Rev.* B **40** 12346.
[542] Portugall O, Yokoi H, Kono J, Miura N, and Bauer G (1993) *Proc. 21st Int. Conf. Physics of Semiconductors*, eds. Jiang P and Zheng H-Z, (World Scientific, 1993) p. 1912.
[543] Matsukura F, Ohno H, Shen A and Sugawara Y (1998) *Phys. Rev.* B **57** R2037.
[544] Dietl T, Ohno H, Matsukura F, Cibert J and Ferrand D (2000) *Science* **287** 1019.
[545] Akai H (1998) *Phys. Rev. Lett.* **81** 3002.
[546] Matsuda YH, Arimoto H, Miura N, Twardowski A, Ohno H, Shen A and Matsuoka F (1998) *Physica* B **256-258** 565.
[547] Munekata H, Ohno H, von Molnar S, Harwit A, Segmüller A and Chang LL (1990) *J. Vac. Sci. Technol. B* **8** 176.
[548] Matsuda YH, Ikaida T, Miura N, Zudov MA, Kono J and Munekata H (2001) *Physica* E **10** 219.
[549] Zudov MA, Kono J, Matsuda YH, Ikaida T, Miura N, Munekata H, Sanders GD, Sun Y and Stanton CJ (2002) *Phys. Rev.* B **66** 161307.
[550] Sanders GD, Sun Y, Kyrychenko FV, Stanton CJ, Khodaparast GA, Zudov MA, Kono J, Matsuda YH, Miura N, and Munekata H (2003) *Phys Rev.* B **68** 165205.
[551] Matsuda YH, Khodaparast GA, Zudov MA, Kono J, Sun Y, Kyrychenko FG, Sanders GD, Stanton CJ, Miura N, Ikeda S, Hashimoto Y, Katsumoto S and Munekata H (2004) *Phys. Rev.* B **70** 195211.
[552] Khodaprast GA, Kono J, Matsuda YH, Ikeda S, Miura N, Wang YJ, T. Slupinski T, A. Oiwa A, H. Munekata H, Y. Sun Y, F. V. Kyrichenko FV, G. D. Sanders GD and C. J. Stanton CJ (2004) *Physica* E **21** 978.
[553] Sanders GD, Sun Y, Stanton CJ, Khodaparast GA, Kono J, Matsuda YH, Miura N, Slupinski T, Oiwa A, and Munekata H (2003) *J. Appl. Phys.* **93** 6897.
[554] Jones H (2003) in [GT-7] Chap. 3.
[555] Progress of superconducting magnets and an interesting story of a world-leading company of superconducting magnets, Oxford Instruments, are found in: Wood A (2001) *Magnetic Venture* (Oxford University Press, 2001).
[556] Herlach F and Miura N (2003) Chapter 1 in [GT-7].
[557] Miura N and Herlach F (2003) Chapter 8 in [GT-7].

[558] Miura N, Goto T, Nakao K, Takeyama S, Sakakiara T, Haruyama T and Kikuchi T (1989) *Physica B* **155** 23.

[559] Foner S (1986) *Appl. Phys. Lett.* **46** 982.

[560] Herlach F, Bogaerts R, Deckers I, Heremance G, Li L, Pitsi G, Vanacken J, van Bockstal L and van Esch A (1994) *Physica B* **201** 542.

[561] Boebinger G, Passner A and Bevk J (1994) *Physica B* **201** 560.

[562] Kindo K (2001) *Physica B.* **294-295** 585.

[563] Takeyama S, Ochimizu H, Sasaki S and Miura N (1992) *Meas. Sci. Technol.* **3** 662.

[564] Motokawa M, Nojiri H and Tokunaga Y (1989) *Physica B* **155** 96.

[565] Gersdorf R, Muller FA and Roeland LW (1965) *Rev. Sci. Instrum.* **36** 1100. Roeland LW, Muller FA and Gersdorf R (1975) in [Pr-3] p. 175.

[566] Grössinger R, Schönhart M, Della Mea M, and Sassik H (2006) *J. Phys.: Conference Series* **51** 623.

[567] Campbell LJ and Shillig JB (2003) Chapter 6 in [GT-7].

[568] Boebinger G, Lacerda AH, Schneider-Muntau H and Sullivan N (2001) *Physica B* **294-295** 512.

[569] Wosnitza J, Bianchi AD, Herrmannsdörfer T, Zherlitsyn S, Zvyagin S (2006) in [RH-8].

[570] Fowler CM, Garn WB and Caird RS (1960) *J. Appl. Phys.* **31** 588.

[571] Pavlovski AI, Dolotenko MI, Kokol'chikov NP, Bykov AI, Tatsenko OM and Bojko BA (1994) *Megagauss Magnetic Field Generation and Pulsed Power Applications*, eds. Cowan M and Spielman RB (Nova Science Pblisher Inc., 1994) p. 141.

[572] Bykov AI, Dolotenko MI, Kolokolchikov NP, Selemir VD and Tatsenko OM (2001) *Physica B* **294-295** 574.

[573] Herlach, F, Knoepfel H and Luppi R (1965) *Proc. Conf. on Megagauss Magnetic Field Generation and Related Experiments (Frascati, 1965)* eds. Knoepfel H and Herlach F (EUR 2750.e - Euratom, Bruxelles 1966).p. 287.

[574] Caird RS, Brownell JH, Erickson DJ, Fowler CM, Freeman BL and Oliphant T (1979) *Megagauss Physics and Technology - Proc. Second International Conference on Megagauss Magnetic Field Generation and Related Topics - Washington, D.C., 1979* ed. Turchi PJ (Plenum Press New York 1980). p. 461,

[575] Solem JC, Fowler CM, Goettee JD, Rickel D, Campbell LJ, Veeser L, Sheppard M, Lacerda AH, King JC, Rodriguez PJ, Bartram DE, Clark RG, Kane BE, Dzurak AS, Facer GR, Miura N, Takamasu T, Nakagawa H, Yokoi H, Brooks JS, Engel LW, Pfeiffer L, West KW, Tatsenko OM, Platonov VV. Bykov AI, Dotenko MI, Maverick AW, Butler LG, Lewis W, Gallegos CH and Marshall B (1998) *Proc. 7th Int. Conf. Megagauss Field Generation and Related Topics*, eds. V. K. Chernyshev VK, V. D. Selemir VD and L. N. Plyashkevich LN, (Sarov, VNIIEF, Russia, 1998) Part 2 p. 756.

[576] Cnare EC (1966) *J. Appl. Phys.* **37** 3812.

[577] Miura N and Nakao K (1990) *Jpn. J. Appl. Phys.* **29** 1580.

[578] Miura N, Kido G, Akihiro M and Chikazumi S (1979) *J. Magn. Magn. Mater.* **11** 275.
[579] Nojiri H, Takamasu T, Todo S, Uchida K, Haruyama T, Goto T, Katori HA and Miura N (1994) *Physica B* **201** 579.
[580] Miura N, Osada T and Takeyama S (2003) *J. Low Temp. Phys.* **133** 139.
[581] Matsuda YH, Herlach F, Ikeda S and Miura N (2002) *Rev. Sci. Instrum.* **73** 4288.
[582] Herlach F and McBroom R (1973) *J. Phys. E* **6** 652.
[583] Nakao K, Herlach F, Goto T, Takeyama S, Sakakibara T and Miura N (1985) (1985) *J. Phys. E, Sci. Instrum.* **18** 1018.
[584] von Ortenberg M, Puhlman N, Stolpe I, Mueller H-U, Kirste A and Portugall O (2001) *Physica B* **294-295** 568.
[585] Mielke CH and McDonald R (2006) in *Proceedings of the 2006 International Conference on Megagauss Magnetic Field Generation and Related Topics (Santa Fe, 2006).*
[586] Kido G, Miura N, Kawauchi K, Oguro I and Chikazumi S (1976) *J. Phys. E, Sci. Instrum.* **9** 587.
[587] Nakagawa H, Takamasu T, Miura N and Enomoto Y (1998) *Physica B* **246-247** 429.
[588] Nakagawa H,Miura N and Enomoto Y (1998) *J. Phys. : Condens. Matter* **10** 11571.
[589] Sakakibara T, Goto T and Miura N (1989) *Rev. Sci. Instrum.* **60** 444.
[590] Sekitani T, Miura N, Ikaida T, Matsuda YH and Shiohara Y (2004) *Physica B* **346-347** 319.
[591] Sekitani T, Matsuda YH, and Miura N (2007) *New Journal of Physics* **9** 47.
[592] Kido G and Miura N (1978) *Appl. Phys. Lett.* **33** 321.

INDEX